Sins Against Science

Sins Against Science

*How Misinformation Affects
Our Lives and Laws*

JUDI NATH

Foreword by Jim Obergefell

McFarland & Company, Inc., Publishers
Jefferson, North Carolina

ISBN (print) 978-1-4766-8639-4
ISBN (ebook) 978-1-4766-4398-4

LIBRARY OF CONGRESS AND BRITISH LIBRARY
CATALOGUING DATA ARE AVAILABLE

Library of Congress Control Number 2021039772

Front cover image: © 2022 Khadi Ganiev/Vitamin444/Shutterstock

Printed in the United States of America

*McFarland & Company, Inc., Publishers
Box 611, Jefferson, North Carolina 28640
www.mcfarlandpub.com*

For my brother, Russ,
who was with me for the beginning of this project but not for the end.
We had great talks about what should be included in this book.
It's so sad that you're not here to read the finished work;
I told you that I was going to miss you like crazy. And I do.

Table of Contents

Acknowledgments

It is with deep appreciation that I acknowledge the following people for their assistance with making this book a reality. First on the list is medical illustrator, healthcare professional, and friend Bill Ober, MD. Bill and I have worked together on anatomy and physiology textbooks for many years; his hand drew every figure in this book, thereby making the complex easier to understand and much more interesting. Bill also read the first draft of this entire book, offering enhanced critiques; I am indebted to his perseverance and skills.

I have also had the distinct pleasure of working many years with medical illustrator, healthcare professional, and friend Claire Ober, RN, who read the first draft and offered excellent suggestions for improvement. In addition to her immense artistic talents, she keeps our team engaged with dog stories and live music; you are truly inspirational.

A mountain of gratitude goes to fellow biologist, Anjali Dogra Gray, Ph.D. Anjali read every line of manuscript and offered her expertise at every turn. In addition to writing the poem found in Chapter 5, she was forthright in telling me when the writing was going off the rails, or where people would want to read more. She trained under the 2018 Nobel laureate in Chemistry, which means she knows her stuff, thus her gentle corrections were duly noted. I am fortunate to call her sister-friend.

Another note of thanks goes to English professor and friend Susan Shelangoskie, Ph.D., who was always just a text or email away, providing a solid sounding board and clarity of prose. Jennifer Sader, Ph.D. was a warmly welcomed reviewer, who read select chapters, offering careful guidance. Excellent feedback was given by long-time friend, Marianne Baker, MAT, who read the pages for fun first and editorial analysis second. I knew I could count on her for honest assessment because anybody who would jump off a moving train in Spain with me definitely had my best interests at heart.

The Supreme Court Case openers were inspired by my lifelong friend and landmark decision plaintiff, Jim Obergefell, who also wrote the Foreword. Thanks for being you and fighting the good fight!

Huge hugs go to my Michigan family, Lily, Paul, and Ethan Krieger, who provided unending support and encouragement for many years. They have a way of making people feel very special.

A warm note of appreciation and much gratitude to my former editor, Cheryl Cechvala, who read pages and chapters and offered rounds of critiques, suggestions, and edits. Thanks for believing that I was on to something here! My life-long friend Martha Wolfe kept the beverages and brownies coming in the midst of a pandemic while providing countless hours of conversation and encouragement. She is the most upbeat person I know. Thanks to all my friends, family, colleagues, and supporters who helped in countless ways when least expected; life is richer because of you.

Rounding out the list is the Love o' My Life and husband, Mike, who never grows tired of reading and re-reading my endless pages of manuscript.

Special acknowledgment is given to McFarland for believing in the project and bringing the printed product to you. Thank you.

Foreword by Jim Obergefell

"How can any rational, thinking person actually think that? Not only believe it but use it as a legal argument in a court case heading to the Supreme Court of the United States?"

I found myself pondering that in reaction to the arguments used by the attorneys general of Ohio, Kentucky, Tennessee, and Michigan in their defense against lawsuits filed by more than 30 other plaintiffs and me demanding the right to marry our same-sex partners in, or to have our lawful out-of-state marriages recognized by, the states we called home.

What were these arguments that I found hard to believe? There were several, but the ones that really caused me to scratch my head were these: allowing same-sex marriage would cause opposite-sex couples to no longer marry, and allowing same-sex couples to marry would negatively impact procreation.

That's right, the states argued in Obergefell v. Hodges that allowing John and me to marry or recognizing that marriage as equal to an opposite-sex marriage would lead to the complete failure of marriage as an institution. Opposite-sex couples would immediately stop marrying because two women said "I do" or two men promised to love, honor, and protect each other until death. Not only that, the future of our species would be at risk because granting same-sex couples the right to marry would cause straight couples to stop procreating.

What was the scientific justification for these arguments, the proof that allowing same-sex marriage would have such detrimental effects on society? No such evidence exists, and as marriage equality court cases proliferated across the United States prior to ours, court after court took defendants to task for making these specious and unsubstantiated claims.

Did I mention that these same arguments were used to justify miscegenation laws that made interracial marriages illegal? Yes, prior to the Loving v. Virginia Supreme Court decision in 1967, two people of different "races" were prevented by law from marrying in many states, because of laws based on these arguments.

Another commonality between Loving v. Virginia and Obergefell v. Hodges is the use of belief, not science or fact, to justify discriminatory laws. Proponents of these laws argued that they were ethical and constitutional because "a supreme being placed people with different skin color on separate continents because they weren't intended to mix" and "the supreme being created only man and woman, and their sole purpose is to procreate."

We *Homo sapiens* tend to have a high opinion of ourselves because we can reason, we can adapt by choice, we have higher cognitive skills than other animals. Yet we still choose to ignore what those cognitive skills and millennia of learning—based in observable and repeatable science—have taught us. Unfortunately, that willful ignorance far too often leads to hatred and discrimination based not on fact but on belief or feelings.

"Homosexuality is a choice" ignores the scientific fact that same-sex attraction and pairing occur in the animal kingdom. "You can be only male or female" ignores the scientific fact that some human beings are born with both male and female genitalia or with differing numbers of sex chromosomes. When I think of what transgender people experience, I wonder how it must feel to wake up every day and see a body in the mirror that does not match who they are inside, especially in a society that is so fixated on the genitals present at birth as the sole determinant of a person's identity and worth.

So much of the animus toward the LGBTQ+ community ignores our ever-evolving scientific knowledge about how gene variances, environmental factors, and more influence who we become as people, who we are as human beings. Another societal cancer, racism, also ignores the clear scientific proof that variation in skin color, a response to environmental factors since the dawn of humanity, is no different than other variations like eye color or handedness. When religious belief, arrogance, and a fear of those who look different collide, racism is born. Our nation and the world continue to grapple with racism—blatant, casual, systemic—for no rational, scientific reason.

Race, sexual orientation, and gender identity are only three areas of society and culture that science can help us better understand. Amid any pandemic, such as COVID-19 in 2020–2021, science should inform every decision made by our nation's leaders, yet anti-science rhetoric hampered every effort to address this health emergency and mitigate the devastation. Leaders pushed unproven and potentially dangerous treatments, regardless of a lack of scientific proof showing efficacy, and undermined the very experts we rely on for guidance—the same experts who were also responsible for developing and testing a vaccine. Humanity has faced pandemics in the past, and science was key to understanding and surviving the threat. New diseases will continue to appear as humanity infringes on the natural world, pushing into animals' habitats, and we must rely on history and science in our responses to these future public health crises.

From health policy to climate change, education to human rights, science and facts should be the basis of every deliberation over laws and policies. The world is better served when science takes precedence over personal feelings and/or religious beliefs. A well-researched book like this one is desperately needed as we attempt to address social issues and other challenges we face.

Humans have gone from living in caves to setting foot on the moon because of observation of the world around us and the knowledge gained through repeatable and provable science. Technology has advanced immeasurably since the invention of the wheel, and we enjoy a life of ease and convenience because of science. Science also helps us understand that we *Homo sapiens* are much more alike than we are different.

Now if only we'd listen.

Jim Obergefell, *a civil rights activist and lead plaintiff in the Supreme Court case of Obergefell v. Hodges, is the coauthor of* Love Wins: The Lovers and Lawyers Who Fought the Landmark Case for Marriage Equality, *the co-founder of Equality Vines, and director of Special Events at Family Equality.*

Preface

Have you ever been fooled by headlines or misled by talking heads? If so, you are not alone. *Sins Against Science* was born out of my frustration with inaccurate accounts and total misrepresentations of science in general and human biology in particular. With a click here and a click there, people post to social media with abandon. Copying, pasting, tweeting, and sending without any regard to accuracy is a widespread practice. Every day, newsfeeds, headlines, and sound bites provide teachable moments by showing the intersection of science and society. For years, I have been fielding questions from every corner and replying with solid science. When the COVID-19 pandemic hit, I simply could not keep up with the queries clogging my screens. Thus, I began to blog. Each blog became a response to what was happening in real time about real issues that required real clear scientific explanations. Sadly, those issues expanded as the Black Lives Matter movement took front and center attention, causing another blog on race. That was followed by a Supreme Court decision on LGBTQ issues, which was followed by another landmark decision on abortion. Moreover, while still in the throes of the pandemic, an unexpected endorsement of same-sex unions came from Pope Francis, the head of the Catholic Church. His statements echoed what most Americans feel: Gay people have the right to be in a family. Yet the pope later said that priests cannot bless such unions. I keep hoping that if we apply solid science and human compassion to so many issues, we'd think differently about them.

I had begun writing this book long before coronavirus became a household name, but each of these aforementioned scientific blog topics were already chapter headings. The blogs and the public responses to them reassured me that people wanted a book on science they could use. Trust me, the haters came out, but their remarks were emotional, not rational. The vast majority thanked me for helping them understand complex issues by writing in a conversational tone. Many people told me their stories, which buoyed my resolve to move forward and get this information out there. Books and blogs from reputable sources are necessary and relevant—even in the age of cyberinformation. As a family member recently said to me, "I just want to know what facts to believe." If I can help people understand the science behind so many contentious issues, the world might be a kinder place.

This book captures our zeitgeist by snatching popular stories, conspiracies, and headlines making the rounds on mainstream media and social networks, and then explaining emotionally-charged topics through the lens of science. Tackling topics like evolution and race, sexual identity, reproduction, vaccines, alternate medicine, and death and dying, fact is separated from fiction. Each chapter walks you through the topic and avoids the clickbait, then teaches you how to be an informed citizen armed with real information rather than an armchair scientist loaded with alternative facts. If I can add to the national conversation, change the way people think, and help others understand the science behind so many heated debates, perhaps we can collectively approach matters from a different perspective. Weaving together history, politics, human biology, and law, you will see how our lives are dependent on understanding the nature of things.

At the heart of much civil unrest is scientific illiteracy. The lack of scientific knowledge among the general population in the United States is dramatic and, as evidenced by numerous events of the last several years, potentially dangerous. The aim of this book is to debunk pseudoscience and deception in our culture of fear mongering by taking current, culturally relevant topics and breaking apart myths, re-assembling them with facts, and presenting the information with interesting backstories and historical relevance.

We can be deceived by headlines, editorials, and news articles. We can also be misled by hearing diametrically opposed commentaries or reports. The reason we can be duped is because the sheer volume of written text and spoken words sounds plausible; but the language is nothing more than rhetoric. The art of effective, persuasive writing and speaking has been perfected by politicians and pseudoscientists to advance self-serving purposes. Now is the time for scientific literacy.

This thoroughly researched book is important because we all need a little applied science to help us understand our society. It's interesting to note that many issues have found their way to landmark Supreme Court cases, which open each chapter. While writing in fall 2020, the United States lost iconic Supreme Court Justice, Ruth Bader Ginsburg, to cancer. Known for her tenacity in advancing social issues, keeping our civil liberties intact, and fighting all forms of discrimination, her decisions and dissents helped us understand the complexity of issues. While one side of the political debate sees her as a "pro-lifer" the other side sees her as a "pro-choicer." I see her as both, for her life's work was as complex as the issues. In 1972 she founded the ACLU Women's Rights Project, and before taking a seat on the Supreme Court bench, she had argued six cases between 1973 and 1979 before the high court. My favorite case to highlight how she can be seen as a pro-lifer and a pro-choicer involves the case of Air Force Captain Susan Struck, a combat nurse in Vietnam. When Struck became pregnant in 1970, the Air Force gave her an either/or option: abortion or discharge.

While this predates 1973's Roe v. Wade decision, abortion was legal on U.S. military bases. Struck did not want an abortion, she wanted to keep both her baby and her military job. As you might have suspected, she was discharged from the army; there were no laws protecting pregnant women from discrimination. That is, until ACLU lawyer Ruth Bader Ginsburg took the case of Struck v. Secretary of Defense and eliminated the unlawful practice of discharging pregnant women officers. Ginsburg also helped Congress draft the Pregnancy Discrimination Act of 1978, which prohibits sex discrimination on the basis of pregnancy. This in itself likely prevented untold numbers of abortions. Everything isn't always as it seems.

We all have opinions on far-ranging subjects, but as informed citizens, it behooves us to have scientific support for those views so that we don't fall victim to falsehoods. This book helps to respectfully and scientifically unravel complex relationships—even when those topics are rooted in religion or faith. You'll be armed with practical information that you can take to your discussion circles, family gatherings, and voting booths. In short, I want to add to the national conversation and change the way people think.

Headlines often perpetuate myths. This is particularly disheartening because the more times a falsehood is stated, the more likely one is to believe it. If people are told over and over by a political leader or figurehead that a truism is false, or if people read a factual statement on social media that a bot cites as fake, this is extraordinarily dangerous because appealing to popular prejudices without any rational argument is how demagoguery gains popular support. We need tools to dismantle such rhetoric.

This book goes behind the scenes and explores the basic science behind hot-button issues, many of which are perceived as controversial, even as scientific evidence indicates otherwise. As science comes to life on every matter, my hope is to reach genuinely curious readers who may simply need a better understanding of the facts to rethink or recast their position.

Why do people so easily believe conspiracy theories? Because they are interesting, dramatic, fun, emotional, and fit what the believer wants to hear. But science and truth can be all of those things—and that's what I strive to do in this book.

It is a challenge to write a contemporary science book within the context of a fickle Internet world. That is why I have chosen subjects that come up year after year in my college courses (human anatomy, physiology, nutrition, and pathology), in interactions with other professionals in my field, in the course of normal conversation, and in countless posts on social media that don't seem to change. Thus, while the list of controversial topics is extensive, this book sticks to human science themes. Each chapter begins with a landmark Supreme Court case relative to the topic and concludes with a wrap-up that serves as the take-home message. I have chosen landmark court cases to show how science comprehension—or lack thereof—can impact public policy and ultimately human lives. Given the condition of our politics, scientific literacy now is as important as it has ever been. Read on.

Introduction

Approach the Bench:
Opening Statements and Newsfeeds

U.S. Landmark Case: New York Times Co. v. Sullivan, 376 U.S. 254 (1964); (9–0 decision)

In the early 1960s, a period of much civil unrest in America, the *New York Times* ran a paid advertisement soliciting donations for a legal defense fund for protesters in the South. The advertisement contained some inaccurate accounts of police actions in Montgomery, Alabama. The city's police commissioner, L.B. Sullivan, sued the *New York Times* for printing an advertisement containing some false statements (defamation). After losing in state courts, the *New York Times* appealed to the U.S. Supreme Court. On March 9, 1964, the court's unanimous decision was announced in favor of the newspaper, stating that the Alabama state law on which the case was based was unconstitutional in that it did not provide safeguards for freedoms of speech and the press as required by the First and Fourteenth Amendments. The court created a standard of "actual malice" whereby a plaintiff had to prove that the defendant acted with an actual intent to do harm. Justice William Brennan noted: "The United States is founded on the profound national commitment to the principle that debate on public issues should be uninhibited, robust, and wide-open, and that it may well include vehement, caustic, and sometimes unpleasantly sharp attacks on government and public officials."

The Necessity of This Book

This landmark case was based on the First and Fourteenth Amendments to the United States Constitution. Passed by Congress on September 25, 1789, and adopted on December 15, 1791, the First Amendment ensures freedom of religion, speech, press, assembly, and petition. The Fourteenth Amendment—one of the most litigated of all the amendments—was adopted on July 9, 1868, and grants all people equal protection of the laws. Thus, regardless of whether that speech is truthful or not, it should be allowed in all forms across all media.

Given our political climate and the hot tempers of the American people, many of us are grappling with this notion of free speech. We're grappling because social media outlets are running amok with purely false statements that people and bots generate to create factions. This is particularly disturbing because the majority of Americans get their news from social media, with Facebook the most commonly used site. In fact, 43 percent get their news from Facebook. Moreover, Facebook poses a major threat to public health. As the 2020 COVID-19 pandemic took hold, people turned to Facebook for health information. However, viewings of the top 10 health misinformation sites greatly outpaced viewing of the top 10 official health institutions in January, March, April, and May. February 2020 was the only month in which people turned to reputable sites for their public health information. When the pandemic was ramping up in April, the top 10 misinformation sites were viewed more than 296 million times, while the official ones only had 71 million hits. The institutions comprising the top ten include leading health institutions in the United States, United Kingdom, Italy, France, and Germany as well as the World Health Organization and the European Centre for Disease Prevention and Control. This means that anti-vaccine communities, conspiracy theorists, and false health gurus delivered bogus information to a greater proportion of people than did bona fide health groups and respected scientific associations.

Political advertisements are extraordinarily guilty of exploiting falsehoods and in October 2019, Twitter's CEO, Jack Dorsey, banned all political ads from the platform. Facebook, however, refuses to ban political ads. Using advertisements to manipulate isn't reserved solely for political gain. Deception is part and parcel of our everyday lives. It can be extremely difficult to tease out fact from fiction as the distinctions among false news, fake news, real news, opinions, disinformation, misinformation, and satire can be subtle. For these reasons, people need to read carefully, fact-check from reputable sources, and think critically about what they see and hear. All too often we have knee-jerk reactions to new information.

The ability to acquire and apply knowledge and skills are hallmarks of intelligence. But are we smarter today than ever before? When I was a child, I thought all adults disseminating information were smart. I thought every person out of high school knew all the answers to everything. Then I became an adult, and over time realized I was wrong. No single person has the correct answers to everything. Of course, there are lots of smart people, but what has changed from my childhood until now is the manner in which people can tout their knowledge. But, if you're reading this and thinking that it does seem as though people are not as smart as in days of yore, you might be right. What we do know is that when it comes to basic scientific knowledge, we are falling a little short. According to a recent study in the *Proceedings of the National Academy of Sciences*, the world's most-cited multidisciplinary scientific journal, when it comes to general science knowledge, people are not scoring high marks. Think about flat-earthers and climate change deniers. People subscribing to the theory that the earth is flat are basing their supposition on their personal experience of walking the earth. Excepting mountainous terrain and bodies of water, the earth does indeed *seem* flat; but satellite imagery and science show otherwise. As for climate change, again, indisputable scientific evidence clearly shows that global warming and climate change are real concerns. Believing that the earth is flat is not harmful; but believing that climate change is a hoax has costly consequences.

This research suggests instead that the decline in understanding basic scientific concepts is the result of many factors including exposure to different intellectual experiences, use of technology, and social media. Thankfully, this research has squelched the dysgenic fertility theory, based on the inaccurate assumption that unintelligent people have more kids than intelligent people, leading to the "dumbing down" of society. Intelligence levels can vary greatly between siblings. Intellectual experiences will always vary because humans vary.

Social media is computer-mediated technology that facilitates our ability to create and share information, ideas, and interests. Social media, a totally intangible thing, exerts profound psychological pressure and emotional influence. Social media is web-based and differs from traditional media platforms like paper-based newspapers, magazines, and television broadcasting. To help you differentiate between traditional media and social media, consider selfies. Selfies are self-portraits, typically taken by holding a digital camera or smartphone at arm's length and snapping digital photographs, with the purpose of sharing on social media platforms such as Facebook, Twitter, and Instagram. People do this a lot, to the extent that the 2013 *Oxford English Dictionary* word of the year was "selfie." So, what do selfies and social media have to do with our discussion? The reason is to show how social media and not thinking are shaping our culture. There's a good chance that you have taken a selfie. There's generally no harm in doing so, unless you're wanted by the FBI (and now they've found you) or you put yourself or others in harm's way. Between October 2011 and November 2017, there were 259 deaths caused by taking selfies. Sadly, these are 259 lives lost to social media. These 259 deaths were preventable. That's more deaths than people who died in avalanche disasters over that same amount of time in North America. People purposefully put themselves in dangerous situations to snap a photo to post for all the world to see. Sometimes, people even do this at zoos, endangering other people too and causing those around the selfie-taker to ponder who actually should be behind the moat. For example, in March 2019, a woman at an Arizona zoo jumped over a concrete barrier to take a selfie with a jaguar. While the jaguar did attack, the woman survived with injuries to her arm. According to Google, in 2015 alone, there were an estimated 24 billion selfies posted to their photo sharing service, Google Photos. Putting oneself in dangerous or risky situations just to snap a photo for other people to see calls into question one's intelligence. But, this shows just how strong a pull social media has on the human experience.

Except in these extreme cases, selfies and selfie idiocy are generally harmless. Being scientifically mindless, however, is harmful; and as this book will show, can have serious consequences to human health and well-being. Why is scientific literacy important? Scientific literacy is important because it provides a context for addressing societal concerns. Higher levels of scientific literacy allow for better decision making across all aspects of our communities. Currently, science knowledge in the United States is just so-so. To illustrate, 15-year-old students in the U.S. currently rank 18th out of 78 countries in science on the 2018 Program for International Student Assessment (PISA), a test that measures their ability to use their reading, mathematics, and science knowledge and skills to meet real-life challenges. This is not just a sad commentary on the state of education in our society, it is potentially dangerous. This book outlines how topics related to human biology are part and parcel of our everyday, consumer-oriented lives, and will prove how important a basic understanding of bioscience really is.

One goal of this book is to reach thinking individuals and provide relevant information so we can all be informed citizens with fact-based positions. Science is inherently interesting, and many people are aware and curious, but uncertain. Others may be more resistant to scientific arguments, but perhaps reading a science-based book written in common language and sprinkled with fun facts will keep everybody reading. Let's keep our collective fingers crossed.

It's the responsibility of scientists and science supporters to provide everyone—parents, students, policymakers, citizens, teachers, healthcare providers—with evidence-based information. This is oftentimes tough to do because pseudoscientific websites and sham articles hijack real scientific terms to appear credible, so even otherwise intelligent people can find it difficult to know which sources to trust. You'll find credible news and scientific sources at the end of this book. Additionally, in every-day life, websites of major hospitals (Cleveland Clinic, Mayo Clinic, and Johns Hopkins to name a few) and government health agencies are good sources of accurate information. When everyone is informed, we are all better for it.

Sussing out science from non-science is difficult. Headlines and Dr. YouTube often perpetuate myths, advance biases, and provide false information. Internet videos and comment sections contain more misinformation than facts, especially when related to health issues. In one study just on the dissemination of misinformation and biased information related to cancer, 77 percent (115 videos) of the hits returned on YouTube for "prostate cancer" contained misinformation and/or biased content within the video itself or within the comments section. This is an example of how false information can be potentially dangerous. These videos were seen by over 6 million viewers. For a point of reference, Los Angeles has a population just above 4 million people. These facts are particularly disheartening because the more times a falsehood is stated, the more likely people are to believe it. It's human nature.

If people are told over and over again by a political leader, talking head, newspaper heading, or social media feed that a truism or fact is fake or false, that fact can actually lose credibility in public conversation, and this state of affairs can have adverse consequences. This is where demagoguery comes into play. A demagogue appeals to prejudices without any rational argument and through this appeal to emotions or so-called common sense, gains popular support— support that eventually overwhelms science, facts, and rationality itself. Americans need to be smarter than this.

Here are two examples from recent headlines to illustrate how average readers are exposed to attention-grabbing phrases that sound scientific:

- Newly-discovered Human Organ May Explain How Cancer Spreads (*New Scientist*, March 27, 2018)
- Are Coffee and Alcohol the Fountain of Youth? (*The People's Pharmacy*, March 5, 2018).

Here are two other examples of politically-charged headlines from lawmakers:

- Legitimate Rape Victims Have Ways to Try and Shut That Whole Thing Down (*Representative Todd Akin, Republican–Missouri, stated August 19, 2012, and again in 2014 in television interviews*)
- House Bill 182 would prohibit health insurers and public employee benefit plans from covering abortion except when necessary to "re-implant the fertilized ovum" or to save a pregnant person's life (*Bill sponsored by Ohio Representative, John Becker in 2019*)

Furthermore, a popular poster-type graphic that has been circulated on Facebook since 2014 lists Republican comments on rape, including those made by Todd Akin. Here is a collection of five other actual statements made by GOP politicians. The remarks sound unbelievable, but they were indeed recited and are not here taken out of context.

1. "Rape is kinda like the weather. If it's inevitable, relax and enjoy it." Stated on March 24, 1990.—Clayton Williams (R–TX)

2. "Rape victims should make the best of a bad situation." Stated on January 20, 2012, on CNN's *Piers Morgan Tonight*.—Rick Santorum (R–TX)

3. "Even when life begins in that horrible situation of rape, it is something that God intended to happen." Stated on October 23, 2012.—Richard Mourdock (R–IN)

4. "In the emergency room they have what's called rape kits, where a woman can get cleaned out." Stated on June 23, 2013.—Jodie Laubenberg (R–TX)

5. "If a woman has [the right to an abortion], why shouldn't a man be free to use his superior strength to force himself on a woman? At least the rapist's pursuing of sexual freedom doesn't [in most cases] result in anyone's death." Stated on February 25, 2014.—Lawrence Lockman (R–ME)

What do you think when you read attention-grabbing headlines like these? Do you believe them? What do you think when you later learn that what you thought was true turns out to be false? How do you know what to believe? The more sensational the headline, the more readers should be on their guard. When you read the first headline, did you really think that we've discovered a new organ in 2018? Was it hiding somewhere over the millennia where no anatomist or surgeon ever found it? Do you honestly think that after all these years of dissecting cadavers (dissection has roots dating back to the third century BCE) and operating on people that a new organ was really discovered? No. What was found was a network of fluid-filled channels with conduits to the lymphatic system. Fluid-filled channels are not organs and this finding was discovered on one patient during a surgery. The "discoverers" titled the chambers "interstitium." The interstitium *is* a real thing: it is a network of fibers and fluid-filled spaces that underlie the skin and surround the intestines, muscles, blood vessels, and lymphatics. However, if you were not well versed in anatomy, you'd be led to think or believe this was something entirely new.

Onto the second headline regarding coffee and alcohol keeping us young. With billions of people drinking both beverages, many on a daily basis, are we living to be 150 years old? No. Yet, both headings were front and center on reputable science publications. The headlines were "click bait" meant to drive web traffic, but they actually mislead.

It's a challenge to unpack Representative Todd Akin's comments above because there are just too many crazy variables. The idea of "legitimate rape" is just a start. Rape is rape and the female body has no such mechanism to shut down any unintended pregnancy. The other comments by Republican politicians are also unfathomable. Unfortunately, many women are impregnated by their rapists, and the body doesn't treat these pregnancies any differently than those conceived in happier circumstances.

As for the last statement regarding the re-implantation of a fertilized ovum to the uterus, it is medically and scientifically impossible to carry out such a feat. Yet, here were erroneous statements from lawmakers who were getting lots of press, whether they actually believed these unfounded ideas or were just cynically trying to sow confusion.

We seem to be at a crossroads between fact and fiction while we're losing our system of checks and balances. Many politicians are getting elected because they're willing to use inflammatory, fact-free rhetoric, while laws are being passed by uninformed legislators. People are moved by emotion and forget to consider sound science. The crux of this book is to show that if people have a better understanding of science, evidence, and facts, then we might be more receptive to counter arguments around key controversial matters of public policy—societal affairs that affect us all.

This book dives deeply into common scientific issues facing us today. As the coronavirus pandemic struck, my inbox and message boards were overwhelmed with questions related to COVID-19. This influx of queries related to the public health crisis was in addition to the myriad questions I routinely receive from students (past and present), family, and friends. At the time of the pandemic I was already writing this book because I saw how science was under attack by "non-believers" and policy makers. I wanted to write a book people would read. And as the pandemic struck, I realized people wanted to read. There are so many important topics crossing our screens and our brains on a daily basis that it became apparent that now is the time to address them. To address them in a non-politicized way that teaches for lifelong use. Each chapter provides guidance on a human biology topic that affects us each day. In order to make sense of this angry world, add a little scientific understanding to calm the rage. Enter this book.

We are continuously being spoon fed heaping helpings of daily nonsense through Twitter feeds, Facebook posts, and news streams. Misinformation has become the diet of the American people. It's all served under the guise of fact, when in actuality many are falsehoods, lies, and piles of inaccurate information deliberately intended to deceive. Sometimes we're all just a little too gullible to know any better. Or maybe we are hearing what we want to hear regardless of truth. Central themes within this book show that if we had a better comprehension of human biology, and allowed the facts to come to the fore, our attitudes about many key controversial and contentious subjects could change. I add "controversial" because it seems as though everything is controversial these days. I never would have thought wearing a face mask to protect humankind would have been controversial. Yet, here we are. The COVID-19 face mask is politicized and tribalized. Let's take a look at the science behind our lives.

Using easy-to-understand narrative, this book walks through human biology topics that have been in the recent public conscious and for which there is bountiful science-based evidence. It just so happens that these topics tend to turn up in the news as displays of public disagreement. While controversy oftentimes surrounds such topics, facts are at their core. You will learn enough about human biology and relevant science to be adequately conversant and to know what makes sense when you encounter headlines, or to know when you need to analyze and seek out reputable sources, or when to call something hogwash. We're living in a culture of unnecessary fear mongering propagated by endless tweets and newsfeeds. In the end, I want people to be skeptical, consider sources, think critically, and have the knowledge to say, "Let me explain that to you so we can make this squabbling go away."

Legislation

Each chapter begins with a "landmark court decision" relative to the topic and concludes with a wrap-up that serves as the take-home message. The phrase "landmark decision" often appears in newspapers, on the news, or in cocktail discussion circles; but what exactly is it? Landmark decisions set new legal principles, or they change the interpretation of existing law in the United States. Landmark court decisions are frequently made by the United States Supreme Court, but they don't have to be. Some landmark decisions are made by Courts of Appeals. Landmark court decisions of the U.S. Supreme Court have been selected to show how understanding science is crucial to shaping policy. People, science, society, lawmakers, and law are inextricably linked.

Now, a little about the Court. The Supreme Court of the United States, often referred to by the acronym SCOTUS, was established in 1789 pursuant to Article III of the U.S. Constitution. Article III establishes the judicial branch of the federal government, which consists of the Supreme Court and lower courts created by Congress. Within each section of Article III, there are three subsections with several clauses. In case you were wondering, cases don't just drop on the Supreme Court desk. They work their way through the lower courts and take time. Sometimes a lot of time. Moreover, getting a case to the Supreme Court is a big deal. A bigger deal is that cases before and decisions by the Supreme Court affect all avenues of American life.

One very important reason I started paying attention to the connection between law and science is because I see legislators and average citizens disregarding the value of scientific facts in favor of public opinion or political party. My front row seat to the Supreme Court process played out with the 2015 case of *Obergefell v. Hodges*. This case opens its relevant section, Chapter 4, Jungle Love: Sexual Orientation. It's also the reason why the foreword to this book is written by my friend, Jim Obergefell. Jim and I have been friends since childhood. Experiencing the hoops he jumped through to marry the love of his life, with whom he'd shared 20 years, was heart-wrenching. You see, Jim was in love with John Arthur, a man who lay dying of amyotrophic lateral sclerosis (ALS), also known as Lou Gehrig's disease. It is one of the most ravaging diseases a person could ever endure. As a progressive disease that destroys motor nerves—those nerves that control every skeletal muscle in your body—it begins robbing a person of life, causing muscle wasting and total paralysis. It's a cruel disease: The person's mental faculties remain totally intact, so they are intimately aware and conscious of what is happening to their degenerating, immovable body. As it progresses, ALS victims gradually lose the ability to talk. They are locked in their heads, unable to move. The average life expectancy for a person with ALS, from diagnosis to death, is two to five years.

Jim and John loved each other as much as any two people could, but they could not marry in their home state of Ohio, because Ohio did not recognize marriage between two men. If they wanted to marry, they had to go to a state that recognized their love: Maryland. But, to make a marriage in Maryland happen, they had to charter a medical jet, transport John on a gurney, fly with an ordained minister (John's Aunt Tootie), and land in Maryland, where the ceremony was performed *in* the jet *on* the tarmac. What a logistical nightmare, especially for two guys who would have thrown one helluva party if they were legally able to get married in Ohio at a time when both men were healthy.

So, Jim and John married in a neighboring state that recognized their union. What transpired after the wedding is as loving as it is inconceivable. Loving because these two epitomized the perfect marriage. Inconceivable, because they learned that when John died—and he was going to die soon—Jim could not be listed on the death certificate as the surviving spouse. Maryland would recognize their marriage license, death certificate, and everything else that came with being a married couple. But they didn't live in Maryland. They lived in Cincinnati, Ohio—a city and a state that would not recognize their marriage nor list either as the surviving spouse on any death certificate. Keep in mind that Ohio was the state where they had lived nearly their entire lives.

Why were two upstanding United States citizens treated as second-class citizens and denied their inalienable rights, which are provided within the Declaration of Independence? That was the question that Jim and many others had and wanted an answer. The answer came via a Supreme Court decision on June 26, 2015. Jim and John's story is eloquently portrayed in *Love Wins: The Lovers and Lawyers Who Fought the Landmark Case for Marriage Equality*, written by Jim and his coauthor, Debbie Cenziper.

Landmark cases typically fall into the following categories: individual rights, criminal law, federalism, First Amendment rights, Second Amendment rights, Third Amendment rights, and the catch-all category of "other." For a fascinating walk through American history, check out Supreme Court Landmarks at www.uscourts.gov. The present book identifies ten key cases pertaining to the chapter topics.

Current Events

People can be easily persuaded by slick talk and pseudoscience. Both are detrimental to democracy. While many are cynical when it comes to politics, there is no denying that the 2016 United States presidential election left many people on both sides of the political aisle wondering "What happened?" No matter your political persuasion, these are the facts: Hillary Clinton surpassed Donald Trump in the national popular vote by nearly 2.9 million votes, but neither candidate won a majority (over 50 percent) of Americans. Across all 50 states and Washington, D.C., Clinton had 2,864,974 votes more than Trump, earning 48.2 percent of all votes cast. Trump earned 46.1 percent. This marked the fifth time in U.S. history that a candidate won the popular vote but lost the election. And this is the largest margin of any losing presidential candidate in U.S. history.

Why bring all this up? One reason is because since 2016, this country seems to be on a deep dive into believing constant, streaming misinformation and media bias. Fake news became a household phrase. But what exactly is fake news? Fake news is the spread of deliberate disinformation, hoaxes, and conspiracy theories via all forms of media transmission. It is akin to "yellow journalism," which is based on sensationalism and dates back to 1895 when an issue of the *New York World* published a cartoon of a child wearing a yellow dress to drum up sales. Adding color to print was not ordinarily done, so this attracted customers. Fast forward to today, and people don't know which news sources to trust because so much is viewed as fake. Yet, there are trustworthy sources. To check the most reliable news sources, turn to Ad Fontes Media at www.adfontesmedia.com and review the interactive Media Bias Chart. Ad Fontes Media derives its name from Latin meaning *to the source*. It was founded in 2018 by Vanessa Otero to analyze content from various media outlets and then rank it for consumers. Other sites that help provide guidance in identifying media bias is AllSides.com and USA Facts, a nonpartisan organization that provides a comprehensive view of data from government sources.

Today, creating news literacy is as important as creating science literacy; but doing so requires practice, comprehension, and critical thinking. It can also be time consuming. While navigating the news landscape can be tricky, finding reliable sources and supporting good journalism contributes to a healthy democracy.

To help steer in the right direction, know that scientific information is reported in published scientific papers in peer-reviewed journals. Here it gets a little complicated, but stick with me. There are three types of scientific papers: case reports, original research articles, and review articles. Finding such papers requires searching primary literature on databases such as PubMed.gov, Medscape, Up To Date, and Google Scholar. But if you are not a scientist, reading these articles will have little meaning, which is why scientific information gets sifted and presented to the average consumer in news sources. This is an excellent reason why we should use reliable sources to depolarize opinions. To assist, here is an alphabetical chart identifying far left-leaning, left-leaning, center, right-leaning, and far right-leaning news biases. Knowing political bents can guide you through the confusion.

News Biases

FAR LEFT-LEANING: Alternet, BuzzFeed News, CNN (online news opinion only), Daily Beast, Democracy Now, HuffPost, The Intercept, Jacobin, Mother Jones, MSNBC, The Nation, The New York Times (opinion), The New Yorker, Slate, Vox

LEFT-LEANING: ABC, Associated Press (politics & fact check), The Atlantic, Bloomberg, CBS, The Economist, The Guardian, NBC, The New York Times (news only), NPR (opinion), Politico, ProPublica, Time, The Washington Post, Yahoo! News

CENTER: Associated Press, Axios, BBC, The Christian Science Monitor, The Hill, The Independent Journal Review, MarketWatch, Newsweek, NPR (online news only), Real Clear Politics, Reuters, USA Today, The Wall Street Journal (news only)

RIGHT-LEANING: The American Conservative, The Dispatch, The Epoch Times, Examiner, Fox News (online news only), New York Post (news only), Newsmax (news only), The Post Millennial, Reason, The Wall Street Journal (opinion), The Washington Times

FAR RIGHT-LEANING: The American Spectator, Blaze, Breitbart, CBN, The Daily Caller, Daily Mail, Daily Wire, The Federalist, Fox News (opinion), National Review, New York Post (opinion), Newsmax (opinion), OAN (One America News)

Many reputable databases are also available to the general public. Scientists and non-scientists alike access these resources. In many ways, it's helpful to know where to look if you want to read the work of researchers and practitioners or simply want data on health issues. Here is a list of databases:

- American Community Survey
- Behavioral Risk Factor Surveillance System (BRFSS)
- Education Resources Information Center (ERIC)
- Google Scholar
- National Library of Medicine's PubMed

Because scientific findings often have political implications, news sources are important. Some lean left, some are centrist, and some are right-leaning. Science should be none of these. Subjects in this book are human biology issues that have definite political ramifications. Fortunately, you won't have to scour the professional literature searching for explanations about human body functioning. Read along and learn some underlying anatomy and physiology as it relates to issues of our society. Using fact-based debate and transparency, you can be better informed about issues and to make intellectually sound decisions.

We are currently living in a time in which our nation is divided, and not just sort of divided, but deeply and profoundly divided, across religious, political, economic, racial, and social lines. Some of this division is because we are like sheep that blindly follow. Some of it is because we are okay being misinformed. Some of it is because we don't believe real experts. Some of it is because we don't trust the government nor like to be told what to do by the government. Yet most of it is because we do not take the time to think critically about topics or we don't know where to go for the answers. Nonetheless, in order to save our democracy, we can no longer blur the lines between fact and opinion, science and snake oil. We don't want historians to look back on this time and label it the Intellectual Abyss Age.

The Intellectual Abyss Age is wrought with civic unrest. Our unrest is rooted in topics with biological underpinnings. These are topics that have been carefully studied and have scientific evidence supporting them, but many people in mainstream society fail to understand them. Prolonged disagreements, forceful views, conspiracy theorists, and legislators are at the forefront of such matters. Unfortunately, the matters seem to also be controversial, and that list of controversies is long. While some still should be debated, many so-called controversies should not exist because there are scientific explanations that make the facts of the situation clear.

Given all the facts, though, some people will still choose opinion over truth. For example, the Earth is an almost spherical planet. This is a fact. Yet there are people today promoting the idea that the Earth is flat. This is an opinion. It seems like a wacky opinion given all the images we have from space, but there are those among us who think of our marble as being flat. Of course, science hasn't answered all the burning questions of our time. Indeed, there is still so much that is not known. However, the tools of science do help us make sense of our natural world, while education enables us to think, question, and find personal answers as to what happens and why it happens while we are here.

Many scientific issues become political agendas which become controversial. The issues and agendas tend to rile our emotions and strike a chord somewhere within our core. Let's suppose, just for a moment or two, that we can live just sitting on our couches and eating chocolate. What a treat that would be. Sitting around, doing nothing, and streaming hours of Netflix. Oh, wait, we all did that during the COVID-19 pandemic. Pandemic aside, that might be quite a life, albeit non-productive and non-contributory. As long as we can live a good life doing it and we aren't adversely affecting society, why should anybody care? If we live in the land of the free, shouldn't we be unencumbered by actions that are meant to oppress? Yet, many policies and laws are viewed as regulating society by oppression. By their nature, they are intended to create civilized actions and impose penalties for offenses. Nowhere are they supposed to oppress personal freedoms that belong solely to that private citizen. But, if that law drains community resources and adversely affects your fellow humans, then that's a problem. And we see this at work on a large scale. Many laws are oppressing people—many are aimed at oppressing women, ethnic minorities, fringe populations, and lesbian, gay, bisexual, transgender and queer/questioning (LGBTQ) individuals. Why? My supposition is that we fail to apply science, stop thinking rationally, advance tribalism, and quit caring as deeply for one another.

So, here we are. Easily deceived, buying in to meaningless rhetoric, but knowing that we can be smarter if we pay attention to our surroundings. From a biological perspective, we're here simply to procreate and carry on the species or, in other words, to have sex and make babies. However, there must be more to life than this, though, because we can't just go around, pretend we're rabbits, and produce offspring. After all, rabbits can produce 100 bunnies per mating season—a feat not possible for humans. If momma rabbit lives 10 years, that's 1000 babies over a lifetime. And that's just from one rabbit. We can't do that. Another reason we can't breed like rabbits is because the bunny mating scene isn't all that terrific. Who really wants a mate to circle, flick a tail, and then urinate on you before commencing with sex that lasts at most 40 seconds? Rabbits also don't have the intellectual capacity of humans. That's where that big brain of ours must be useful for something. Turns out, it is useful—but only if we engage it. Another aim of this book is to show the intersection between thinking and science in framing our world.

As a society, we've been duped a lot in the past and in the present. We call the deception a hoax. The term *hoax* is likely derived from a contraction of the word *hocus*. You've heard it before when magicians are performing, as in, hocus-pocus, abracadabra, presto! Magicians use magic phrases to distract us from the sleight of hand. People today use their digital devices for trickery by advancing untruths. Now's the time to use that brain of ours, especially since it's using about 10 percent of the calories we eat!

A Popular Press Science Book Is Born

I have spent nearly 30 years teaching science and writing biology textbooks. What I discovered in my personal and professional life was that regardless of the course I taught or the people I met, people are curious. Some folks may not be interested in physics or chemistry, but if you tell somebody that you are an anatomy and physiology professor and textbook author, the litany of questions begins. Everyone—and I mean everyone—has questions about the human body and how it works. Along with those questions are lots of misconceptions. That's where this book comes in.

This book has been percolating for years. As I watched the media landscape become dominated by "fake news" and "alternate facts," it became increasingly apparent that this book's time had come. Once coronavirus took center stage, my inbox and message boards filled, further underscoring the necessity of getting easy-to-understand, non-political scientific information into the hands of the American people. Thus, I set out to write a straightforward book using clean, concise language from a scientific perspective. I know it may be naive to believe that people really will read every page and walk away with an informed view, but that is my hope. I still hope this despite knowing that research shows that knowledge of truth is no match for illusion of truth. Please approach the topics, apply reason, and become knowledgeable about scientific matters that are part and parcel of our everyday lives.

Tropism

Read a Facebook post, Twitter feed, or magazine headline and you'll see recurring, significant stereotypes. Stereotypes are widely held, but over-simplified ideas about a person or thing. Sexual and racial stereotypes never seem to go away. Much like stereotypes, we also have tropes, which are words or expressions used repeatedly, that appear in those same platforms. If you want people to believe what you say or write, you merely need to repeat it a few times, until it starts to seem true. This is a trope. What's oddly interesting is that our rating of something as being truthful increases the more we see it or hear it, even if it is total nonsense. If you hear day in and day out that all cows are brown—especially if you've never seen a cow—then you will be more apt to believe that all cows are brown. These repeated ideas become rooted, and then persist even in the face of new information. If you have been convinced that all cows are brown and then one day somebody tells you that this was a myth and that some cows are black and white and even shows you a picture, you are still much more likely to believe and state that all cows are brown, because it will be hard to overcome your deep-seated belief. If you hear repeatedly during presidential campaigns and printed in quotes from speakers that Mexicans are criminals and rapists, you may come to believe it. However, this is a racist trope as old as colonialism. In fact, data from Texas, the state with the second largest Latino population, shows that both legal and illegal immigrants have a significantly lower criminal conviction rate than native Texans. So, we can see why fake news is a problem and why we need tools to help us fight it. We need to use these tools and engage our brains because research shows that while we may be resistant, humans are capable of using evidence-based approaches to address problems. See? Stating again that this book can be helpful will start to reinforce the idea, and you'll come to believe it.

Sometimes tropes do emerge around facts. Unlike the cow example, such tropes are not about convincing people of something that is untrue but rather used to rally people. They are a cultural shorthand: useful for quickly contextualizing a topic of discussion, but imprecise. A trope of this type that has been repeated ad nauseum is related to "the one-percent." This phrase comes up during election seasons, running a constant loop among the political talking heads. Ever since I've heard this number with its percentage, I've tried hard to figure out *what exactly* are they talking about. Does it refer to how much the one-percenters earn? Does it mean how much accumulated wealth they've amassed? Does it mean something else? It's said over and over and over again, so you'd think the average person would have an answer, right? Do you know? It's difficult to know who politicians are referring to when they continually recite "the 1%." Determining the one-percenters is quite complicated. Commonsense tells you that it's the ultra-wealthy, the uber-rich, the super-elite. It's these so-called one-percenters that led to the Occupy Movement, an international progressive, socio-political movement opposed to social and economic inequality. Quite a valiant effort. But how do you oppose something that isn't easily defined?

Where does one find the answer? One place is from the Economic Policy Institute (EPI). They know who these one-percenters are. The EPI is a nonprofit, nonpartisan think tank created in 1986 to include the needs of low- and middle-income workers in economic policy discussions. According to a July 19, 2018, report by the EPI, income inequality has risen in every state since the 1970s and the incomes of the top 1 percent grew faster than the incomes of the bottom 99 percent in 43 states and Washington, D.C. Ah, so the one-percenters are the top income earners, right? But I need some real numbers to wrap my brain around. According to the EPI, the average annual income of the top 1 percent in the United States is $1,316,985. The average income of the bottom 99 percent is $50,107. Knowing actual dollar amounts helps me put facts to a trope. If you are super curious, you can look at a nice table that shows income inequality by state, metropolitan area, and county. But what number do I need in order to reach the *threshold number* to be considered part of the top 1 percent—not the average? That national minimum dollar amount is $421,926. Got it. The bare minimum to reach the top is nearly half a million dollars. If you're only interested in reaching the top 1 percent in your own state, note that considerable variation exists among the 50 states. For example, if you want to crack into the top 1 percent in my state, Ohio, you need $334,979 of annual income. The state with the highest threshold is Connecticut at $700,800 while Mississippi is last at $254,362. Knowing who comprise the 1 percent is relevant and important, because it helps us to elect public officials, who in turn, shape policy and the tax codes. We need to know more than just some mantra if we want to seriously address income inequality and not just build resentment. Think about this example the next time you hear or read a trope.

Intellectual Curiosity and Skepticism

Intellectual curiosity and critical thinking are underpinnings of this book. As such, here is an example to underscore the significance of critical thinking. What if you had a broken arm? Your friend who just finished a course in medical transcription tells you that the new treatment for fractured bones is to do nothing and let nature heal the disorder. This seems like sound advice to you because you've heard that nature works wonders. You put your arm in a sling you had lying around the house, take a couple aspirin, and decide to give it some time to heal. Your arm still hurts, so you increase the dosage of aspirin and the pain subsides. As a savvy consumer, you think, well, I'll check the Internet to see what the wise world of the web tells me. Within the global network of information found there, you read over and over and over that the new treatment for broken bones is to do nothing and let nature heal the fracture. Sounds good to you, because by now you've spent hours searching and you keep coming up with mostly this same information. You're satisfied with keeping your arm immobile. Then again, there's a nagging feeling that maybe you should see a physician because you've known people with a broken arm who were hospitalized and released wearing a cast. Your neighbor, who is a physical therapist, tells you that you really need to see a physician because doing nothing is not the best way to heal a broken bone. But, your other neighbor, who just finished a course in healthcare administration, and thus must know everything, tells you that the new way to fix a broken arm is to do nothing. So now you have two opinions that agree and one that disagrees. By the next day, your arm is swollen, so you apply some ice. Since you have the day off, you go back to the trusty Internet to get some more information. This time, the Internet algorithm shifted, and you find some sources that say to seek medical treatment with an orthopedic surgeon while other sources say that nature will heal the bones. Now, you have a conundrum. Your arm still hurts, and it is swollen, but you know things take time to heal. Do you stay the course or seek medical treatment? You pride yourself on being smart, so you decide to seek medical treatment. The doctor tells you that you have a bone fracture that will require surgery. She admits you to the hospital, sets the broken bone, casts the fracture, and tells you that it should be healed in about a month.

You might be thinking that this is a ludicrous example; however, this scenario illustrates the point about believing fake news. The physician treated the broken bone by following evidence-based medicine. Had you gone to reputable sites on the Internet to obtain your knowledge, you likely would not have treated the problem yourself. Knowing which sites are reputable isn't always easy, though. Especially when the topic is something with which you are not familiar. But you did have one clue here: a person you knew who had a broken arm didn't treat it herself.

As I write this, the COVID-19 pandemic has created an avalanche of Internet misinformation related to treating the virus. Some of it is very dangerous, causing leading health experts, established bastions of science, and manufacturers to release "do's and don't's" statements. When faced with stuff you read on the Internet, ask yourself if what you are reading really seems plausible. Think. Be a fact checker. Ask yourself why this would be true. Do you know it to be true? Do you think it's true because you've encountered it over and over before? Do you know of any evidence that refutes the claim? Familiar claims without supporting evidence can be dangerous. The scientific issues (some described as controversies) in this book are all rooted in familiar claims that are simply propaganda or nefariousness. These sorts of things threaten our lives and our society.

Here is another widely held belief: Emergency rooms/emergency departments have more visits on nights with a full moon. Have you heard of this? Research has documented that the so-called full moon madness is a myth. Emergency rooms can be busy places with lots of activity on most any night, but people may pay more attention to the chaotic atmosphere when they are aware there is a full moon and just experience heightened awareness. Heightened awareness is not the same as busier emergency rooms. What's quite astonishing is that 40 percent of medical staffs are convinced that lunar phases can affect human behavior, so more education and training doesn't necessarily make people resistant to myth. When scientific studies are done to test myths, we get real scientific data. In this case, scientific analyses have shown that moon phases, zodiac signs, and Friday the 13th have no influence on emergency room frequency or surgical blood loss.

Doing research in the face of uncertainty may seem obvious, but how do you know what you don't know? That's a tough question to answer. Scientists have a term for such self-awareness and self-understanding. It's called metacognition. It's also a very difficult thing to teach. Lack of metacognition often displays itself in test-taking. To illustrate, in the academic world, more students tend to visit professors' offices *after* taking an exam rather than *before* taking the exam. The reason? Poor-performing students seek help on topics after they've received a low test score. The overwhelming comment from students is "I thought I knew the material." Many factors could be at play here, including not knowing how to study, not being good at taking exams, or more likely, students—and others—have a hard time determining what they don't know. When students study, they tend to study the stuff they know really well because that's much easier to do and it boosts confidence. In reality, they should study the stuff that makes them uncomfortable, because that's where true learning comes in. Then they'll begin to grasp what it is that they *don't* know. Hence, they'll gain steps toward metacognition.

Outside of the academic world, how do you know what you don't know? If the topic relates to science, you can read reputable sources, listen to trusted scientists, be questioning, and approach works to build your knowledge.

One obstacle to self-education is the Dunning-Kruger Effect. David Dunning and Justin Kruger are cognitive psychologists who discovered that people of low ability have illusions that they really are doing a superior job. This illusion is a cognitive bias in which the person's perception of their performance is contrary to their actual performance. That is, they think they are doing a great job when in reality, they are performing at a mediocre level. Going back to the example of students performing badly on an exam, low performing students may *tell* you that they knew the material inside out, forwards, and backwards. However, when you ask them to explain something to you, they can't do it. Outside the classroom, this same tactic can be used. Ask somebody to explain why they believe something to be true. Doing so will reveal much. Many topics in this book tend to be controversial because opinions are made without knowing the actual science. Hope is not lost though, because once you know facts, controversies can fade.

The Dunning-Kruger Effect is evident everywhere in our daily lives whenever truly incompetent people display profound confidence. To combat the Dunning-Kruger Effect, we have to adopt an attitude of intellectual humility. None of us can be an expert on everything, so when a topic is outside our field or comfort level, we may need help to explore and approach some topics knowing that some truths are inconvenient. We have to look at reputable sources, admit to ourselves what it is we really don't know, and think critically. Like the brown cow or Mexican immigrant examples, just because we have repeatedly heard arguments or compelling stories doesn't mean they are correct. Lies and misinformation can spread as quickly as your gullible friend can hit the send button.

Learn to be skeptical. History is rich with hoaxes that even fool educated people. Examples include Sokal's hoax and the Dr. Fox Effect. In 1994, mathematical physicist Alan Sokal, submitted a sham article to the cultural studies journal *Social Text*. He knew it was a sham, wrote with tongue-in-cheek prose, peppered the article with meaningless discourse about postmodernism, and submitted his article. The editors did not pick up on his nonsensical approach and published the article in the spring/summer 1996 issue. It was satire.

In another example, which became known as the Dr. Fox Effect, two different speakers—one an actor and the other an actual scientist—gave lectures on irrelevant topics to a classroom of psychiatrists and psychologists, all of whom had MDs or PhDs. The actor was Michael Fox (not the *Back to the Future* film actor), who took on the identity Dr. Myron L. Fox during the test. The topic was "Mathematical Game Theory as Applied to Physician Education"—a topic that nobody in the audience had any background knowledge of. The actor was coached to use excessive double talk, non sequiturs, contradictory statements, jargon, and neologisms (newly coined words). He was also coached to be lively, humorous, and to demonstrate warmth toward his audience. The other speaker was a dry, no-nonsense academic who delivered a truthful lecture. After the lectures, the audience rated the professors. Students—the psychiatrists and psychologists—rated the sham professor higher than the genuine professor, thus demonstrating that charisma affects student ratings more than knowledge. Studies such as these have been replicated with the same results: nonverbal behaviors mask meaningless presentations. However, the studies have also shown that while the evaluation of the actor-professor may be higher, achievements on actual exams covering the subject matter were poor. Which totally makes sense: if you are lectured over nonsense, you can't do well on exams covering real stuff.

Similar to the Dunning-Kruger Effect is the halo effect, which is the belief that somebody is smarter than they are because of strengths in an unrelated area. Basically, the impression of a person we create in one area influences our opinion of that person in another area. This is why athletes and movie stars are often used to promote products that have nothing to do with sports or acting. It's a *cognitive bias* and it happens across life and in professional fields, including medicine. In medical practice, the halo effect is usually beneficial; the manner, attention, and care of a provider during a medical encounter, regardless of which medical procedure or services the encounter involves, has positive consequences. Numerous other studies have shown that personality, charisma, popularity, lecture fluency, non-verbal behavior, attire, appearance, and voice quality all affect our perceptions of a person—regardless of what they say or how knowledgeable they are. Sound familiar? As you'll see in Chapter 6 on vaccines, the halo effect has had great influence on the anti-vaccine movement.

We're also drawn to conspiracy theories, those beliefs that some covert but influential organization is responsible for a circumstance or event. We're so enamored with them that conspiracy theorist Jerome Corsi has two *New York Times* best-selling books: *Unfit for Command* (2004) (co-authored with John O'Neill) and *The Obama Nation: Leftist Politics and the Cult of Personality* (2008). Both books are loaded with inaccuracies. Now, if you read Corsi's educational pedigree, you'd want to believe him. Especially since he holds a B.A. from Case Western Reserve University and two degrees from Harvard University (MA and PhD). In 2018, Corsi was subpoenaed by the Mueller Special Counsel Investigation related to his association with Donald Trump and Roger Stone. Corsi's *Unfit for Command* (published by Regnery) is a book that details falsehoods about then presidential candidate John Kerry and his naval service. Although Kerry's fellow sailors released point-by-point rebuttals to Corsi's claims, the damage was done, and Kerry lost the election. The 2008 book *The Obama Nation* (published by Threshold Editions) followed the same format as *Unfit for Command* by discrediting Barack Obama's fitness for office by stating inaccuracies and making unsubstantiated false claims. Obama's team was able to counter the falsehoods before the election cycle and the damage wasn't enough to keep him from being elected.

Given Corsi's academic credentials, how is the average person supposed to trust what they read? First, if something sounds outrageous, it generally is. Second, it sometimes is wise to do a little homework and check the publishing source. Per their website, "Regnery Publishing is America's leading publisher of conservative books, serving conservative audiences with bold, bestselling books for over 70 years." The mission of Threshold Editions, founded in 2006, is to "provide a forum for the creative people, bedrock principles, and innovative ideas of contemporary conservatism." When authors search for book publishers, they seek out those that best match their style, catering to their target audience. This happens across all political and scientific ideology. After all, when you write a book, you want it to sell. Here is a chart identifying key publishers with their political bent, which are historically labeled as left, center, and right. Centrist and independent publishers (which are different from the publisher with a similar-sounding name, Independent Publisher) are not listed as they are too numerous to cite; however, they also dip into either sway because money is money.

Publisher Matched with Political Leaning

Liberal (Left-Leaning)	Conservative (Right-Leaning)
Haymarket Books	Crown Publishing
Independent Publisher	Liberty Hill Publishing
Liberalis Books	Regnery Publishing
Seven Stories Press	Sentinel
Verso Books	Threshold Editions

Before the Mueller investigation there was Pizzagate, the fake news conspiracy rumors manufactured by Trump campaigners and Twitter bots, that 2016 presidential candidate, Hillary Clinton, ran a pizza restaurant as a front for a child sex ring. Though no reputable news sources carried the story, Edgar Welch believed it and took his AR-15 rifle to the Comet Ping Pong Pizzeria in Washington, D.C., and fired shots. Fortunately, nobody was hurt, but in March 2017, the 28-year-old North Carolina man pled guilty to a federal charge of interstate transportation of a firearm with intent to commit an offense and a local charge of assault with a dangerous weapon. For many people, the news source does not matter. Nevertheless, news sources and rational thought must matter.

Science

For this book, I chose topics that come up repeatedly in the popular media. Many are the so-called "hot button topics" that strike emotional chords. Non-coincidentally, most of these topics are also the very same ones at the fore of politics and lawmaking. Because of that, it became apparent that the average person really was not aware that there are scientific answers to many debates. And lawmakers are either not very well-versed in science or refuse to accept facts.

To understand most every topic covered, we need to place each within the context of what science is and how science works. At its core, science is the intellectual discovery and systematic observation of the natural world. Sounds impressive. It involves observation and experimentation, both of which are removed from bias and opinion. That last bit is important. No bias. No opinion. Just an evaluation of the actual findings.

Let's look at how science is done. All science "works" in the same way, in that the basic principles applied to some areas should be applied to others. If something is found to be true—for example, the sun rises in the east—then this concept is true until something demonstrates it to be not true. Scientific explanations come about through the scientific method, through carefully designed experiments, and through repeatable results. A big difference between science and pseudoscience is that with science, the methodology is sound and well-documented, and the same results are obtained again and again no matter who performs the experiment. We explore this topic in greater detail in Chapter 1.

When thinking about how science is done and how controversies persist, it's worth mentioning confirmation bias. Confirmation bias is the tendency to interpret new evidence as confirmation of one's existing beliefs, theories, inclinations, or prejudices. Another form of confirmation bias exists when people avoid exposure to information that doesn't fit their biases. To test your confirmation bias, watch several different news sources and record how you feel after viewing each. At the same time, record some "facts" from each and check reputable sources to see if what was stated is actually true, false, or somewhere in between. To help you, reputable websites are found at the end of this book.

Tomes have been written about confirmation bias, and the topic has been studied and explored at great length among evolutionary biologists, sociologists, and cognitive scientists. Two recent books exploring why people continue to believe erroneous information and ways to correct wrong thinking include (1) *The Enigma of Reason, the Knowledge Illusion: Why We Never Think Alone* and (2) *Denying to the Grave: Why We Ignore the Facts That Will Save Us*. It's even more important to explore it here because this book is intended to provide facts to shape and inform your opinions. However, changing somebody's mind is difficult to do, despite mountains of information. This is where I ask you to consider the scientific method.

The scientific method removes personal bias. Given the manner in which scientific studies are done, the results are reproducible and done by researchers with no motives. Independent researchers the world over submit their findings, the studies are peer reviewed, others perform the experiments, researchers squabble, but the methodology is sound. Science prevails and society is advanced.

As in the case of Sokal's hoax, some studies with falsified data or that were influenced by funding sources, do slip through. This means that even in the best of circumstances, some bad or false information gets published, as in the case of vaccines causing autism (they do not). Generally, the scientific community quickly comes to the rescue and dubious articles get retracted. But once bogus information is out there, it's difficult to get it back—especially if you never see or hear about the retracted article. Science is an evolving process, building on previous studies, so as more is known, more will be published—even if it refutes previous information.

We're living in a time when all we seem to need to solidify our knowledge base is a friend or two who agrees with us, a person tweeting something, or strong convictions that we're unwilling to change. There must be a reason why humans will hold fast to false beliefs, even when they are dangerous. The example that comes to mind here is not vaccinating children against diseases.

Evolutionary biologists are seeking answers for our propensity toward holding on to misinformation or dogma. This resistance to new information must have some sort of adaptive function. Why don't facts change our minds? The answer may be related to our social instinct for cooperation. When our ancestors lived as hunter-gatherers in small groups, there was no advantage to reasoning. Reasoning took time and discourse. However, there was great advantage to winning arguments or convincing others of our beliefs, or in agreeing with a charismatic leader. In doing so, we immediately got what we wanted. Our lives were short-lived, so long-term consequences didn't matter. But we've come a long way since those days. Or, have we? Tribalism seems to be making a resurgence. That's where science must prevail.

Throughout time, we have relied on the expertise of others. After all, how many of us can really explain how some things we use on a daily basis actually work? Stop and think. Could you explain how electricity is generated at the power source and ends up in your house? Do you know the mechanism behind your flushing toilet? Examples are endless. So, it's now time to put emotion aside and pay attention to scientific facts, instead of confirmation bias, because not doing so is dangerous. Cognitive scientists Hugo Mercier and Dan Sperber give this scenario to bring home the point: Imagine if a mouse thought the way a human did. If a mouse was "bent on confirming its belief that there are no cats around," then it would soon be dinner. To dismiss evidence is threatening to our survival.

We humans can be peculiar creatures. Sometimes we accept a particular scientific teaching in one area, but don't accept it in others. Why do we accept science in some areas but reject it in others? Only you can answer this question for yourself. As each topic is presented, keep an open mind. We live in a society in which scientific advancements continue to shape our everyday lives. Think about going through your day without using a refrigerator or a smart phone or reaching into the medicine cabinet for an aspirin or prescription drug. Unless you live totally off the grid and are a total recluse, it is probably pretty hard to imagine going through the day without some of these scientifically-produced objects.

Here's another example of science affecting our daily lives. It's also an example of how scientific discoveries evolve. Can you remember not having a microwave oven? Perry Spencer, an engineer for Raytheon, is credited with inventing the microwave oven. The first commercial microwave oven hit the market in 1947, the first domestic microwave was available in 1955, and by 1975 over a million microwaves were sold annually. Today, 90 percent of American homes have a microwave oven. Phenomenal? Perhaps. But science was involved. Science also advanced microwave technology, and currently, microwave technology is used to diagnose and treat cancer. Cancer is an emotional, physical, and economic burden. With over 1.6 million cases diagnosed annually with nearly 600,000 people dying yearly of the disease, we welcome cancer research and any cancer advancements. This only happens through science and the scientific method. With each scientific move forward in cancer investigation, human science, biotechnology, or medicine, we are left with more things to think about. But, think we must.

TAKE-HOME MESSAGE

When you don't know something, look it up in *reputable* resources. Nurture your inner curiosity, be skeptical, and don't perpetuate misinformation. Aim to be scientifically literate.

Get your facts first, and then you can distort them as much as you please.
                    —Mark Twain (1835–1910); American writer, humorist

1

Blinded by Science

The Discipline of Facts

U.S. Landmark Case: Daubert v. Merrell Dow Pharmaceuticals, Inc., 509 U.S. 279 (1993); (9–0 decision)

In the late 1980s two parents of children born with birth defects sued Merrell Dow Pharmaceuticals, Inc. The parents (plaintiffs, with Daubert as the lead plaintiff) claimed that the mothers had used Dow's medicine Bendectin to overcome morning sickness and that this medicine in turn caused the birth defects. The plaintiffs presented unpublished, unreviewed evidence based on animal studies. Merrell Dow presented extensive scientific statistical studies which showed no such ill effects from taking this drug. The lower court did not allow the plaintiffs' expert testimony and directed a verdict for Merrell Dow. The appeals courts held likewise. Both courts based their ruling on a long-standing general acceptance test called the Frye standard, which is used for the admissibility of scientific evidence. Based on the Frye standard, expert opinion must be based on scientific technique, and the evidence must be generally accepted as reliable in the relevant scientific community (*Frye v. United States, 293 F. 1013 [D.C. Cir. 1923]*).

In 1993, the Supreme Court took the plaintiffs' appeal in *Daubert* and ruled unanimously that the lower courts erred in using the Frye standard. They remanded (sent back) the case to the lower courts to reconsider the disputed evidence using the guidelines of the Federal Rules of Evidence, Rule 702 (which had been written in the 1980s for federal cases) and further said that judges must act as gatekeepers and think like scientists using the same criteria the scientific community uses. Here are the four criteria: 1. Is the claim falsifiable and has it been tested? 2. Was the claim peer reviewed? 3. Has an error rate been determined? 4. Has the underlying science won general acceptance?

Eventually the plaintiffs' expert testimony was not allowed, based on the new standards, and judgment was granted for Merrell Dow. The major result of this case is that judges are granted more flexibility in determining what scientific evidence to allow into court to help the jury and the judge to make decisions. The key point of this case is that, in plain terms, the judge and jury can get all the relevant facts and information to make a legal decision, whether experts are brought in by the plaintiff, defendant or the judge herself.

Scientific Literacy

In this landmark case, understanding the scientific method was germane to the proceedings. While this case was about the morning sickness prevention drug Bendectin, it evokes memories of another drug, thalidomide, taken by mothers to treat morning sickness. Thalidomide (discussed in Chapter 8) was marketed between 1957 and 1962 and resulted in babies being born without arms, partially developed and twisted limbs, undeveloped organs, and other deformities that resulted in death. Thalidomide had ties to Nazi concentration camps where human trials of the drug were carried out. Although the drug was sold worldwide, the FDA never approved the drug for use in the United States, thanks to Dr. Frances Oldham Kelsey, a general practitioner who in 1960 began a long and distinguished career at the Food and Drug Administration. She had concerns over the lack of clinical trials and testing on pregnant animals, and insufficient evidence of adverse effects. She withheld approval despite pressure from the drug's manufacturer. This is an FDA success story, because scientists were carefully watching infant deformities increase around the globe in women who used thalidomide, thus there were only about 17 cases reported in the United States.

Reaching a level of scientific literacy requires reading appropriate sources, discussing, and thinking. If the COVID-19 pandemic has taught us anything, it's that armchair scientists equipped with the most recent YouTube video or viral "scientific paper" made their voices known, causing great distress for a population already experiencing extreme anxiety. It's dangerous when people think that the natural world can be explained by personal experience or belief systems that directly clash with science. At its core, science aims to explain the natural world through observation and repeated experimentation.

Solid science is not fake, and fake science is not real. Despite the word "fake" being used over and over again within mainstream news cycles or as a rallying cry for something that doesn't fit with a current belief, scientific information is something we can count on. We get to scientific knowledge through science education. Think of science education as inoculation against ignorance, and such inoculation has been known for centuries. Consider the work of Francis Bacon (1561–1626), an English statesman and philosopher who advocated for the inductive method in his notable work *Novum Organum Scientiarum*. The Latin title means *new instrument of science*. Written in 1620, Bacon identified various idols—Idols of the Tribe, Idols of the Cave, Idols of the Marketplace, and Idols of the Theatre—that clouded logical thought. From his perspective, these idols were tendencies that got in the way of logical thinking. His remedy was the scientific method.

Science education and scientific literacy are critical to our times. According to the National Center for Education Statistics, the 2016–2017 (the latest date for which data are available) adjusted cohort graduation rate (ACGR) for public high school students was 85 percent. ACGR is an indicator that examines the percentage of students who graduate on time. This is good news because this is the highest it has been since ACGR was first measured in 2010–2011. The states with the highest ACGR are Iowa and New Jersey at 91 percent, while the lowest is New Mexico at 71 percent. Compare these statistics with 2007 research done at Michigan State University that showed that approximately 28 percent of American adults qualified as scientifically literate. Another study done at that same time showed that 70 percent of Americans could not read and understand the science section of the *New York Times*.

Fast forward a few years to more recent studies conducted between 2015 and 2018 and reported by the Pew Research Center in March 2019 which show that science knowledge remains strongly tied to education, as those with higher education levels are more likely to answer science questions accurately. That seems like a no-brainer, but we see differences associated with ethnicity, gender and political party even when education is the same. For example, white people score higher in scientific literacy than do black people or Hispanic people; men score higher than women; and Democrats and Republicans score about the same, but conservative Republicans and liberal Democrats tend to score higher than their more moderate counterparts. This may not sound concerning, but if we look at many of the contentious or controversial issues that have found their way into politics, many of these so-called issues shouldn't be issues because they have answers embedded in science. That is, there is no need for controversy because science has already offered an explanation. However, we're currently facing a political threat to the role of science in American society. In June 2019, the White House issued an Executive Order slashing the number of federal advisory committees, including those that inform decisions on public health, thereby reducing the amount of evidence-based information available to government officials.

What is scientific literacy, and why is it so important? Scientific literacy has several definitions, ranging from having pure knowledge and understanding of science to having the ability to find answers to being able to use scientific facts in making personal choices and informed decisions. Did you get all that? Basically, it's finding the facts and being able to use those facts as we swagger through life.

In this book, cases are made for scientific literacy and scientific knowledge. Yet, we are all humans who have a difficult time setting aside personal beliefs, feelings, and agendas. To illustrate, research has shown only a modest positive relationship between science knowledge and people's support for science. The issue of climate change starkly illustrates the importance of scientific literacy and knowledge. On the left side of the aisle, alignment with what science knows—that climate change is real and mostly caused by human activity—is aligned with respondents' level of scientific knowledge. To illustrate, 93 percent of Democrats with a high level of scientific knowledge said climate change was mostly due to human activity while only 49 percent of Democrats with a low level of scientific knowledge stated climate change was due to human activity. But the other side of the aisle shows that beliefs can beat science. Scientific knowledge level had no effect on Republican beliefs: Republicans with a high level of scientific knowledge were no different than Republicans with a low level of scientific knowledge. Both groups stated overwhelmingly that climate change was *not* the result of human activity. That means that scientific literacy does not necessarily have a direct relationship between scientific knowledge and personal positions.

At the time of writing this book, a Pew Research study found that the average score on scientific literacy in the United States was 7.9 out of 12—that's about 66 percent. On the common A, B, C, D, F grading scale, that is a solid D. If C is average (70 percent), we are below average. That's a little embarrassing for a highly-developed, industrialized country that boasts some of the greatest universities.

Here's another startling statistic: Scores obtained from the 2018 Program for International Student Assessment (PISA) found that 15-year-old students in the United States rank 37th out of 79 countries in mathematics, 18th out of 78 in science, and 13th out of 78 in reading. The PISA is a cross-national test that is given every three years to measure math and science literacy, reading ability, and other key skills among 15-year-old students in developed countries. Here's an interesting breakdown: China ranked first in math, science, and reading; the Dominican Republic ranked last in math and science; and the Philippines ranked last in reading. Ranking above the United States in science were the following countries: China (four provinces), Singapore, Macau (China), Estonia, Japan, Finland, South Korea, Hong Kong (China), Taiwan, Poland,

New Zealand, Slovenia, U.K., Australia (tie), German (tie), and Netherlands (tie). The next five countries ranking *after* the United States were Sweden (tie), Belgium (tie), Czech Republic, Ireland, and Switzerland. If you are an American, how does this make you feel?

As an educator, I want my students to have the skills to think critically about statements and determine whether the content is true. As an average person, I want my fellow citizens to be able to do the same. If we as a society cannot determine whether statements have merit, then we should be equipped with the know-how to seek the answers from reputable sources. This ability is important to scientific literacy: we can't know everything about every topic, but we should all possess the skills to question dogma. We should all be able to know real news from fake news. We should all be able to legitimately defend our thinking. This is where science and the scientific method come in.

There's much to be gained from independent thinking when we don't subscribe to a political party's belief system. While writing this book, I came across a sentence in a work of fiction, *Cutting for Stone*, that sums up the human mindset. Here it is: "Ignorance was just as dynamic as knowledge, and it grew in the same proportions." As you read this book, please consider that passage.

How Science Works

It's important to know how science works so that we can better inform ourselves. To that end, let's talk about the scientific method. It is not magic. It does not involve a supernatural being. There is no crystal ball. When dealing with studying the natural world, scientists take two basic approaches: (1) observation and (2) experimentation. The ability to notice things and to notice significant details about something in order to gain information is key to observation. After careful observation and discovery, inductive conclusions can be made. Inductive conclusions are those reached on the basis of evidence and reasoning. Observations lead to questions, which then lead to more observations and experimentation to answers those questions. This latter approach leads to scientific experimentation and the utilization of the scientific method, which underlies science and the answers derived from science.

The scientific method is a time-valued, essential process. It is a series of procedures consisting of:

- observation
- measurement and experimentation
- forming hypotheses
- making predictions
- testing
- obtaining results/data
- analyzing the results/data
- being able to reproduce the results

This is what is commonly referred to as "doing science." Science is "done" so that we can find answers to questions in our physical world. There is a lot to it.

Scientists formulate hypotheses and conduct experiments. Hypotheses are conjectures to confirm or refute the experiment. At every step, explanations are based on the evidence. In the common vernacular, the word hypothesis has a very different meaning than it does in science. Using everyday language, "hypothesis" is often misinterpreted or misrepresented as an assumption—something that is understood to be true without proof—or speculation—guessing without firm evidence, which is the available body of facts that show whether a belief is true or valid.

But these are not scientific definitions of hypotheses. This is an important point to underscore because it devalues the process of science. These are definitions of a postulate, which is something that is taken to be true without proof.

Hypotheses, on the other hand, are tentative answers. Because they are tentative, they must be further investigated with more study and research. Research is a rigorous process that seeks to find truth and answers. Sometimes the research goes nowhere; sometimes it takes the scientist on a totally different course; and other times it leads to more research along a common thread. When it leads to further research along the same line of questioning, conclusions can be drawn.

Just because one researcher or a team of researchers comes up with something, it doesn't necessarily mean that their findings will become dogma or that these findings are considered incontrovertibly true. Instead, these findings are subjected to another test called peer review. Peer review is the process of evaluating research proposals, grant proposals, publication manuscripts, and abstracts for presenting at scientific meetings. Part of the peer review process is determining if something is valid, reliable, and statistically significant. When conducting research, results have internal validity if they are interpreted accurately, and results have external validity if they can be generalized to the population. Reliability is the consistency of the research and the extent to which it can be replicated, meaning that utilizing the same methods and conditions, other researchers will arrive at the same results. In order for any study to be valid, it must be reliable first. When researchers are analyzing results, they look for statistical significance. Simply put, statistical significance means that the results occurred by something other than chance; for example, a treatment really had an effect. Findings from research should be made available to the scientific community at large. The scientific community crosses country borders and peoples. The research proposals, grant proposals, manuscripts, and abstracts are judged for technical and scientific merit by other scientists and professionals working in the same field. For example, you want people who are well versed and educated on a topic to ensure that crackpot scientific ideas do not come to the fore and become the standard in science. Remember, we want to be sure that the information is well vetted. Our goal is for the intellectual, systemic study of structure, function, and behavior in the natural world through observation and experimentation. We don't want an opinion that is not based on fact to inform our decision-making.

Keep in mind that the scientific method allows for continuing research and broadening knowledge. During the COVID-19 pandemic, there was a constant information whiplash: no need to wear masks, wear masks, only some need to wear masks, everybody wear masks. This was fueled in part because pre-press papers were viewed on the Internet by the general population and those armchair scientists who were reaching premature conclusions. Pre-press articles are published online ahead of the peer-review process so that a greater scientific audience has access to the information. However, until the papers have undergone the peer-review process, they cannot be relied upon for the best information. While some papers may come to the top and indeed be valid, we don't know that until they have been through the tried-and-true process. Another issue at stake during the pandemic stemmed from the coronavirus being "novel." The phrase "novel coronavirus" was stated so often that people forgot that "novel" meant "new," and scientifically-speaking, this meant that not much was known about it. So, any time a discovery was made, it was new. While coronaviruses have existed for quite some time, this *particular* coronavirus, SARS-CoV-2, had not been around very long. In fact, the acronym COVID-19 stands for **co**rona**vi**rus **d**isease, and the 19 indicates the year it was found, 2019. Using our evidence-based approach utilizes relevant information from peer-reviewed research to address specific issues. That is, the simple rules of science and peer-review, not opinion, must be applied to determine the validity of the information.

At some point within the scientific method, a theory *may* be developed. Again, this is another term that has entered mainstream language and has come to be used colloquially as meaning an idea to explain or to justify something. Scientists refer to this as an "everyday theory." Everyday theories are based on personal experience or personal knowledge. Everyday theories haven't been proved. This is not the definition of a scientific theory.

Scientific theories are backed by a substantial amount of scientific evidence. An example of an everyday theory was found in the Introduction's discussion on emergency room (ER) visits. Many theorize that there are more ER visits during a full moon than on other nights. Although many people believe this to be true, scientific studies have found no direct correlation between the lunar phase when the moon is full than another night in the ER. The ER is often a hectic place, and people may just pay particular attention to the activity on nights of a full moon. However, it is an incorrect assumption to make that the moon's phase affects medical crises. By the way, the full moon is also not associated with witchcraft, werewolves, or increased crime. The moon has no mystical powers, but it and the sun do cause ocean tides.

A scientific theory is a reasoned explanation drawn from the facts of robust study. Scientific theories have been replicated through experimentation so often and the results obtained consistently and reliably that scientists can agree on a theory for a physical phenomenon. Thousands of experiments support the hypothesis and this hypothesis has never been disproved. An example from human biology is the cell theory, which states that all living things are made of cells. Another example is the germ theory of disease.

Scientists are not married to their hypotheses, though. If new evidence comes along that refutes the hypothesis, that changes the whole game and the process repeats itself. Like life itself, science is dynamic. Scientists have no problem changing previous "facts" if new evidence comes to the fore. Many people don't trust science and scientists because they hear one thing one time and another thing some time later. Besides wearing face masks, another case in point is eggs. You might have thought eggs were bad for you and then you read another study that says eggs are good for you. The egg dilemma has its roots in nutrition science, which is a relatively new field of study. As such, new findings come to the headlines quite often. Nutrition is also an area in which we know individual biology has to be considered. Stay tuned on that topic.

Let's look at the germ theory of disease. Today, we know that infectious diseases, like common colds, are caused by microorganisms, also called microbes. Microorganisms are germs that are too small to be seen without a microscope. This theory didn't happen overnight. It was first proposed in 1546 by the Italian physician and scholar, Girolamo Fracastoro. It was further advanced by Austrian physician, Marcus von Plenciz in 1762. That's right, 216 years later! However, it wasn't until around 1850, building on the work of Louis Pasteur and Robert Koch, that the germ theory of disease was accepted. Pasteur (1822–1895) was a French chemist and microbiologist and Koch (1843–1910) was a German bacteriologist. Note the international effort. Their work helped shape the science and provide the compelling evidence that infectious disease was caused by germs—not by supernatural beings—and thus the theory was born. Incidentally, Koch received the Nobel Prize in Physiology or Medicine in 1905. Even acknowledgment took a while.

In science, there is a progression of research, and at times laws are developed. A *scientific law* is a statement of fact that details a sequence or relation of physical phenomena that does not vary under a set of conditions. These laws are based on empirical data and many laws are encapsulated as mathematical equations. Both scientific theories and scientific laws are generated using the scientific method. Building on our previous bullet-point list, here's a flowchart demonstrating the scientific method, beginning with observation and previous experimentation and ending with scientific law:

Observation and Previous Experimentation
↓
Hypothesis
↓
Experiment
↓
Data Collection
↓
Data Analysis
↓
Determine Data Biases

   —If yes, then redesign experiment
   —If no, then move on to next step

Refine Hypothesis
↓
Repeat Experiments

   —If experiments are repeatable, then move on to next step
   —If experiments are not repeatable, then propose alternate hypothesis

↓
Accept as Theory
↓
Time, More Experimentation, and High Level of Confidence
↓
Accept as Law

Below are some examples of theories and laws you likely have heard of. The first and the last items, which are related to evolution and natural selection, are often considered controversial because they are contrary to some religious teachings. Evolution and natural selection will be discussed in Chapter 3.

- Big Bang Theory—explains how the universe formed through a rapid expansion of matter from a state of extremely high density and temperature that marks the origin of the universe. In essence, a fireball of radiation at extremely high temperature and density occupying a very small volume formed about 13.7 billion years ago. This fireball expanded and cooled, and its subatomic particles condensed into matter that accumulated and formed the galaxies and stars. In present time, these galaxies are still retreating. The residual radiation cooled and can be detected as background microwave radiation.
- Universal Law of Gravitation—any two objects, despite their mass, will exert a gravitational force toward one another. Gravity is what keeps us grounded. Literally.

When we jump off a diving board toward the swimming pool, we can be assured that we will land in the water and not up in a tree.

- Newton's Three Laws of Motion—an object in motion stays in motion; there is a connection between an object's mass and its acceleration; for every action there is an equal and opposite reaction. You can experience this if you are riding a bicycle and slam on the brakes. The bike stops, but you topple forward.
- Archimedes' Buoyancy Principle—the force acting on a submerged object equals the weight of the liquid that the object displaces. This explains how huge lake freighters and cruise ships can stay afloat.
- Theory of Evolution and Natural Selection—evolution through natural selection accounts for the diversity of life. The concept of evolution has been around for a very long time. First proposed by ancient Greek thinkers, it was rejected in Europe because it was contrary to religious teachings. Jean-Baptiste Lamarck (1744–1829) was a French naturalist and early evolution proponent who stated that organisms transformed in response to their environment. Sort of like putting on a coat when you get cold. Through his studies, he knew something was going on, but he couldn't explain it.

Lamarck's work was advanced through scientific study by Scottish geologist Sir Charles Lyell (1797–1875). Once again, note the international contingent picking up the scientific ball and running with it. As a geologist, Lyell thought that the Earth's features, mountains, valleys, waters, geological deposits, and the like, were shaped over long periods of time by elemental forces that changed the face of the planet. Think of this as early plate tectonics, which is the theory that explains the structure of the earth's crust and many associated phenomena resulting from the interaction of lithospheric plates (rigid outer part of the earth) that move slowly over the underlying mantle (region of the earth's interior between the crust and the core). The foundational work by these scientists helped shape Charles Darwin's theory of evolution.

Darwin (1809–1882) was an English naturalist who proposed a theory of evolution by natural selection. Darwin's formal training was in religion, and he was expected to become a vicar. His hobbies, however, were of the natural world and included such activities as beetle collecting and geology. His unpaid position aboard the *Beagle*, a ship whose purpose was to circumnavigate the globe, was as a "gentleman naturalist." His duties were to collect things of scientific interest and send them back to John Stevens Henslow (his friend and mentor), Lyell, and other luminaries of the scientific community. He was also supposed to be a good conversationalist for the captain.

Darwin proposed that evolution was gradual and was shaped by natural selection, meaning that some organisms were better adapted to an environment and thus more likely to reproduce. Remember the opening section about our purpose in life being to produce descendants? With time and advances in cell biology, molecular biology, and genetics, his theory has been modified and shaped and serves as one of the unifying concepts in biological science.

Something we take for granted, like germs making us sick, is rooted in science. In order to arrive at this conclusion, experiments were done. Experimentation began with observation and making a hypothesis. Then a further line of questioning was developed, and methods were outlined to conduct experiments. The results from those experiments were obtained and analyzed, and finally, conclusions were drawn. In human biology, these are the criteria for analyzing the world we live in and for determining whether something is true or not. In the field of medicine, it is known as evidence-based practice.

Science, Religion and Life

Sometimes it is helpful to know where words come from or to break down words into their word parts. To illustrate, the word *science* is derived from the Latin term *scientia*, which means "to know." Knowing something is not the same as *believing* something. Make no mistake, science is not a belief system. To believe something generally means you have faith in something or strong feelings for something; many belief systems are generally linked with religion and feelings are linked with experience.

Although science and religion may not seem closely associated, social scientists have been studying the origins of religion for quite some time. Specifically, scientists have explored the reasons why humans ascribe supernatural powers to natural events. As humans evolved, they started to attribute meaning to things in the world. If something couldn't be explained rationally or through science, then some supernatural power must have been at work. Many ponder the meaning of life. We naturally assume that we and everything around us acts for a purpose. Some scientists propose that the tendency to explain the natural world through the existence of supernatural powers is the foundation for religious beliefs. Others propose that religious beliefs are an adaptive trait of humans. Regardless of its origins, religion is still with us today and is shaping policy in America and the rest of the world.

Consider this statement: More people have been killed in the name of religion than for any other reason. Is this true or false? Remember, you are supposed to think. In *Encyclopedia of Wars* by Charles Phillips and Alan Axelrod, they list 1,763 wars. Good grief, that is a huge number of battles. Yet, of those 1,763 wars, only 123 had a religious cause. That's less than 7 percent of all wars and less than 2 percent of all people killed in war. Finding that hard to believe? Are you thinking, oh, that can't be? The Crusades, aimed at recovering the Holy Land from Muslim rule, were a series of holy wars organized by Christians. The Spanish Inquisition, directed against converts from Judaism, Islam, and Protestants, was a holy war established by Catholic monarchs. The conflict in Northern Ireland was rooted in religion. The partition of India and Pakistan in 1947 was based on religion, and close to a million people were killed. The attacks of 9/11 were definitely motivated by religion. The list goes on. However, the total number of lives lost in the aforementioned religious wars is less than the 35 million soldiers and civilians who died in World War I. These numbers are still atrocious, but these may be examples in which you might have heard something so often that you believe it to be true. If you stop to think, though, can you name *all* the wars? Of those wars, would you be able to name all those fought in the name of religion? Point taken? Most wars are caused by greed, unbalanced power, and hatred. It's time to peace out, people.

Religion is part of a complex web of life. Given that it has so many roles, there is a reason that religion—and not science—is the opiate of the people. Religion provides many with a purpose in life and a feel-good sentiment—despite the fact that the Christian Bible chronicles numerous wars. It gives hope in troubled times and comfort for sadness. Religion helps many people understand that we're part of a greater universe. Sometimes, though, it is used as a weapon of oppression. Because people are, well, people, some folks have an uncanny ability to tease out take-home messages and decide what is historical truth or spiritual balm. There are also those who believe everything in life is part of God's plan. Others subscribe to religion and religious freedom, but only if it is *their* religion. A more worldly view may help. For example, in *Why I Am a Hindu,* author Shashi Tharoor writes, "I can honour the sanctity of other faiths without feeling I am betraying my own." As a writer of science and a person who spent her academic career working at a Catholic institution, I think it is important to show that there is space in our big wide wonderful world for both scientific inquiry and religious thought, as long as neither one is oppressive, each seeks the truth, and both approach humanity with kindness.

Science is connected to our everyday life. This book focuses on scientific issues pertaining to human biology. As stated in the opening, many public controversies or contentious issues exist because *opinion* is oftentimes confused with *facts*. Science is powerful because it depends on evidence that comes from quantifiable observations that can be replicated repeatedly by scientists in different parts of the world. While many controversies have religious connections, we have to step back and realize that religion is not science because religion is not testable. It is a belief system based on faith. Although many people may hold the same religious views, we are still unable to test these views or assumptions using the scientific method. For these reasons, we have to separate what we *believe* to be true from what we *know* to be true.

With respect to religion, some research has found that students in countries with higher levels of religiosity perform lower in science and math. So, if a population is highly religious, that group does worse on science and math tests. These studies used aggregate data. Yet, we all know plenty of highly-intelligent people who are both scientific and religious. In fact, many quality institutions of higher education are sponsored by religious orders.

It is interesting to note that when the Pew Research Center polled Americans on religion and public life, they found that 53 percent of adults in the United States say that religion is important to their lives, 24 percent stated it was somewhat important, 11 percent not too important, 11 percent not important at all, and 1 percent just didn't know. While volumes of publications have been written about the role religion plays in our lives, it's not the focus of this book. Religion and science can coexist peacefully, and lots of good scientific research and education advancement is done at religious-affiliated schools and universities.

However, we must be concerned about the low level of scientific literacy in this country because it has consequences for the health of our citizens and the economic impact of our country. Here is a real story that underscores this last statement. According to the news site *Axios*, during Trump's presidency, he suggested that nuclear bombing could weaken hurricanes. Nuking a hurricane would not work because hurricanes produce much more energy than a single nuclear bomb. There is an energy mismatch. But, a world leader—one with codes to nuclear weapons—made this assertion. This was fact checked, and according to the Hurricane Research Division of the National Oceanic & Atmospheric Administration, hurricanes release 50 trillion to 200 trillion watts of heat energy. This is equivalent to a 10-megaton nuclear bomb exploding every 20 minutes. Moreover, this is a terrible idea because trade winds would carry the nuclear radiation across the globe, devastating and annihilating life.

Science is a part of life and can be applied to cinema. *Breakthrough* is a 2019 faith-based movie about a boy who drowned in icy water and was revived by prayer. The film is an adaptation of the book *The Impossible: The Miraculous Story of a Mother's Faith and Her Child's Resurrection* in which the mom prayed for her child's recovery and he was resurrected through prayer. What likely caused the boy to recover is something known as the diving reflex, a physiological event that occurs when a body has been submersed in cold water. This topic is discussed in Chapter 9.

Science can also be applied to prayer. Let's explore the concept of intercessory prayer, which is the act of praying to a god on behalf of another person. Does this work for both good and evil? A few scientific studies have been done to test the effect of intercessory prayer. In 2001, a double-blind study conducted at the Mayo Clinic randomized 799 discharged coronary surgery patients into a control group and an intercessory prayer group. Those patients in the intercessory prayer group received prayers at least once a week from 5 intercessors per patient. The study took place over 26 weeks and measured "primary end points" such as death, cardiac arrest, and rehospitalization. At the end of the study, the researchers concluded that intercessory prayer had no significant effect on medical outcomes. A 2007 systemic review of 17 intercessory prayer studies found "small, but significant, effect sizes for the use of intercessory prayer" in 7 studies, but "prayer was unassociated with positive improvement in the condition of client" in the other 10. The researchers concluded that intercessory prayer must be considered an experimental intervention.

But let's be clear. Both religion and science have enhanced human life. It seems that religion has been around for nearly as long as humans have existed. The measurable effects of religion show that those people who attend regular religious services and events benefit from strong social networks. This is powerful. Byproducts of strong social networks include less loneliness, depression, and drug use and greater life expectancy. In short, belonging to a group of like-minded individuals has positive health benefits. It may not make you smarter, but it might make you live longer. Likewise, scientific advances have also contributed to human health. We live longer now because we can thwart many diseases as a result of medicine, which is rooted in science.

Advances have also raised ethical issues, which oftentimes bump up against religious beliefs. As a society, we cannot discount either scientific advances or religious beliefs. We are all people living together, sharing the same resources, and pursuing many common goals.

Here are some basic foundations for all of us to consider: (1) Nobody can control where they are born; (2) Just because we *can* do something, doesn't mean we *should* do it. What we should do as a society is gather all the scientific evidence, think about societal implications, and move forward with ethical reasoning. We need good science, sound ethical practices, and an educated populace so that public policy can be shaped in the most positive manner for *all*.

The subsequent chapters outline controversies, contentious issues, and thought-provoking topics that may be at odds with beliefs. Because public opinion is often divergent from reality, please approach the upcoming chapters with an open, objective, unprejudiced mind. Then, think and make informed decisions based on the best available information. Read on to explore the science behind the headlines.

Take-Home Message

Hands-on approaches to problem solving and learning are often used by people to make decisions. Examples of such approaches are using educated guesses, stereotyping, and intuitive judgments to arrive at conclusions. Science uses observation, experimentation, and replication to arrive at conclusions.

> *If pundits tell you something is true, and you don't have the critical skills to evaluate their statements, then you are putty in their hands, as they shape your world view to their will.*
>
> —Neil deGrasse Tyson (1958–); American astrophysicist

2

Hello Dolly

Genetics

U.S. Landmark Case: Association for Molecular Pathology v. Myriad Genetics, 569 U.S. 576 (2013); (9–0 decision)

This case involved a company, Myriad Genetics, that identified the precise location and sequence of two genes, BRCA1 and BRCA2. Mutations of these genes dramatically increase the risk of breast and ovarian cancer. With this knowledge, Myriad Genetics could screen patients and offer insight as to their risk for developing breast or ovarian cancer. Myriad Genetics also created synthetic complementary DNA (cDNA). Moving forward, the company obtained patents on the discovery and isolation of BRCA1 and BRCA2 genes as well the cDNA they created, preventing others from using their findings and information. These patents caused immense controversy in the scientific community and among healthcare organizations, physicians, and individual patients. Finally, a lawsuit was filed arguing that naturally occurring DNA cannot be patented and that the cDNA was not actually synthetic. After lower courts' decisions favored Myriad, the case made it to the U.S. Supreme Court. Their ruling was that a naturally occurring DNA segment is a product of nature, thus it is not patent eligible, and merely isolating it does not change this basic fact. However, cDNA is patent eligible because it is not naturally occurring.

Principles of Genetics

It's often difficult to determine who owns what when it comes to something as personal as our own genes. So, when a company uses cells from real people to identify the sequence of genes, identifying "ownership" gets a bit messy. For these reasons, understanding baseline genetics is essential, especially as science advances in the name of humanity.

But when it comes to genes and genetic modification, the general public perceives genetic manipulation as something that results in designer babies. Or genetically-modified food is deemed less safe than conventional food. Many people feel that genes in humans and plants should not be altered, while others see the value in editing a few genes. In addition to scientific endeavors, genetic fiddling also occurs naturally. Nature itself modifies both animal and plant genes. Because scientific manipulation of genes allows for enhancements in medical treatment, food supply, and drug development, it remains a viable option for life on our planet. Since opposing views exist, it makes sense to start with describing what genetics really is in order to dig deeper into the morass.

Genetics is the branch of biology concerned with the study of genes. Famous geneticists include names you probably know like Gregor Mendel, James Watson, Francis Crick, and Barbara McClintock. As a scientific study it can be fascinating because it searches for information that makes each of us who we are. That genetic code influences our traits such as hair color, eye color, and disease risk, among other things. It also explains why we differ from each other. Central to genetics are our genes—those "things" we inherited from our biological parents. Genes are made of DNA, so at its core, genetics is about DNA. This means that any discussion of genetics has to begin with the genes and DNA. Textbooks, tomes, and courses centering on genetics abound, but all we need is a rudimentary understanding.

Genes are discrete segments of DNA molecules that make us who we are and what we are. We're not talking alternate forms like Lee, Levi, or Calvin Klein. We're talking about little sections of DNA that influence traits like height, weight, skin color, personality, and some intelligence. Detectable effects such as these are controlled by gene expression, and genes express themselves in a variety of ways. For example, there are dominant genes, recessive genes, codominant genes, and sex-linked genes. Before continuing, it's wise to have a little vocabulary lesson. Brace yourself, it's a little rough, but in order to appreciate the gravity of genetics, we need some common terminology to frame the discussion. Keep in mind, you can always look back to the definitions. Even Neil deGrasse Tyson had to use the words of the discipline when explaining astrophysics. Roll up your sleeves for a concept list and vocab check:

- Think of genes as basic words of life's instruction manual. The word *gene* comes from the Latin term *genos*, meaning *birth*. For example, various genetic features like black hair and green eyes are encoded in parental cells and transferred to their offspring. Think: mom and dad pass down some traits that their kids inherit.
- Heredity is the genetic characteristic passed from one generation to the next within genes. The term *heredity* is derived from the Latin word *hereditas*, which means inheritance. So, we can use the terms heredity and inheritance interchangeably.
- DNA is the abbreviation for **d**eoxyribo**n**ucleic **a**cid, a self-replicating molecule that carries genetic instructions. It is found packaged and organized in our chromosomes (defined next). Each DNA molecule consists of a double-stranded helix that is shaped like a spiral staircase. Each step of the staircase is made up of chemical bases called nucleotide base pairs. These bases are adenine (A), thymine (T), cytosine (C) and guanine (G), and they combine in specific pairs so that one sequence on one strand of the helix complements the other strand. For example, A combines with T and C combines with G. The specific sequences in which they combine make up our genetic information.
- Chromosomes are thread-like structures made of protein and DNA. The number of chromosomes in an organism varies from species to species; humans have 23 pairs of chromosomes, but bacteria and viruses have a single chromosome. On the other end of the spectrum, hermit crabs have 127 pairs. Our beloved dogs have 39 pairs while our cats have 19 pairs. Look at Figure 2.1 to see the relationship between DNA and chromosomes.
- An allele is version of a gene that a person receives from each parent. Think of alleles as alternate versions of a gene.
- Dominant genes override other genes, even if only one parent has the gene. It's like

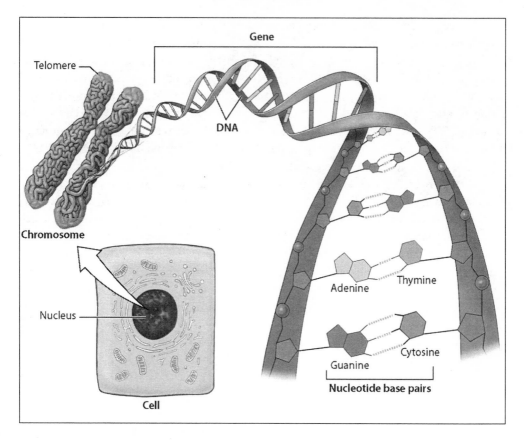

Figure 2.1: Relationship Between DNA and Chromosomes.

having two siblings in which one is quiet and the other is loud. The loud sibling is heard while the quiet sibling is there but still. Traits like dark hair color can be transmitted to offspring even if only one parent has the gene.

- Recessive genes can only be expressed if both parents have the gene. Genes for blonde hair are recessive and can only be expressed in offspring if both parents have the gene.

- With codominant genes, two versions of a gene are expressed and thus both show their associated traits. Having blood type AB is an example of a codominant trait if a person inherits allele A from one parent and allele B from the other parent.

- Sex-linked genes are carried on sex chromosomes, X and Y. Females are genetically XX and males are genetically XY, although there can be variation here, as we'll learn about later in this book. Almost all sex-linked genes are recessive genes and are present on the X chromosome. Since males do not have a second X chromosome, the sex-linked genes express themselves like dominant genes.

- Here's a quick synopsis: chromosomes are found within the nucleus of a cell → each chromosome contains DNA → genes are found along strands of DNA → genes carry traits that you inherit → in total, the genes make up the complete set of genes in us known as the human genome, our total genetic instruction manual.

Mapping human genes was a high point in science. In order to do this, the scientific world came together, pooled resources, and sequenced the human genome as part of the international Human Genome Project consortium. Research centers and universities from across the globe, including the United States, the United Kingdom, France, Germany, Japan, and China collaborated to sequence the genome. We worked *together* to make this happen.

In order for us to understand what genetics, genetic engineering, and genetically-modified foods are about, a basic understanding of how genes operate is important. For the most part, genes direct protein synthesis. Most part? What does that mean? Not that long ago when I started my career in biology, I learned and taught students that genes are stretches of instruction-containing DNA. The instructions are copied into ribonucleic acid (RNA) and then turned into proteins. Today, we know that this isn't always the case. Yep, there are exceptions to the rules.

The word *protein* is derived from the Greek word *protos*, which means first. This should give you an idea that proteins are pretty important because being first is pretty important. In fact, three-fourths of the dry weight of a cell is protein; and proteins are involved in structures, hormones, enzymes, muscle contraction, immunity, and other essential life functions. Proteins are made up of amino acids, which make up the genetic code. Are you starting to see the connections?

DNA (found in the cell's nucleus) is necessary for life because it encodes for these proteins and by extension, everything necessary to sustain life itself. But, some organisms, like microscopic single-celled bacteria and cyanobacteria, do not have a distinct nucleus, so their DNA is found in that single chromosome in a region called the nucleoid.

Bacteria and cyanobacteria are frequently in the news, especially as we learn more about how important they are to life on earth. Bacteria are found in soil, air, and water and in or on plant and animal tissues. We're learning more and more about the significance of bacteria to our microbiome, the microbial environment that protects us against germs, breaks down food, and produces vitamins in the intestines. However, bacteria are also essential to global well-being as they cause organic decay and nitrogen fixation (chemical conversion of atmospheric nitrogen into organic compounds)—both essential to our survival. Bacteria are frequently used in genetic studies because they grow easily and reproduce rapidly in laboratory cultures.

Cyanobacteria, also called blue-green algae, are related to bacteria, but they are capable of photosynthesis and represent the earliest known form of life on earth. Recall from your school biology class that photosynthesis is the process by which green plants use sunlight to synthesize food from carbon dioxide and water. Generating oxygen is a byproduct of photosynthesis, and we know that this vital gas is essential to our survival.

Chromosomes have little caps on their ends called telomeres, and telomeres are noncoding parts of DNA. Refer back to Figure 2.1 to see the telomeres capping chromosomes. Telomeres are like aglets—the plastic tips glued around the ends of your shoelaces. Each time our cells divide, the telomeres on the tips of chromosomes shorten. You might be thinking, so what. But wait, telomere shortening is important, because it is involved with aging at the cellular level. In fact, one theory of aging rests on the premise that telomere shortening is what causes us to age. When the telomeres become too short, the cell stops dividing and it becomes senescent (old and inactive) or it dies. In addition to aging, telomere shortening is also associated with increasing cancer risk. Now I have your attention.

In 2009, Americans Elizabeth Blackburn, Carol Greider, and Jack Szostak shared the Nobel Prize in Physiology or Medicine for their discovery that telomeres protect chromosome ends and telomerase is the enzyme that causes elongation. Add a few drops of telomerase to the caps of chromosomes, and they'll grow. Telomerase is quite active in cancer cells, but it is either absent or present in very low quantities in our somatic cells (those are all the cells in our bodies except the sex cells called gametes). Said another way, telomerase is active in eggs and sperm. These findings are important to studying cancer, aging, and stem cells. What we know is that stress and inflammation affect telomeres causing them to shrink. Conversely, novel research has shown that exercise helps increase their length while mindfulness, meditation and a Mediterranean diet can also positively affect telomere length and lifespan.

There's also another way to increase telomere length, but it involves some travel. If you want to increase your telomere length, head into space. That's right. It appears that space travel caused astronaut Scott Kelly's telomeres to lengthen. The lengthening was short lived and within two days of returning to Earth, a majority of his telomeres shortened. In an ongoing and fascinating study, NASA researchers have been comparing data points between Arizona twin brothers, astronauts Scott Kelly and Mark Kelly. You may also know the brothers through politics: Mark Kelly was elected to the United States Senate and is married to former United States Representative and gun control advocate, Gabby Giffords. Scott Kelly spent a year in space while his twin brother stayed on Earth. Since the two are identical twins, their DNA is also identical. (Twinning is discussed later in this chapter.) However, it's important to point out that while genetically identical twins have identical DNA, gene *expression* is not identical. Gene expression refers to how active a snippet of DNA is. DNA produces messenger RNA, colloquially known as mRNA. Messenger RNA is what your cells use for body functions, while how much and which mRNA your genes produce is gene expression. This means that identical twin DNA stays the same, but how it behaves can be different. It's a classic example of the environment affecting genes. We'll also encounter a few more examples in this chapter showcasing how the environment affects genetic expression.

Genetics as a Science

People studied heredity long before DNA and genes were discovered. Scientists who study genetics are known as geneticists, and the most notable early geneticist was Johann Gregor Mendel (1822–1884).

Mendel was an Austrian monk who spent years studying pea plants in the monastery gardens. He conducted countless experiments, crossing various strains, and cataloguing his findings. Through his research, he discovered the fundamental laws of heredity and earned the title, "Father of Genetics." Mendel painstakingly charted the mathematical patterns of heredity in pea plants from generation to generation to generation to generation. He understood the principles that governed heredity in pea plants, and today we know these findings by their catchy title, Mendel's Laws of Heredity. These laws, which determine characteristics passed to offspring, are the Law of Segregation, Law of Independent Assortment, and Law of Dominance.

Mendel's contemporary, Charles Darwin (1809–1882), published *On the Origin of Species* in 1859 as Mendel was conducting his experiments. The full title is actually *On the Origin of Species by Means of Natural Selection, or the Preservation of Favoured Races in the Struggle for Life*. This quintessential work is considered to be the foundation of evolutionary biology. Darwin was an English naturalist who proposed evolution by natural selection. Remember that we encountered Darwin and these terms in Chapter 1 when we discussed the foundations of science. Interestingly, because Mendel's work was published in an obscure German journal, Darwin never knew about it. In fact, Mendel's work wasn't rediscovered until many years after the publication of Darwin's *On the Origin of Species*. It is too bad that Mendel's work wasn't known, because the mechanisms by which natural selection was occurring was the one thing missing from Darwin's book. Between the work of Mendel and Darwin, it was established that gene mutations are the source of variation and that statistical methods could be used to analyze new mutations. Ever so slowly, experimental evolutionists were morphing into field geneticists.

Although the science of genetics is much more complicated now and fields such as cellular genetics and molecular genetics exist, Darwin and Mendel paved the way. Their work demonstrates the long history we've had with looking at offspring and figuring out what's happening. Here's a sampling that charts historical findings within the field, beginning at the turn of the last century. In the early 1900s, recessive inherited disorders were described, including albinism (1903), congenital cataracts (1906), Huntington disease (1913), Duchenne muscular dystrophy (1913), red-green color blindness (1914), and hemophilia (1916). Fast forward to today and according to the Genetic Disease Foundation, there are over 6,000 genetic diseases caused by inheriting an altered gene.

Manipulating Genes in Animals and People

It might be surprising to know that the principles of genetics predate Mendel and Darwin. Ancient peoples—we're talking 10,000 to 12,000 years ago—used their understanding of heredity to domesticate and selectively breed plants and animals for desired traits. Very early domesticated plants included barley, wheat, and peas; and domesticated animals included dogs, goats, horses, camels, and sheep. Wheat, the most widely produced and consumed grain in the world, actually originated from three grass-like species. Yes, your bowl of breakfast cereal began as a grass about 75,000 years ago. Techniques were used to enhance yield, increase size, vary color, intensify taste, and alter the ripening season.

What is wrong with any of this? Several of today's greatest controversies center on genetically-modified foods, biotechnology, and cloning. To many people, these terms conjure up images of Frankenstein, the titular character in Mary Shelley's novel, subtitled *The Modern Prometheus*. In Greek mythology, Prometheus is a demigod. As the story goes, Prometheus stole fire by trickery from Zeus, who was hiding it from humans. Prometheus took this fire and returned it to earth. As punishment, Zeus chained Prometheus to a rock where an eagle pecked away at his liver, which re-grew at night. (Fortunately for Prometheus, he was rescued by Zeus.) What's interesting about this story, though, is the notion of the liver re-growing, because the liver can indeed regenerate. However, the regeneration isn't indefinite, so if the liver is injured beyond its regenerative capabilities, a liver transplant is the only course of action.

Another fictional breeding experiment, Audrey II, is the blood-requiring plant that looks like a cross between a Venus fly-trap and a butterwort that was created by a florist in the musical *Little Shop of Horrors*. Are Frankenstein and Audrey II just fictitious examples of genetics gone bad or are they worthy of serious consideration?

Cloning and Twinning

People are afraid of cloning. In some instances, the fear is justified. Nobody wants a rogue scientist to clone humans or attempt to create a super-human species. This is a reason why understanding the process is important and why ethics panels should always be involved. It's just as important to know that cloning occurs in nature without any scientific intervention. Let's begin with some definitions and some natural examples.

Cloning describes the biological process in which genetically identical copies are made. It happens naturally and through genetic manipulation. Examples of natural cloning include asexual reproduction (cell division) in bacteria and sexual reproduction in humans. Bacteria produce genetically identical offspring from a copy of a single cell from the parent bacterium. This type of cloning enables bacteria to divide rapidly. To illustrate, the common bacteria that live in our intestines, *Escherichia coli*, can divide every 20 minutes. The growth rate is exponential: If one bacterium divides every 20 minutes, within seven hours 2,097,152 bacteria will be produced! You can see why eradicating a bacterial infection can be challenging. Humans, on the other hand, reproduce through sexual reproduction so the numbers are much smaller.

To begin, some definitions are in order. Oocytes (-oo = egg + -cyte = cell) are cells that can be fertilized by male sperm. Oocyte is a three-syllable word that is fun to say, so give it a try: oh-oh-site. Colloquially, we call oocytes eggs or egg cells. If an oocyte is fertilized, it becomes a zygote. The word *zygote* is an apt term derived from the Greek word for *yolk*, and a yolk sac forms early in embryonic development. Zygotes are future embryos. We'll explore zygotes in greater detail in Chapter 5 Birds and Bees: Human Reproduction. And, a quick primer on counting: one = mono, two = di, and tri = three. You get the picture.

Now back to cloning. Whether you realize it or not, you are already familiar with natural clones in humans: They are called identical twins. You may also know them as maternal twins. Humans are not alone in twinning as identical twinning occurs in other mammals, too. In scientific circles, identical twins are known as monozygotic twins. When not in scientific circles, you can still use scientific terminology to impress your friends and family.

In the case of twinning, monozygotic twins result when one zygote separates into two independently growing cell clusters at an early developmental stage. The two-cell cluster then becomes two separate embryos of the same sex and identical genetic makeup. In simple terms, a single embryo split in two. Scientists still do not know why this occurs, but most identical twins have a mirror face of the other, meaning defining characteristics are on opposite sides.

Fraternal twins—those two humans that do not look exactly alike—are called dizygotic twins. Fraternal twins are not clones, but it's interesting and worth learning about the process, nonetheless. With dizygotic twins, there are two egg cells that were fertilized by two separate sperm cells. This usually occurs through hyper-ovulation, which means that the mother released two egg cells during her menstrual cycle. Thus, these twins can be male and female, male and male, or female and female and look no more alike than other siblings with the same mom and dad. Although specific genes linked to hyper-ovulation have not been located, one thought is that as women get older, they release more eggs—nature's way to ensure continuation of the species.

Other types of cloning, called quasi-cloning, also occur in other kinds of twins known as sesquizygotic twins. Breaking this five-syllable word apart, the prefix, *sesqui-* means ½. Clear as mud, right? Sesquizygotic twinning is an intermediate result between monozygotic and dizygotic twinning. These sesquizygotic twins are identical on the mother's side but fraternal on the father's side, making them semi-identical. With this twinning scenario, one egg is fertilized by two sperm, forming a triploid, which then splits in two—hence the ½. While this type of twinning is fascinating, it is an extremely rare event. Yet, it's important to be aware of this situation because it's something to consider when we visit issues surrounding abortion in Chapter 5. In cases where sesquizygotic twinning does occur, the embryos contain too much genetic information to enable them to successfully complete cell division, so in nearly every single case, the newly-formed embryos die.

There are cellular mechanisms in place for favoring singular zygotes. One mechanism ensures *monospermy*, meaning that only one sperm is allowed entry to the one oocyte. It's a one-for-one ordeal. Monospermy happens through a process called *oocyte membrane block*, when one sperm has successfully docked with the oocyte receptor. Once the sperm has docked, all other cellular receptors become unavailable. Another mechanism called the *cortical reaction* clears other sperm from the area. Yes, this is a very simplified explanation for the events, but what's important to know is that human oocytes block polyspermy, thus only one sperm can fertilize an oocyte. And, yet, polyspermy *can* happen. In February 2019, scientists confirmed semi-identical twins in Australia. This was only the second case confirmed; the first set was identified in 2007 and reported in *Nature* magazine. The genetic makeup of the 2007 sesquizygotic twins was only discovered because one of the twins had genital ambiguity. Genital ambiguity refers to incomplete development of the sex organs. It's important to bring this up here because it shows that in nature, people can be born without a clearly defined sex, a topic further explored in Chapter 4.

While still on the topic of twinning, you're probably wondering how conjoined twins fit into this scenario. Are they clones? It depends. Conjoined twins are monozygotic twins that are physically joined at birth. Sometimes these twins share organs and sometimes they have the full complement of organs but are joined otherwise. The extent of the union and degrees of duplication vary. Sometimes these twins can be separated by surgery, depending on the degree of union. A very dated term—and one that should no longer be used—for conjoined twins is Siamese twins. In the early 19th century, two men, Chang and Eng Bunker (1811–1874), were a much-publicized pair of conjoined twins born in Siam (present-day Thailand). They traveled the country and were known as the Siamese twins, a term that stuck for many years to mean conjoined. These brothers were joined at the torso by a band of flesh, cartilage, and their fused livers. Today, surgery could have separated the two.

Is it possible to naturally manipulate the probability of twinning (cloning) in humans? Perhaps. Perhaps you've heard that "twinning runs in families"? Let's look at twinning and genetics. The biology behind twinning is complex and is likely determined by multiple genetic and environmental factors. When we look at the statistics of twinning, monozygotic twins occur in 3 to 4 births per 1,000 births worldwide. Dizygotic twins are about twice as common as monozygotic twins. Succinctly, it appears as though having identical twins is not determined by genes; but genetics can play a role in having fraternal twins. Thus, fraternal twins can run in families, but only the mother's genes matter. Stick with me here. Fraternal twins happen only when the mother releases two egg cells. Usually, a female releases one egg cell per menstrual cycle. On the other hand, males release thousands of sperm cells with each ejaculation. Sperm release has no effect on how many egg cells the mother will release. With monozygotic twins, a single zygote/future embryo splits in two and this zygote splitting is a single random event that occurs by chance. Chance doesn't run in families. With dizygotic twins, there are two zygotes that develop into two individuals. Some women have alleles (different versions of genes) that make them more likely to release more than one egg cell, an event termed hyper-ovulation. These genes are passed down through the familial lineage. For this reason, fraternal twins *can* run in families.

But, the family background for determining if twinning runs in families only matters for the mother's side. Remember, only females ovulate, not males.

Twinning is a fascinating phenomenon. In fact, there are events devoted to twins. Every August, the city of Twinsburg, Ohio, hosts Twinsburg's Twins Days Festival. Touted as the largest annual gathering of twins and multiples in the world, it is a total ogle fest. In no other setting is it as accepted to virtually stare and gawk at people.

Twinning and cloning also piques student interest. Case in point: one night (everything seems to happen in night classes), I was taken aback when one of my Nigerian students told me that in her village, twinning was very common and could be attributed to eating yams. As you can imagine, this started some lively discussion, and the phones and fingers were fast at work checking to see what this was all about. As we soon discovered, there may be some truth to this phenomenon. The fraternal twinning rate in Western Nigeria is quite high at a rate of 45–50 per million pregnancies. The high incidence might be attributed to diets high in estrogen-like substances. Guess what vegetable contains estrogen-like substances? Yams. And what role does estrogen play? Estrogen is a hormone involved in ovulation. This is an example of the environment influencing gene expression. It's also worth noting that in 1943, researchers were looking for a treatment to help women suffering from gynecological disorders such as painful menstruation when Penn State chemist Dr. Russell Marker, first extracted a plant sex hormone, diosgenin, from a wild Mexican yam. His extraction method became known as the Marker Degradation and paved the way for developing synthetic sex hormones such as estrogen and progesterone.

If you've ever known identical twins or spent time with identical twins, you know that they are not completely the same in every way. For example, they do not have the same fingerprints because those whorls on the tips of fingers are influenced by amniotic fluid flowing past developing fingers. It's like water lapping the shore and causing swirl marks in the beach sand.

While the Kelly twins and space travel studies is a fascinating area of research, experiments can also be done here on Earth. Let's look at how the environment influences genes and why some identical twins don't look exactly alike. Physiologically-stressful situations can influence gene expression as in the case of increased red blood cell production. To illustrate this event, we'll place a set of identical twins in two different environments with Twin #1 on a high mountain and Twin #2 at lower sea level. Twin #1 lives in the Andes, the longest continuous mountain range in the world and also an area of high altitude. Twin #2 stays living near sea level at a much lower altitude. The lower level of oxygen in the Andes will cause an increased production of red blood cells. The Andes environment will influence gene expression, causing more red blood cells to be produced. Why is this important? Red blood cells carry oxygen—the more red blood cells that are produced, the more oxygen that can be latched onto for survival.

We also know that nutrition has an impact on the development and health of identical twins. Scientific investigations into why identical twins have physical differences point to dietary factors that influence genes. When genes change because of environmental modification instead of genetic code alteration this is referred to as epigenetics.

Types and Purposes of Cloning

There are several types of cloning: gene cloning, reproductive cloning, and therapeutic cloning. Let's take a look at these types of cloning to gain a better understanding of the process.

Gene cloning involves inserting genes or segments of DNA into a cell, which can grow colonies of identical cells *in vitro*—in a lab outside a living organism. Reproductive cloning produces copies of whole animals. Therapeutic cloning produces embryonic stem cells that can be used experimentally for treating diseases or replacing injured or diseased tissues.

When you read the phrase *embryonic stem cells*, what do you think? In laboratory settings, embryonic stem cells are derived from embryos grown in an *in vitro* fertilization clinic. The egg was fertilized *in vitro* and donated for research with informed consent of the donors. (Informed consent is discussed in Chapter 3.) These embryos were not grown in a woman's body; the cells were grown in a lab to create a cell culture in a dish. It's important to use stem cells because these are very early cells that are unlike any other cells in the body. Stem cells are capable of dividing and renewing themselves, they are unspecialized (meaning they haven't "decided" what to become yet), and they give rise to specialized cell types. Stem cells are precursor cells that have the ability to differentiate into other cell types.

Why don't we just use adult stem cells? Adult stem cells can renew themselves and differentiate into the same type of cell in which it was found. Therefore, femur adult stem cells can only become a femur. Locating stem cells can be challenging. Some stem cells have been found in the bone marrow, allowing them to be used for generating bone, cartilage, fat cells, and blood. There is some evidence that they can be found in the brain and the heart.

Again, this can get very complicated and we can write tomes on genetics, but there are different things that can be done with the cloned cells of organisms, many to help humans. For example, the transfer of human genes into sheep and cows has led to the production of clotting factor IX, which can be used to treat the bleeding disorder hemophilia. For people deficient in factor IX, this ability to manufacture the clotting factor synthetically is a boon. Cloning has also enabled the production of alpha-1-antitrypsin, a protein used to treat emphysema. People who are deficient in alpha-1-antitrypsin are predisposed to pulmonary emphysema, a dreadful lung disease. Biosynthetic insulin, created using recombinant DNA technology, is a mainstay of insulin therapy for both type 1 and type 2 diabetes. Without using cloning technologies, this life-saving treatment for people with diabetes would not exist. To put this into perspective, the Centers for Disease Control and Prevention (CDC) notes that in 2020, 34.2 million Americans have diabetes. That's about 1 in every 10 people.

When it comes to cloning mammals, Dolly and Louise Brown come to mind. Dolly was a sheep who was cloned from an adult cell in 1996. Named after the singer Dolly Parton, she lived to the age of 6. You might be wondering what Dolly Parton has to do with this. Well, researchers used a mammary gland cell to create Dolly the Sheep and Dolly Parton is recognized for her large breasts. While Dolly the Sheep is the most famous cloned animal, cloning has been around for a really long time. Recall that you can see cloning in nature whenever bacteria reproduce. Before Dolly, frogs and mice had been cloned in the lab. What's remarkable about Dolly was that she was the first mammal to be cloned from an adult cell line. The important piece of this is that the DNA of adult cells are already specialized for particular cell types; had embryonic cells been used, it would have been easier.

Through *in vitro* fertilization and a surrogate mother, Dolly was born. Dolly lived out her days at the Roslin Institute in Scotland, produced offspring naturally, and was euthanized in 2003. The average life expectancy of sheep raised outdoors is about 11–12 years. Dolly was raised indoors and contracted a virus-induced lung tumor, which is common in sheep who are raised in such a manner. She also had some arthritis in her hind leg. Since Dolly was already "old" to begin with because her cell line began with an adult cell, this might explain why she developed arthritis, which is typically an age-related disorder.

Louise Joy Brown, born in 1978, was the world's first test-tube baby. "Test-tube baby" refers to a type of *in vitro* fertilization (IVF), a process in which multiple oocytes are placed in a medium (nutritive broth in a test tube) and then sperm are added for fertilization to create a subsequent zygote. The zygote is then removed from the test tube and placed into a woman's uterus where it undergoes normal developmental stages, forming an embryo and then a fetus, and grows like a usual pregnancy until birth. Brown was born July 25, 1978, in England, and in 2015 published her autobiography, *My Life as the World's First Test-Tube Baby*.

Eugenics

The topic of cloning invariably leads to a discussion of eugenics. Eugenics is the pseudo-science aimed at improving the human population by controlled breeding to increase the occurrence of desirable heritable characteristics. Basically, it is the practice of trying to create a better human stock. It was largely developed by Sir Francis Galton, a cousin of Charles Darwin. Galton (1822–1911) was an English scientist credited with founding and introducing methods of measuring mental and physical abilities. He also pioneered the Galton system of classification of fingerprints as a means for identification. His fingerprinting work was based on the variations in patterns of ridges, arches, loops, and whorls on individual fingertips. Recall that fingerprints were discussed earlier in this chapter with respect to identical twins having different fingerprints. His later work is positively noteworthy while his former work in eugenics is abhorrent.

Eugenics involves practices that are directed toward enhancing the human race by selecting desirable traits and breeding for them. Since this topic involves genes, a detailed description is given here, but it will be discussed again in Chapter 3 because such practices are firmly rooted in racism.

A leading principle of eugenics is racial hygiene. Galton's racial hygiene attempt to breed a "superior race" by selective breeding was passed on to the Nazis. Wiping away "unfit humans" and creating a "master race" were tenets of Nazi ideology. But, don't think that eugenics was contained to Nazi Germany. Eugenics and sterilization were practiced in the United States. In fact, the effectiveness of the sterilization program in the United States inspired the Nazis.

Mendel's work involved breeding peas for desirable traits. Unfortunately, Mendel's laws were also of particular interest to Nazis and eugenicists keen on breeding humans with selected characteristics. The Nazi Party's platform promoted racist ideas and right-wing governmental rule. It flourished under the reign of dictator Adolf Hitler. Nazis held true to anti–Semitism eugenics and pure "Aryan" Germans, meaning Caucasians who were not Jewish. Keep in mind that scholars have rejected the notion of an "Aryan race," which refers to a group of people descended from Indo-Europeans. After the end of World War II, the Nazi Party was outlawed in Germany. However, the American Nazi Party, founded in 1959 by George Lincoln Rockwell with headquarters in Arlington, Virginia, is alive and well in the United States. Their purpose is stated in fourteen words: "We must secure the existence of our people and a future for White children."

Eugenics played a role in forced sterilization. Forced sterilization, also known as coerced sterilization and compulsory sterilization, is a shameful part of American history. Throughout the 20th century (1929–1974), federally-funded sterilization programs took place in 32 states. It was used as a means of controlling "undesirable populations." Who were these undesirables? They were immigrants, people of color, poor people, unmarried mothers, disabled individuals, and those deemed mentally ill. These sterilization programs were part of the eugenics movement in the United States and were driven by prejudice, pseudoscience, and social control—and they informed policies on immigration and segregation. If this sounds scary, it should. According to researcher Alexandra Stern, "Many sterilization advocates viewed reproductive surgery as a necessary public health intervention that would protect society from the deleterious genes and the social and economic costs of managing 'degenerative stock.'"

In North Carolina, the dehumanized victims were disproportionately black, female, and poor. North Carolina sterilized 7,600 people. It's also important to note that in 1927, the U.S. Supreme Court upheld a forced sterilization law for the supposed good of society. This is another example of public policy intersecting with science. Presently, North Carolina is giving compensation to some victims of forced sterilization, though it's impossible for material compensation to make up for such an atrocity.

We can see echoes of eugenics and its racially charged events by recalling the "Unite the Right Rally" in Charlottesville, Virginia, in August 2017. The purpose of that white supremacist rally was to unify the American white nationalist movement while chanting Nazi slogans. For a history refresher, the United States remained neutral for more than two years after World War II broke out in September 1939. Then, on December 7, 1941, the Japanese attacked the U.S. naval base in Pearl Harbor, Hawaii, forcing President Franklin Roosevelt to declare war on Japan. Shortly after that, Nazi Germany declared war on the United States, and America entered the war, fighting against the Axis powers. When Americans were posed with the question, "Should President Roosevelt have declared war on Germany, as well as on Japan?," Gallup public opinion polling from December 12–17, 1941, showed that 91 percent of Americans said "yes." This was the strongest support for the war since it had started. During World War II, the U.S. military lost 416,800 soldiers, but the United States emerged victorious as a global superpower. These people lost their lives fighting Nazism. We dishonor every American soldier and civilian who sacrificed to wipe out the Nazi regime every time we tolerate white nationalism.

Modifying Genes and Genetic Screening

Today, many in the mainstream shun genetically-modified anything because they think gene manipulation is "unnatural" or "harmful" or is an attempt to breed super-humans or to create designer babies. Perhaps much of this fear or reluctance to embrace genetically-modified products is because it reminds people of the eugenics movement. Superficially, that is understandable. Yet, there is benefit to genetic modification. Recall that nature modifies genes without any scientific help. Some animals and plants that exist today are not like the fauna and flora of yore because of genetic modification. Many have undergone genetic alteration to increase disease resistance, nutritional value, and yield. That said, the ethics of gene editing must always be considered.

Are you familiar with the Green Revolution? This is not the same as "going green" or the "Green New Deal," the trend toward being more environmentally conscious and legislation that supports environmental responsibility. The Green Revolution, also known as the Third

Agricultural Revolution, was the application of research and genetics to increase crop production in developing countries between 1930 and 1960. Examples include high-yield crops, dwarf wheat, new methods of cultivation, fertilization, mechanization, and hybridized seeds. A key innovator in the Green Revolution was American geneticist, agronomist, and humanitarian, Norman Borlaug (1914–2009). His work in developing semi-dwarf, high-yield, disease-resistant wheat varieties improved food availability in Mexico, Pakistan, and India. For his achievements, he was awarded the Nobel Peace Prize (1970), the Presidential Medal of Freedom (1977), and the Congressional Gold Medal (2006). This is an example in which molecular genetic methods were used to secure food for swaths of people across several countries and thwart hunger. Genetically-engineered apples, corn, soybeans, and tomatoes are an important part of our current food supply. We humans are also genetically engineered through thousands of years of evolution and natural selection. For the most part, there's nothing you can do about it. You are a product of your environment.

Genetic modification is a part of our everyday lives. Many of the 6,000 inherited diseases that make up the human genome are fatal or debilitating. Genetic tests are available to screen for some genetic diseases, but not all genetic disorders have a screening test available. Different types of screening and testing include newborn screening, diagnostic testing, carrier testing, prenatal testing, preimplantation testing, predictive and pre-symptomatic testing, and forensic testing. Whew! To begin, let's look at one test that is performed on infants just after birth: the test for PKU.

Soon after entering the world, newborns are screened for the inherited disease phenylketonuria (PKU). It's important to identify PKU right out of the gate so that babies can be treated. Phenylketonuria is a recessive inherited disorder in which the infant cannot metabolize phenylalanine. As a recessive disorder, it means that both copies of the gene—one from mom and one from dad—have mutations. Phenylalanine is an amino acid that is found in proteins and some artificial sweeteners, like aspartame sold under the brand names Equal and NutraSweet. Aspartame is a derivative of aspartic acid and phenylalanine. Individuals with PKU lack the enzyme, phenylalanine hydroxylase, that metabolizes the phenylalanine. If phenylalanine can't be broken down, it builds up in the blood, leading to brain damage, nerve damage, intellectual impairment, behavioral problems, and in some cases heart problems and abnormally small head size. Treatment is lifelong and involves specific dietary guidelines to reduce certain proteins. It is detected at birth by a heal prick blood test, known as the Guthrie test. Federal regulations require that any food containing aspartame must have this on the label: "Phenylketonurics: Contains phenylalanine." Check it out the next time you reach for a food that contains artificial sweetener.

Screening for some other genetic diseases occurs during pregnancy. Amniocentesis and chorionic villus sampling are two such screens. Amniocentesis is the transabdominal aspiration of amniotic fluid from the amniotic sac, which surrounds a developing fetus. Using ultrasound, a long needle pierces the abdomen and is inserted into the amniotic sac to withdraw amniotic fluid. Amniotic fluid bathes the fetus and contains fetal cells and other chemicals produced by the fetus. In addition to other diagnostic tests, genetic amniocentesis is usually done between week 15 and 20 of pregnancy and tests for chromosomal abnormalities and sex determination. Common disorders identified by amniocentesis include Down syndrome (trisomy 21), Edward syndrome (trisomy 18), Patau syndrome (trisomy 13), Turner syndrome (XO syndrome or Monosomy-Turner Syndrome), Klinefelter syndrome (XXY syndrome), and other sex chromosome anomalies. Some of these conditions are discussed in Chapter 4.

Chorionic villus sampling (CVS) is another test that provides information relative to genetic makeup and chromosomal condition. This test, which can be done sooner than an amniocentesis, involves biopsy of the tiny vascular processes of the placenta through the vagina. Done between weeks 10 and 13 of pregnancy, CVS can be used to test for conditions such as Down syndrome, cystic fibrosis (CF), sickle cell anemia, thalassemia, and muscular dystrophy (MD).

These are routine tests offered to pregnant women so that they can be well-informed about their pregnancies. Women choose whether or not to have genetic analyses and must meet with genetic counselors before the procedure and after the procedure when the test results are revealed.

The only similarity among Down, Edward, and Patau syndromes is the tripling of chromosomes. Humans typically have 23 pairs of chromosomes for a total of 46. Twenty-three come from the mother and 23 come from the father. The 23rd pair, designated X and Y, are called the sex chromosomes. Individuals who are XY are designated male and individuals who are XX are designated female. Down syndrome is marked by an additional chromosome 21, Edward has three number 18 chromosomes, and Patau is characterized by three chromosomes 13.

With Turner syndrome, there is a loss of the second sex chromosome, so it affects development in females only. The condition affects about 1 in 2,500 female newborns worldwide, but it is more common in pregnancies that do not reach full term and result in miscarriages and still-births. Females with Turner syndrome are infertile, but intelligence is not affected.

Individuals with Klinefelter syndrome have two X chromosomes and one Y chromosome. Since the Y chromosome is present, it is considered a male disorder. It affects 1 in 500 to 1 in 1,000 newborn males. Males are infertile, and many have developmental disabilities.

Why spend so much time talking about these disorders? The reason is because people are born with differing chromosomes and these differences can affect them where gender issues are at the fore. Think about regulations and laws that seek to discriminate against people who are unlike the majority. Does this really seem just? Turner syndrome and Klinefelter syndrome will be discussed again in Chapter 4.

Genetic Engineering

Fundamental knowledge of genetics is important so that you can make informed decisions and form educated opinions about the role of genes and genetic manipulation in everyday life. An example of something good brought about through genetic engineering is drug development. Genetically-engineered mice are used in the process of developing new drugs and studying various human diseases/conditions like cancer and obesity. Scientists have been using genetically-modified mouse models since the 1980s to test drugs because *in vivo* (live) animal studies are crucial to drug development and preclinical trials.

The United States Food and Drug Administration (FDA) has very stringent policies for bringing a drug to market and 80 to 90 percent of potential therapeutic drugs do not get approval. Drug development is costly, and the hurdles are many, so cost-effective tools such as using mice are necessary. This is merely the beginning of drug development. Just because everything seems to go fine with mice, new drugs still require several trials in humans before a new drug can be brought to market. Hence, it makes sense to use genetically manipulated mammal models for testing and research before "trying it out" on humans. According to recent studies, the cost of launching a new drug exceeds $1 billion. Common drugs you might already know about that utilized genetically-engineered mice include Allegra, Celexa, Claritin, Epogen, Paxil, Premarin, Prevacid, Prilosec, Procrit, Prozac, Zoloft, and Zyrtec. These drugs treat a variety of ailments ranging from depression and anxiety, to allergies, to blood production and hormonal issues.

A new tool that scientists can potentially use is CRISPR, a recent technology for genomic editing. Pronounced "crisper," CRISPR is an acronym for Clustered Regularly Interspaced Short Palindromic Repeats, and is basically a family of DNA sequences in bacteria. What this means is that scientists can use CRISPR to specifically engineer animal models to be used in research fields such as cancer, human development, immunology, neuroscience, and physiology. Teams led by American biochemist Jennifer Doudna (1964–) and French professor Emmanuelle Charpentier (1968–) discovered and developed the groundbreaking CRISPR genome editing technology. Moreover, in October 2020, these scientists were jointly awarded the Nobel Prize in Chemistry "for the development of a method for genome editing."

Why is CRISPR such groundbreaking technology? With this technology, scientists can use CRISPR to target specific stretches of genetic code and edit DNA. Doing so will enable permanent gene modification. With CRISPR genome editing, cell and animal models can be created quickly for researching various diseases. This can be quite exciting and useful because CRISPR can be used to correct mutations at exact locations in the genome with the end goal being the treatment of genetic diseases.

There's also a potential dark side to CRISPR. Because you can edit individual genes, specific targeting of certain traits may also be possible. In 2018 a rogue Chinese scientist, He Jiankui, revealed at the Second International Summit on Human Genome Editing that he had edited the genes of human embryos using CRISPR. He edited the genes of twin embryos to help them resist HIV infection. This was the first instance of DNA editing of human babies. In July 2019, the World Health Organization asked that all countries stop experiments using CRISPR to genetically edit the human germ line. Egg and sperm cells that join to form an embryo are germ cells, and the germ line DNA is the source of DNA for all other cells in the body. Human germ line editing is essentially banned in the United States because of a law that prevents the United States Food and Drug Administration from reviewing clinical trial applications if they involve genetically modifying embryos.

Mice aren't the only animals used for genetics research. Viruses are used as well. In Chapter 3 we discuss cystic fibrosis (CF), the congenital lung disease that causes abnormal mucus production. Researchers are currently using gene therapy for treating CF. Using virus-based vectors, they are able to replace faulty genes with "good" genes that restore faulty proteins in pigs who have cystic fibrosis. It sounds simple, right? Replace the CF-causing gene with a normal gene and everybody lives happily ever after. However, it wasn't until 2008 that a suitable animal model had been developed to test the new virus-delivered genes. Mice worked well for initial investigations, but the pig model allowed researchers to take it a step further.

More recently (2017), genetically tweaked viruses saved a cystic fibrosis patient who had received a double lung transplant. This patient was suffering from a bacterial infection originating from the suture sites. Antibiotics were not working and there were no other measures left to try except possibly a bacteriophage. A bacteriophage (phage for short) is a virus that preys on and parasitizes a bacterium by infecting it and then reproducing inside it. Microbiologist

Graham Hatfull at the University of Pittsburgh genetically engineered a phage for use in the patient and represents the first-ever engineered phage to help a human patient.

TAKE-HOME MESSAGE

The field of genetics as a science is quite old, and genetic modifications have allowed advances for human civilization. An understanding of genetics helps the average person comprehend how genes are expressed and can be altered. Furthermore, it aids in correcting misperceptions.

> *Nature uses only the longest threads to weave her patterns, so each small piece of her fabric reveals the organization of the entire tapestry.*
> —Richard Feynman (1918–1988), American theoretical physicist

3

Time Travel

Evolution and Race

U.S. Landmark Case: Edwards v. Aguillard, 482 U.S. 578 (1987); (7–2 decision)

In the early 1980s, Louisiana passed its "Creationism Act" (actually titled the "Balanced Treatment for Creation-Science and Evolution-Science Act"), which forbad the teaching of the theory of evolution in public elementary and secondary schools unless accompanied by instruction in the theory of "creation science." After being struck down by the Federal District Court, Louisiana appealed to the U.S. Supreme Court. The Court's decision (7–2) held that the Louisiana act unconstitutionally infringed on the Establishment Clause of the First Amendment in that it did not have a legitimate secular purpose and had the intent of promoting religion. The Court also found that science education is undermined if teachers cannot teach evolution or if creationism must also be taught. In 2005, a subsequent attempt to use the term "Intelligent Design" in a Pennsylvania school district was also declared by federal courts, in the Kizmiller v. Dover Area School District case, to be religious teaching.

Dinner Party Topics

These landmark and federal cases delve into the topic of evolution and its opposite counterpart, creationism. Creationism is also known by its updated term, intelligent design. This chapter covers some touchy topics, so let's dive right in. Evolution is a scientific principle that belongs to the science classroom, while creationism and intelligent design are religious principles that belong in the theology classroom. There are many reasons for this, notably that science teachers typically have no formal education in religion or theology, and generally are not certified to teach topics reaching into religious beliefs. Conversely, religious studies teachers generally are not trained or certified in science, so they stick to theological topics in their courses. Regardless of the course, you want the most qualified people teaching the content. Imagine being trained as a sociologist and having to teach calculus or being trained as a mathematician and placed in an art course. While each is a worthy discipline of study, we want to be educated by those best equipped to handle the topics.

The mere mention of these words, evolution, intelligent design, creationism, speciation, survival of the fittest, and natural selection likely cause a little anxiety. And rightly so, because these are topics with heated discussions centered around politics and religion, two topics considered taboo at dinner parties and family reunions. Before going further, please know that religion is compatible with science.

This chapter focuses on evolution as it relates to skin color, race, and racism. While 2020 will be known for the COVID-19 pandemic, it will also be known for its protests for racial justice. An aim of this chapter is to help you understand how we wound up with different skin colors in the first place, and to then comprehend where we are today. Merely thinking about these two topics will conjure emotions. But why talk about evolution and race together? The reason is because you can't discuss the scope of each without considering the biological correlates that underlie them.

Many believe that evolution doesn't exist or is contrary to religious teachings, that race is a biological category, and that people of different skin coloration are different from one another. These beliefs set the stage for mainstream controversy. From a scientific perspective, evolution is a process and genes shaped by evolution determine skin color. Let's dig a little deeper.

While there is no such thing as "various races," there are various skin colors. For the purposes of this discussion, we must agree that there is one human race. So, regardless of where you live or what color your skin is, you belong to *the* human race. That means you belong to the same human genus and species (*Homo sapiens*) and you can mate and produce offspring with other human beings. From a pure biological perspective, reproducing (procreating)—producing offspring—is a purpose for being on the planet. Biology needs us to carry on our species; otherwise, we fall to extinction. But when we talk about skin color, an automatic leap often occurs, and we begin grouping people by some contrived catalogue. To understand the faulty reasoning here, let's frame our discussion around evolution. As humans, we fit our environment. We're able to fit our environment because we adapt to situations, changing to meet new conditions. At the biological level, adaptation is a result of evolution by natural selection. Natural selection doesn't happen overnight—it is a process and it randomly favors the best of what is out there. The false narrative is that natural selection turns this organism or that organism into some super creature. It does not.

The underlying premise of natural selection is that organisms—those that are like us and those not like us—adapt to their environment. Adaptation enhances our potential to produce offspring. Those that are better adapted tend to survive, and by extension are able to produce more offspring. Over time, as our environment changes, natural selection causes characteristics to change and thus evolution occurs. At its core, evolution is the source of diversity from teeny tiny microbes to entire ecosystems. Ecosystems are biological communities of interacting organisms and their physical environments. Think of an ecosystem as a complex network with lots of interconnections.

At this point you're likely recalling information from an introductory biology course that also mentioned survival of the fittest. Survival of the fittest is linked to natural selection in that it refers to the continued existence of organisms that are best adapted to their environment. It's also likely that when you hear the term "fittest" you are thinking about physical strength, good health, and great endurance. This definition does describe a person who is fit. However, from a biological perspective, fitness refers to an individual's ability to reproduce and for those offspring to survive and ultimately reproduce. The more an organism can reproduce and have viable offspring that reproduce, the greater the fitness. Yet, if an asteroid fell on you or you got hit by a car—either of which killed you—survival of the fittest plays no role. If something catastrophic happened, like an earthquake, and wiped out hundreds of people, the human species would not go extinct because there are millions more who are able to reproduce. A modern-day example of survival of the fittest is discussed later in this chapter with the topic of sickle cell disease and its relationship to malaria. People with the sickle cell trait are more fit to survive malaria than are those without the trait.

Before your eyes glaze over or you think these topics clash with belief systems, please read on with an open mind. Also be fully willing to accept that you can still hold on to your religious system. Science does—and does not—do many things. Science doesn't ask you to suspend religious belief, but science asks you to look at the evidence. Science doesn't tell you what to believe. Science points to the known. Science is a system based on examining facts, while religion is a system of faith. Science will never tell you to give up your faith, because science can only study the physical world while religion is founded in the metaphysical world. Science and religion can coexist peacefully; oftentimes science strengthens personal belief systems through its inability to explain everything or to know everything. Lots of people—from all walks of life, profession, political persuasion, and geographic location—consider themselves to be religious. This chapter is not written to dissuade you from religion. It is written to give some solid examples of facts, answers, and truths already well established in the scientific literature.

Evolution

Let's go further into evolution and the acceptance of this scientific premise. Did you know that the United States lags behind every other Western country (and Japan), in accepting evolution? That's right. Other developed countries and most of the world are okay wrapping their brains around the topic. There is evidence for evolution; and when evidence exists, it makes sense that we accept it.

Evolution is not a new topic and has deep roots. Charles Darwin, the father of evolution, first announced his theory of evolution through natural selection in 1859. That's 160 years ago! Virtually every scientist and 62 percent of Americans accept this to be true. However, about 25 percent of Americans believe evolution was guided by a supreme being, while another 34 percent reject that idea entirely. The terms "believe" and "idea" are being used quite loosely, because first and foremost, evolution and the science behind it are not belief systems or abstract ideas. Evolution is the dominant unifying concept of modern biology.

By its very nature, science is fact based. The reason for the rejection of evolution as a basic scientific principle lies in religion. For believers of a higher power, evolution smacks up against religious beliefs with an underlying premise that we were all created by a divine superpower. It doesn't smack up against *all* religious beliefs, for on October 27, 2014, Pope Francis, the leader of the Catholic Church, proclaimed at the Pontifical Academy of Sciences that "Evolution in nature is not inconsistent with the notion of creation." Granted, the pontiff could have avoided using a double negative (*not* and *in-*) to make this easier to understand, but the message is clear that evolution can coexist with creation. Even for believers of a supreme being, logical thinking warrants that this supreme being would create *something* or *some system* where organisms are capable of adapting and revamping within an ever-changing environment.

Patterns of evolution can be seen in modern times. Such evolutionary evidence resides in vestigial organs. Vestigial organs are structures that have degenerated over time and have become functionless in the course of evolution. Common examples of such organs are wisdom teeth and body hair. Wisdom teeth once had a function when human mouths were bigger, and our diets revolved around eating tough, fibrous foods. The bigger mouths gave us room for the extra two molars on each side, which also replaced teeth lost from chewing hardy foodstuffs. We have much less body hair than our distant ancestors, but little muscles, called arrector pili, still exist at the base of our hair follicles. Contractions of these tiny muscles cause our hair to stand on end. If we were hairier, the hair would stand up, causing us to look bigger and more menacing when we were scared. You see these muscles in action in your dog or cat when they encounter a threat: the hair along the ridge of their backs stands on end. Today, we experience goosebumps instead.

Other rarer examples of rudimental structures are vestigial tails and preauricular sinuses. In humans, we see an example in our rudimentary tailbone, the coccyx. The coccyx forms from 3–5 fused vertebrae. During the fifth to sixth week of embryological development, human embryos have a tail with 10–12 vertebrate, but by week 8, this human tail disappears as the vertebrae fuse, forming the sacrum and coccyx. However, sometimes during embryological development, the vestigial tail persists, and the infant is born with a tail. The infant can be born with just a stub of a tail or one that measures about 5 inches long. When this occurs, the tail is surgically removed shortly after birth. Some people are born with preauricular sinuses, tiny holes in front of the ears, where the ear attaches at the face. Other names for them are ear pits or auricular fistulas. These pits develop during the first two months of embryological development when the ear is developing and are evolutionary remnants of fish gills. They appear at birth and a person usually only has one, but both ears could be affected. In many cases, they are relatively unnoticeable and cause no harm. In other cases, they accompany other genetic conditions such as branchio-oto-renal syndrome (disorder of kidneys with hearing loss) and Beckwith-Wiedemann syndrome (abnormal earlobes, enlarged tongue, with liver and kidney problems).

Creationism is the belief that the universe, matter, and all living organisms originate from the specific acts of a divine creation. This system adheres to a biblical account of natural processes rather than natural processes taking place through evolution. One of the largest testaments to creationism is found in the United States. In Petersburg, Kentucky, stands the 75,000 square foot Creation Museum, which opened its doors in 2007. It now boasts a life-size Noah's ark. According to the museum's website, it is a Christian evangelistic outreach that shows God's infallible word. Scientists in Kentucky, Indiana, and Ohio—the three states closest to the museum—signed a statement calling the exhibits "scientifically inaccurate materials."

According to an ABC news poll conducted in the summer of 2017, 83 percent of Americans identify as Christians, 13 percent as no religion, and 4 percent as non–Christian. If we look worldwide, 52 percent are non–Christian while only 33 percent are Christian. Countless books have been written about the religions of the world, and a dissertation on the topic is not appropriate for this book. The purpose in citing these statistics is to provide personal reference for you in our world.

Why do people trust science when it comes to so many things in their lives, yet reject it when it comes to topics such as evolution and vaccines (which are discussed in Chapter 6)? On August 21, 2017, we witnessed firsthand the droves of people who rearranged schedules and traveled great distances to experience a phenomenal event. That event was the solar eclipse, an occurrence in which the sun is obscured by the moon and was predicted by astrophysicists years before it happened. The earth experienced several total solar eclipses between 2020 and 2021, but the next one to pass through the United States will be April 8, 2024. Using the tools of science, exact dates and times for eclipses can be predicted.

We use science daily without giving it much thought. Every day, people will get into cars (designed by mechanical engineers), take chemotherapy for cancer (developed by research scientists), rely on transportation by airplanes (built by aeronautical engineers), and use phones and electricity without questioning the science behind them. Yet, many people reject evolution even though it is backed by science and continues to shape our lives.

Maybe the concept of evolution is just not clear in our minds. Evolution is a change from one state, condition, or form to another over a long period of time. A VERY long period of time. It has been proposed that perhaps many Americans can accept evolution, but they have difficulty accepting *speciation*. Speciation is the evolutionary process by which diverse species of animals or plants are formed from a common ancestral stock. It can be thought of as forming new or distinct species in the course of evolution. When discussed among "evolution non-believers" the first thing that is often blurted out is something along the line of "I don't believe that humans came from apes." However, the scientific phrase is much different from this statement. Scientists will never say "humans came from apes." What scientists will say is that there is evidence for human evolution and that humans and apes share a common ancestor. Biologists refer to the evolutionary development and diversification of a species or group of organisms as phylogenesis. This word comes to us from the Greek term *phulon* meaning *tribe* plus *genesis* meaning *formation of*.

The fossil record is rich with evolutionary evidence. Fossils are the remains or impressions of prehistoric organisms that are preserved in petrified form or as a mold or cast in rock. Fossils can be dated using several techniques based on principles of chemistry, geology, and physics. Studying human fossils enables scientists to study many things, for example, changes in brain and body size, methods of moving, various diets, and occupations of early humans over the past six million years or so. For example, even in modern day humans—you and me—skeletal changes will provide clues to our occupations or whether we are left-handed or right-handed. X-ray imaging that shows exaggerated curving in the cervical (neck) spine—like bowing your head and putting your chin on your chest—is indicative of jobs requiring a great deal of head bending as in working at a desk all day, while bone thickness in one hand over the other hand demonstrates handedness. You see, as more stress is placed on bones, they grow thicker. If you use one hand more than the other, the bones in that hand will be a little thicker, thereby suggesting hand dominance.

We know lots of this cool stuff because scientists such as anthropologists make careers out of studying it. Anthropologists are experts who study human societies, their behaviors, cultures, and development. Implements of human behavior, like stone tools and paintings, found in the fossil record, give anthropological evidence of how early humans lived; while genetics, the study of heredity and the variation of inherited characteristics, show how we are related to one another as well as to how we are related to other primates. Findings thus mark our evolution.

We're likely most interested in us: primates. By definition, primates are mammals with hands, hand-like feet, and forward-facing eyes. Besides humans, examples of other primates are lemurs, bushbabies, tarsiers, marmosets, monkeys, and apes. Apes do not have tails and ape species include gorillas, chimpanzees, and orangutans. In case you were drawing a blank on bushbabies, they are those small, night-time tree-dwelling creatures with big eyes that live in Africa. Through the use of exploration, genetics, and fossil dating practices, scientists were able to develop the human family tree that shows the various species and lineages constituting modern day humans.

Tracing Our Ancestry

With a brief walk through time, we can trace our own ancestry, starting with groups. There were four main groups: *Ardipithecus, Australopithecus, Paranthropus,* and *Homo.* No need to worry about pronunciations. Just be aware that these groups existed. Each of these groups had between 3 and 4 different species. Species are a group of living organisms who are similar and can interbreed. Interbreeding is discussed in just a bit. If this were a biology course, we'd discuss all the species. However, our aim is to outline a simple timeline. Today's humans can trace their ancestry back 6 million years ago to the earliest human ancestors who made up the *Ardipithecus* group. It's hard to imagine what 6 million years is like because 6 million (6,000,000) is an incredibly large number and our brains have cognitive limitations. (That's thanks to evolution, too.) To comprehend what 6 million of anything looks like, 6 million paperclips would fill a little more than half a rail car. If we look at 6 million in terms of people, there were 6 million Jews killed in the Holocaust. If we look at generations of people over the course of 6 million years, and we assume 18 as the age at which each generation had a child, then 6 million years spans roughly 333,333 generations!

Back to *Ardipithecus* and us. *Ardipithecus* was a group that evolved out of Ethiopia (horn of Africa) who took the first steps *toward* walking, but they did not walk. The next group, *Australopithecus,* emerged about 3 million years ago. This group walked upright but still climbed trees. About 2 million years ago, the *Paranthropus* group emerged. The term (Paranthropus comes to us from the Greek word part, *anthropos* meaning *man,* and a recently introduced term, *anthropologist,* also comes from this word part.) These early humans walked upright and had a varied diet. Our closest ancestors were members of the *Homo* group, which evolved about 1 million years ago. This group had large brains, used tools, and migrated out of Africa. This group moved and roamed, which becomes an important point in our next topic of race, for it is our ability to roam that led to various skin colors.

What is race? Dictionaries define races as major divisions of human kind, each with distinct physical characteristics—namely, skin color. We often hear the term "race" in the convenient phrase, "people of all races, colors, and religions," as if these are three separate entities. But what does race really mean?

From a biological perspective, race is a social and political construct, not a scientific categorization. A social construct is something invented by people as a means for grouping people. There are many ways to group people ranging from height and weight to socioeconomic status and skin color. Height, weight, and socioeconomic variables are all measurable. However, skin color is not truly measurable because its hue covers the spectrum with individual variation occurring even between and among family members. Yet, skin color remains one of the biggest issues in terms of categorizing or classifying people. In fact, judging people by their degree of skin tone within a group is known as *shadeism* or *colorism* and is considered a form of intraracial discrimination. Note, this is not interracial discrimination, which occurs outside of a group. Social life is complicated, and explaining this is best left to sociologists and others in related fields studying this. My goal is to show the origins of skin color so you will be able to look at all humans through the lens of scientific relevance in an effort to eradicate any contrived hatred.

Classifying Living Things

Scientists do, however, categorize living things using a classification system. This system is a family tree with many branches. In grade school you might have learned this mnemonic for scientific taxonomy: **K**ing **P**hillip **C**ame **O**ver **F**or **G**ood **S**oup. This is often used to remember the breakdown of species classification **K**ingdom, **P**hylum, **C**lass, **O**rder, **F**amily, **G**enus, and **S**pecies. Placing humans into the classification scheme we get:

Kingdom: Animalia
Phylum: Chordata
Class: Mammalia
Order: Primates
Family: Hominidae
Genus: Homo
Species: sapiens

Outside of biology books and scientific papers, we usually just use the genus and species. A genus is a broad, systematic classification of living organisms, while a species is a smaller biological division within a genus. For example, your pet dog is a member of the genus *Canis* and the species *familiaris*—familiar canine (dog). So, to scientists, the domestic dog is *Canis familiaris*. The wolf belongs to *Canis lupus*—wolf dog. Fido the domestic dog does quite well living with us humans, though. In fact, humans have selectively bred the dog species for various traits so that man's (and woman's) best friend suits us quite well, too. This selective breeding is part of evolution! Our pal Fido provides comfort, companionship, and aid; and Fido can be trained for myriad roles to enrich our lives. As members of this particular group, dogs can mate with each other and produce offspring. This is why that little smooth-haired Chihuahua can breed with the neighbor's fluffy large German shepherd dog and produce stylish mutts.

Homo sapiens

Humans, on the other hand, belong to the scientific category known as *Homo sapiens*. The terms are derived from Latin and literally mean "wise (*sapiens*) man (*Homo*)." *Homo* is the genus of primates—just discussed—that includes humans. All humans. Humans, regardless of skin color, belong to the same "race."

Whenever we look at another person, we make quick judgments relative to that person's height, weight, age, and "race." Somebody can be tall or short, fat or skinny, young or old, and…. And, what? Black, white, brown, red, and olive? From social and cultural perspectives, race equals skin color. But skin color does not equal a biological category. Skin color is determined by evolution and genes, those tiny packets of information inherited from our parents.

If skin color is determined by evolution and genes, skin color *differences* also result from genes. Biology explains why we have different skin colors. Skin color is a polygenic trait. This means that many (poly) genes (genic) at multiple sites on our chromosomes determine our skin color. Chromosomes are microscopic (very, very tiny) threadlike structures found in the nuclei (singular = nucleus) of our cells. Chromosomes contain DNA, our genetic material. Every chromosome contains our DNA, a double helix with a linear sequence of these genes. All cells in our body, except red blood cells, have a nucleus, and that nucleus contains our genes. To reiterate, cells have nuclei, which contain chromosomes, which have our DNA.

Coding for Skin Color

Those chromosomes within the cell nucleus contain many genes that code for everything that makes you, well, you. Human skin color varies from very pale (extreme "white") to very dark (extreme "black"). This color variation is due to pigment production by specialized cells. That dark pigment is called melanin, those cells are called melanocytes, and the amount of melanin produced is determined by genes.

Beginning at the cellular level, the term *melanocyte* literally means "pigment producing cell." The pigment produced by melanocytes is called *melanin*. We all have melanocytes in the epidermis, which is the outer layer of skin. This covering layer is about 0.1 mm thick—the thickness of a single sheet of paper. Regardless of our skin color, genes dictate *how much* melanin we produce. You are no doubt quite familiar with melanin and melanocytes if you have ever gotten a suntan or a sunburn. Sunlight causes melanin production to increase; thus, dark-skinned people turn darker brown and lighter skinned people tend to turn tan or pinker. However, too much sunlight can also cause dark and fair skin to burn as well.

This means that the genes we inherited from our mom and dad signal how much melanin is produced by our melanocytes. Basically, very dark-skinned people produce more melanin than do very light-skinned people. However, regardless of *how much* melanin your cells produce, the actual *number* of melanocytes is about the same for *all people*, regardless of skin color. Said another way: black people have the same number of melanocytes as white people.

Melanocyte number is true even for people who have the inherited disorder known as *albinism*, a term derived from the word part *albo* meaning white. Albinism is characterized by a lack of pigment in the skin, hair, and irises (colored parts of the eyes). People with albinism do have melanocytes, but these cells do not produce pigment, resulting in white skin, white hair, and pink eyes. The eyes appear pink as a result of the underlying blood vessels, which can show through more readily. So, black, white, yellow, and tan-skinned people can all have children with albinism. Albinism occurs in all parts of the world, and the term dates back to early 18th century when Portuguese and Spanish people used it to denote albinism among African black people.

Why do we have various skin colors? Melanin is produced to protect our cells from the deleterious effects of ultraviolet (UV) radiation in the form of sunlight. Too much sunlight is harmful. Too much sunlight causes premature skin aging and can cause skin cancer. This is why if you map skin colors across the globe, you'll find that people who live in regions with lots of direct sunlight—such as Africa and equatorial zones—have the darkest skin, while people living in more northern climes have paler skin.

How did skin coloration happen? The answer lies in evolution. Evolutionary pressures influenced our genetic production of melanin. As an evolutionary process, the variation did not happen overnight. We humans acquired dark skins about 1.5 million years ago. It was during this time that our bodies were losing the protective hair covering and required something else to protect our cells from the beating sun. Underneath that furry exterior that we lost, the skin was likely pale. This is based on the fact that if you look at today's chimpanzees and gorillas—two close "cousins" of humans—they have light skin beneath their fur. Something was needed to protect cells below the light skin.

Think of melanin production as a darkening shield that protects our underlying cells. In particular, the shield is protecting the DNA in our cellular nuclei. The ozone layer prevents a good amount of the sun's UV radiation from reaching us on the Earth's surface. While a little sunlight is good because it enables vital vitamin D production in the skin, too much can lead to cancer, genetic mutations, and folic acid destruction. We can do without cancer and bad genetic mutations, but folic acid, a vitamin of the B complex, is necessary for the normal production of red blood cells. Specifically, folic acid is needed to form heme, the iron-containing part of hemoglobin in red blood cells.

Stick with me here. Hemoglobin carries oxygen in our blood. And we all know that we need oxygen to survive. To prove this point, hold your breath for as long as possible. Are you doing it? You won't be able to hold it very long before the carbon dioxide (CO_2) builds up in the blood, triggering the respiratory centers in your brain, which in turn cause you to stop that breath-holding and take a breath. Proof that our cells don't tolerate the buildup of carbon dioxide and that they require oxygen. If you don't get enough folic acid in your diet, red blood cells cannot mature and carry oxygen, and this results in anemia. Anemia causes our skin and gums to look pale, and we feel very tired.

Folic acid is also really important prenatally because folic acid deficiency during pregnancy leads to birth defects, an increased risk of preterm delivery, and low birth weight. This is an example of an environmental pressure: If babies are born prematurely, they have a low chance of survival. If they do not survive, they cannot live to reproductive age and carry on the human species. Something had to change over thousands of years to protect our cells. One such change was a darkening of skin color.

As a species, humans are nomadic by nature. We just can't stay home. When humans migrated out of Africa toward northern latitudes, they needed more sunlight to penetrate the skin to enable a chemical reaction for vitamin D synthesis. Vitamin D is essential for calcium absorption. You may hear and read about how important calcium is to building strong bones. Yet, without vitamin D, your body cannot absorb calcium. The chemical reaction whereby sunlight is converted to vitamin D is a little complicated; in essence, when skin is exposed to sunlight, cells in the skin's outermost layer (epidermis) convert a cholesterol-related compound into cholecalciferol. It is pronounced koh-lee-cal-SIH-fer-ol. Practice saying this six-syllable word out loud so you'll have some fun while reading. There are several forms of vitamin D. Cholecalciferol is vitamin D_3, an active form of vitamin D. Vitamin D_3 deficiency affects calcium levels, causing rickets (bowed legs) in children and osteomalacia (bone softening) in adults. In children, the long bones of the legs bend due to the weight they are bearing. In adults, the bones can also bend. Sometimes osteomalacia is called adult rickets. The term rickets comes from the word *wrick* that means *to twist*.

To recap: just the right amount of folic acid is necessary to produce critically important red blood cells, and being able to produce vitamin D is necessary for calcium absorption to build strong bones. Red blood cell production, vitamin D synthesis, and rugged bones are all required for healthy humans. Dark-skinned people lived in areas with high UV-light. The increased melanin production protected folic acid from destruction while the high UV-light intensity allowed sufficient vitamin D production, and reproduction was successful. Lighter-skinned individuals in low UV-light environments reproduced successfully because folic acid is not destroyed in low UV environments and the lower melanin levels enabled sufficient vitamin D production. Each of these processes is interdependent on skin.

For these reasons, we have various skin colors. It is basic biology. When we view skin color through a biological lens, it makes no sense to stereotype people according to color shades. When populations are studied, we know that there is more genetic variability between two people than there is amongst an entire population. Said another way, individuals from different populations, for example sub–Saharan Africans and Europeans, can be more genetically similar than are individuals from the same population. Even identical twins—called monozygotic twins in scientific vernacular because they developed from one zygote (union of egg and sperm)—show genetic variation. Why? Because mutations and gene-copying variations occur during development.

Fraternal twins—siblings developed from two zygotes—can also have different colored skin. Marcia and Millie Biggs, who were born July 3, 2008, are living proof that complex genes for determining skin color are at play. These twins living in England are the product of a white mom and a black dad. The Biggs sisters are not the only known fraternal twins to appear as black and white, yet the odds of this occurring are around one in a million. Stories of twins born with different skin colors are easily found, as are stories of siblings having darker skin shades than the other. In Lori L. Tharps' book *Same Family Different Colors: Confronting Colorism in America's Diverse Families*, she outlines the real-life ramifications to having darker skin, even within the same family. Her book also looks at other stereotypes, such as meeting Enrique Martinez, a young man born in Mexico who had white skin and red hair—such characteristics are not often ascribed to Mexican heritage. Color bias creeps into politics, family life, society, and nearly every aspect of being human. But why? That's the quintessential question given what we know about genes.

Color, colorism, shadeism, mixed-race, biracial, multi-racial, and every other term used to categorize people by skin tone is pure and simple racism. Like many other characteristics, skin pigmentation is not a binary trait with only two possibilities. We all have genes coding for pigment that exist along a spectrum. Science writer Elizabeth Kolbert states it eloquently: "the visible differences among peoples are accidents of history resulting from mutations, migrations, natural selection, the isolation of some populations, and interbreeding among others." This point is further underscored by American geneticist, Craig Venter, who was involved with the second sequencing of the human genome: "the very concept of race has no genetic or scientific basis." Yet, the perception of race continues to divide us.

Human Zoos

The 1860s were a troubled period in American history, but an important document that would shape the course of the nation was being drafted by President Abraham Lincoln. On September 22, 1862, while the North and the South were engaged in the Civil War, Lincoln announced that all black slaves within the rebellious states were emancipated. Its effect was to change the course of the war to one against slavery. The actual Emancipation Proclamation went into effect a few months later on January 1, 1863. Although Lincoln didn't live long enough to see the end of slavery, his proclamation paved the way for the abolishment of slavery by the 13th Amendment of the Constitution. After ratification, the 13th Amendment was officially adopted on December 18, 1865. The 13th Amendment ensures that "neither slavery nor involuntary servitude shall exist within the United States, or any place subject to their jurisdiction." While slavery had been officially abolished, the aftermath persists.

After the abolition of slavery, white people were still fascinated with people who looked differently. To support the fascination, entrepreneurs (collectors of "human rarities") who were looking to make money and anthropologists, who claimed to be supporting science, would travel the globe looking for "other" people. Basically, they would capture people in other lands, bring them back to the United States or Europe, and put them on display. The so-called scientific name was "ethnological exhibitions," but in reality, they were human zoos.

One of the most egregious acts was conducted by William McGee (1853–1912), an American anthropologist from Iowa, who created one of these human zoos at the 1904 World's Fair in St. Louis, Missouri. McGee's theory was that there were several races of people with each race demonstrating a different stage of evolution with white Europeans demonstrating the pinnacle of evolution and superiority. He believed African Pygmies, who were dark-skinned and short in stature, were at the bottom of the evolutionary tree. To showcase so-called differing stages of human evolution, McGee paid explorer Samuel P. Verner, to travel the world (South America, Africa, and Asia), capture people, and bring them back to the United States to be shown as human specimens at the World's Fair. People of varying cultures would be on display, wearing their native clothes and going about their tasks of daily living all for the viewing pleasure of onlookers. Sometimes this meant engaging in practices that were not necessarily routine but would certainly awe or shock the audience. Some people actually froze to death because they did not have clothing suitable for the vastly different climate. Although racial superiority could not be demonstrated, human zoos were quite popular and one of America's greatest shameful acts of human exploitation. America wasn't alone, though. Europe was also peppered with such exploits.

Human zoos also paved the way for the eugenics movement described in Chapter 2. Without the benefit of modern science and scientific literacy, men like lawyer and self-proclaimed anthropologist Madison Grant (1865–1937) were able to advance the notion that whites were indeed the superior race. In 1918, Grant published *The Passing of the Great Race, or the Racial Basis of European History*. Racists, including Adolf Hitler, were fans of the book.

The human zoos were simply human exhibitions—areas to exploit native cultures and advance the notion that differing ways of living equate to humans of lesser value. To date there is still no scientific evidence to support different human species.

The Tuskegee Study

The Tuskegee Study is a very early study in racism perpetrated by the United States government. From 1932 to 1972, the U.S. Public Health Service conducted a clinical study called "The Tuskegee Study of Untreated Syphilis in the Negro Male." Its purpose was to study the natural course of untreated syphilis. The 600 African American men who took part in the study were impoverished sharecroppers from Macon County, Alabama. About two thirds of the men had contracted syphilis before the study began, and the others did not have the disease. Why did these men participate? They were led to believe they were receiving free health care from the United States government. The men were told that the study was only going to last six months and they would be given meals, free medical care, and free burial insurance if they took part in the study. They were also told they were being treated for "bad blood" which was the colloquial term used for syphilis infection. Figure 3.1 shows a blood draw from a patient in the study.

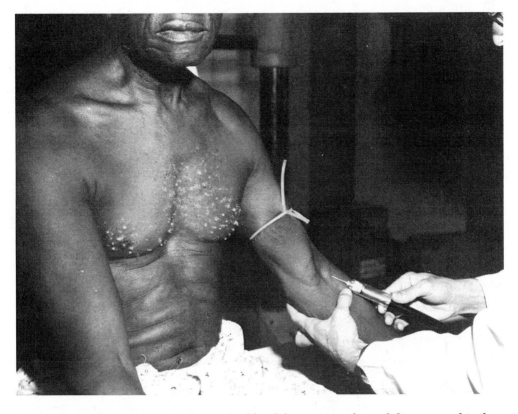

Figure 3.1: Photograph of a man having his blood drawn. Note the syphilis sores on his chest (National Archives at Atlanta).

The Tuskegee study ranks as one of the most horrific, racially-motivated, unethical studies ever performed. None of the men were ever told whether they had the disease and those that did have it were not treated, despite the fact that penicillin was known as an effective treatment against syphilis since 1947. Because of these practices, 40 wives contracted the disease and 19 children were born with congenital syphilis. A whistleblower, Peter Buxtum, eventually revealed the atrocity. He was a former employee of the U.S. Public Health Service who leaked the story to the press.

Healthcare providers withheld treatment and information. We currently have laws in the United States in which healthcare and reproductive information can be withheld from women seeking abortions. The gravity of withholding information, facts, data, and guidance cannot be measured. Knowledge is needed so that informed decisions can be made.

Experiments done on humans should have protocols in place to ensure protection of the participants. Studies done today require informed consent, the opportunity for participants to leave the study at any time, and the dissemination of information. Outside of Nazi Germany, the Tuskegee incident is the most egregious abuse of human subjects.

Today, we have federal laws and regulations requiring Institutional Review Boards (IRBs) to protect human subjects because ethical oversight is essential when conducting any sort of experiment or study on humans. Before any experimentation on people can be started, the study must be approved by an independent ethics committee who determines if the study can move forward. It reviews appropriate steps and measures to protect the rights and welfare of human subjects and to ensure that the study is appropriate and ethical, safeguarding humans so that atrocities like the Nazi doctors performed do not occur.

The genesis of such oversight began with the National Research Act of 1974, which created the National Commission for the Protection of Human Subjects of Biomedical and Behavioral Research. It ensures that people are treated ethically and are respected. The guiding principles are respect for persons, beneficence, and justice. For research to be conducted at academic institutions, the proper paperwork must be filed with the IRB who must approve the research. After much discussion and intense deliberation, the Commission published *The Belmont Report* in 1976. This report "identifies basic ethical principles and guidelines that address ethical issues arising from the conduct of research with human subjects." This document is so important that the Office for Human Research Protections of the U.S. Department of Health & Human Services has a website devoted to it so researchers and participants alike can access important resources.

Racism in Medicine

Myths about physical and physiological differences between black people and white people were popular during the times of slavery and many are still with us today. That black people are innately different than white people is a preposterous assumption, but to "prove" the point, medical experiments were performed on slaves, the likes of which rival those conducted by the Nazis on Jewish people held hostage in concentration camps. We know of these experiments because one fugitive slave, John Brown (c. 1810–1876), lived to tell and write of his horrific tale at the hands of Dr. Thomas Hamilton, who used Brown as a human guinea pig.

In telling his story, Brown said that Dr. Hamilton was fixated on showing that black skin was thicker than white skin. In trying to prove his point, Hamilton blistered the skin on his hands, legs, and feet repeatedly leaving permanent scarring. The thinking at the time was that black bodies were much different than white bodies and that black people had larger sex organs, smaller skulls, and higher pain tolerance than white people. This was also accompanied by a belief that black people had weak lungs, that could only be strengthened by hard labor. Of course, this is not true, but it fit into the racist ideology. See Figure 3.2 of escaped slave Gordon, also known as "Whipped Peter," showing his scarred back at a medical examination in Baton Rouge, Louisiana, in 1863.

Dr. Hamilton wasn't the only physician promoting higher pain tolerance pseudoscience. Dr. J. Marion Sims (1813–1883), who is considered to be the father of modern gynecology, experimented on black women by cutting their genitals repeatedly in an effort to perfect his surgical precision in repairing vesico-vaginal fistulas. His practicing was done throughout 1845–1849 without informed consent (of course) and without anesthesia, causing women to endure excruciating pain and disfigurement. Moreover, he wrote about this in his autobiography, *The Story of My Life*, with exacting detail, describing the pain and agony these women experienced.

Thomas Jefferson also wrote of unproven theories regarding perceived racial differences between blacks and whites in *Notes on the State of Virginia*. Then, there was Dr. Samuel Cartwright (1793–1863) of the University of Louisiana, which is now Tulane University. In 1851, he published a paper, "Report on the Diseases and Physical Peculiarities of the Negro Race," in the May issue of *The New Orleans Medical and Surgical Journal*. In addition to advancing the weaker lungs fallacy, he also purported that enslaved people were also prone to "drapetomania," a made-up name for a disease of the mind that caused enslaved people to run away. The preventive measure for the disorder was for slave masters to "whip the devil out of them." This paper was widely circulated, read, and believed.

Racial bias in pain management, including assessment, perception, and treatment, persists to this day. In a 2016 study of 222 white medical students and residents, researchers found that half the respondents endorsed at least one health myth regarding black people. These healthcare providers and providers-in-training believed that black people feel less pain than white people, that black skin was thicker than white skin (so much for that gross anatomy course), and that the nerve endings of black people were less sensitive than whites. These findings may explain why black people are prescribed less pain medication than white people as well as identify racial disparities in our present-day healthcare system. A quick literature search

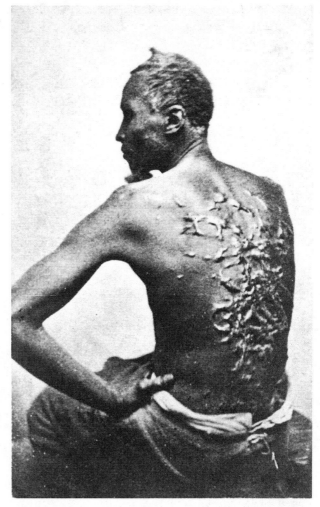

Figure 3.2: Scarred back of escaped slave, photographed by McPherson & Oliver. Baton Rouge, Louisiana, 1863 (Library of Congress).

shows that the information regarding racial disparities is out there—that is, thousands of studies have been conducted and published in peer-reviewed journals identifying racial bias in medicine.

The COVID-19 pandemic has exposed racial health disparities, particularly with Black Americans, who were hit hard. Black Americans with diabetes were especially at risk of dying from the novel coronavirus. Sadly, black patients also lose limbs at triple the rate of other patients, largely as a result of healthcare inequity. While medicine is making great strides in treating diabetes mellitus, the amputation rate across the country increased by 50 percent between 2009 and 2015. According to research gathered by *ProPublica*, 130,000 amputations are performed on diabetic patients, often in low-income and underinsured neighborhoods. Oftentimes, critical preventive care or tests are not covered by insurance, thus patients become chronically sick, and the end result is amputation.

Dr. Foluso Fakorede, a cardiologist in Bolivar County, Mississippi, is working to end this Black American amputation epidemic. Using maps identifying average annual amputations in 2007–2009 and enslaved populations in 1860, Fakorede discovered that amputations from vascular disease—the underlying reason for amputation—closely tracked with slavery. He realized that amputations are a current form of racial oppression. Since the article first published in May 2020, the American Diabetes Association (ADA) has included an initiative to prevent unnecessary amputations as part of a plan to reduce racial disparities in diabetes care. The ADA wrote the Health Equity Bill of Rights to help lawmakers draft policies addressing racial disparities.

After I wrote a blog about skin color, I received quite a few questions and positive comments. The questions I received mirrored those that I field from students in the classroom. All are legitimate, for it shows just how much our society really does not understand basic human biology; and it is refreshing to see that people really are keenly interested and curious. At the top of the list are the queries regarding Kenyan, Ethiopian, and Jamaican runners: Why are they so good? Do they have different musculature and bone structure? Would it be factually correct (but obviously not politically correct) to equate various human races with various dog breeds? To unpack these questions and assuage my own curiosity, I delved into the medical literature just to see how many scientific studies had been done on these runners. Google Scholar revealed 5,980 results for "Kenyan runners," 6,820 results for "Kenyan and Ethiopian runners," and 3,170 results for "Jamaican runners." It's no surprise that there are so many citations because exercise physiology is an exciting field of study and regardless of the sport, elite athletes are continuously studied to figure out competitive edges.

The professional literature is ripe with research, no doubt. But what the average person wants to know is whether these dark-skinned people have something in their genetic composition that makes them stand apart from white-skinned people. Basically, black people must obviously be different than white people or else black runners would not dominate sprinting and distance running. The findings show that there is nothing unique regarding genetic or physiological characteristics. These people are good at what they do through a combination of environmental factors, psychological motivation, and training regimens. Sports, race, and ethnicity are complex issues entangled in anthropology, gender, history, philosophy, politics, racism, science, sociology, and globalization. Rigorous training of select muscle groups will indeed change the way they and the underlying skeleton develop, but it will not change a person into something else. Recall the days when people speculated that there was something profound about cyclist Lance Armstrong's physiology that enabled him to win race after race including seven Tour de France titles. It was later discovered that he cheated by using erythropoietin (EPO), blood transfusions (blood doping), testosterone, and corticosteroids—all of which enhanced performance. Elite, dark-skinned runners and white cyclists are not a different species or subspecies with superhuman powers. They are still human, still *Homo sapiens*.

Another question I am often asked centers around Pygmies. As in, are Pygmies human? The answer is yes. Pygmies are peoples, short in stature, with dark skin who live a nomadic, hunter-gatherer life in tropical rainforests. Their average height is approximately 4 feet 11 inches. Biological and anthropological studies find that the small body size confers selective advantages for humans living this type of existence.

We're More Alike Than We Are Different

As genetic beings, we simply follow the laws of nature. Humans across the planet are 99.9 percent genetically identical to each other, regardless of where they live. Stop to ponder that. With all the people on the planet, we are nearly all the same: Only .1 percent of our genes make us different. According to the most recent research, there are only 8 genetic variants in 4 regions of our entire human genome that have a strong influence on skin color. To put this in perspective, there are around 24,000 genes. Some code for proteins and others do not. We look different because our physical appearance has been shaped by adaptations to our environments throughout the past 200,000 years. Skin color evolved as a response to environmental UV levels and represents a trade-off reminiscent of Goldilocks and the three bears: too much sunlight destroys folic acid, too little sunlight inhibits vitamin D production, but some sunlight is just right.

Humans today can trace a portion of their genetic ancestry back to Mitochondrial Eve. Who is Mitochondrial Eve? She lived 200,000 to 150,000 years ago in Africa and is the common ancestor to us all. She represents an unbroken line purely through mothers. Mitochondrial DNA—the genetic blueprint found in the powerhouses of our cells, is passed down the matrilineal line. Mitochondria exist outside the cell's nucleus. This unique DNA is passed down from females to their offspring of both sexes and is separate from the rest of our DNA. Think of it as a little package in the mitochondria partitioned from the other DNA. The matrilineal line is the kinship each of us has to our maternal line and is passed down from generation to generation. This makes Mitochondrial Eve our theoretical ancestor, linking all of us together.

While we do have genetic variation, the differences are more than skin deep. But there are some interesting points to consider about medical conditions linked to our skin and by extension, our *various populations.* While genes do not show racial differences, we do know that there are some disorders linked to populations, and these populations just happen to have the same skin color. Keep in mind what we just learned about population genetics and hemoglobin in red blood cells before reading the next sentences.

Examples of disorders often linked to populations include sickle cell disease and thalassemia. Sickle cell disease is frequently associated with black people, while the condition thalassemia is connected to Mediterranean people. There's good reason for that, but it has nothing to do with race and everything to do with location. Let's break it down. Both of these conditions are inherited hemoglobin disorders.

With sickle cell disease, an abnormal form of hemoglobin is produced causing the red blood cells to curve into a semicircular, sickle shape. Normal red blood cells are nice, round, biconcave disks that can travel freely within the confines of our tubular blood vessels. Sickled red blood cells are problematic because they can't travel freely, and they stack up like dinner plates in the blood vessels, which cause blockages. If oxygen-rich blood can't get to our cells, the cells die. If this occurs over a long enough period of time to enough vital organs, the person dies.

Thalassemia is a blood disorder of hemoglobin metabolism in which the globulin portion of the hemoglobin molecule is impaired, causing the body to have less hemoglobin than normal. With less hemoglobin, there is a reduced ability to carry oxygen, resulting in anemia, leaving the person very tired. Both sickle cell disease and thalassemia are inherited. And both evolved to protect populations from malaria. Areas where malaria is endemic are equatorial zones, subtropical areas, and tropical regions—areas where people had darker skin. Malaria is caused by the *Plasmodium* protozoa that invades red blood cells. The disease is characterized by intermittent and fluctuating fever. It is transmitted to humans by the bite of an infected female mosquito that previously sucked the blood from a person with malaria.

Both sickle cell and thalassemia traits evolved to help these populations resist a potentially fatal disease. To understand, we need to know another term: allele. This term is pronounced as "uh-LEAL," in which "leal" rhymes with feel. An allele is one of two or more alternative forms of a gene that arise by mutation. They are found on the same place on a chromosome. For every gene, we get one allele from our mother and one allele from our father. The allele for sickle cell disease is pleiotropic. Pleiotropy is a genetic term and refers to one gene influencing multiple, seemingly unrelated effects. If a person is homozygous, referring to having identical alleles (Hb^AHb^A) for normal hemoglobin, then that person does not have sickle disease but is susceptible to malaria. If another person is heterozygous, having different alleles (Hb^SHb^A) for hemoglobin, then that person does not have sickle cell disease and is immune to malaria. If a third person is homozygous for sickled hemoglobin (Hb^SHb^S), then that person has sickle cell disease and is immune to malaria. If individuals are carriers for either sickle cell disease or thalassemia, the protozoa has difficulty entering the red blood cells due to their size or surviving due to their structures. This is also can example of the *founder effect*.

The Founder Effect

The founder effect is responsible for a high frequency of a gene in a particular population derived from a small set of ancestors. Much like environmental pressures influencing skin pigmentation, a founding ancestor carried a mutation to a specific region. Other examples of "founder effect" conditions include Tay-Sachs disease and cystic fibrosis (CF). Tay-Sachs disease is found among Ashkenazi Jews whose DNA links them to people from the Middle Ages, and cystic fibrosis is found among people of northern European descent. Individuals who are heterozygous for cystic fibrosis have an advantage to surviving cholera, those with Tay-Sachs disease are resistant to tuberculosis (TB).

Tay-Sachs disease is an inherited metabolic disorder in which certain fats accumulate in the brain leading to blindness, muscle spasticity, seizures, and death in childhood. Crowded living conditions can lead to tuberculosis; the high prevalence of Tay-Sachs disease and low prevalence of *Mycobacterium tuberculosis* among Ashkenazi Jews suggests a protective effect of the disorder. Cystic fibrosis causes the production of abnormally thick mucus that blocks pancreatic ducts, intestines, and the lung passages. There was a time when cholera, caused by the bacterium *Vibrio cholerae*, decimated populations by dehydration. The bacterium caused profuse watery diarrhea, extreme fluid and electrolyte loss, which led to dehydration and total collapse. In individuals with CF, cells have a defective chloride channel that causes thick, clogging mucus. Individuals who are carriers for the CF gene, but do not have CF, may be protected by the configuration of the chloride channel by preventing the dehydrating effects of the bacterium. If people could survive the profound dehydration, they could survive cholera.

Did skin color have a role in attaining or thwarting these diseases? No. Did evolution have a role in protecting the species from annihilation? Yes. Science now understands that some genetic conditions, not skin color, may confer resistance to other infectious diseases. Melanin is only skin deep. If you remove the skin, every organ, organ system, and body structure look the same regardless of where the person originated. Do a few cadaver dissections for yourself and you'll quickly see the point. Or look at a few X-rays and try to tell the race of a person. If neither of these is an option, ask a surgeon or an anatomist.

To this day, skin color is significant as it drives racism. Racism is the prejudice, discrimination, or antagonism directed against someone with a different skin color. The history of the United States is rich with racial discrimination and slavery. Although the 13th Amendment to the U.S. Constitution abolished slavery in 1865, people of color are still underrepresented in positions of power and prestige. The 117th Congress represents the most racially and ethnically diverse to date. That's good news. Yet, as a whole, it remains disproportionately white when compared with the U.S. population. As humans travel more and have more children with people who look less like them, perhaps the concept of race will disappear. When we start looking at each other and treating each other as human, we can wipe away invented notions of superiority and inferiority based on skin color.

To conclude, I have dissected countless cadavers and human body organs, and I can assure you that once you remove the skin, we all look pretty much the same. Which is why I am left wondering: How can something as thin as a sheet of paper cause so much hatred in the world?

TAKE-HOME MESSAGE

Evolution occurs through natural selection and the resulting cumulative changes occur in a population. Skin color is determined by genes, and racial categories are not shown in our human genome.

> *I have a dream that my four little children will one day live in a nation where they will not be judged by the color of their skin, but by the content of their character.*
> —Martin Luther King, Jr. (1929–1968); American minister,
> Civil Rights activist, Nobel Peace Prize winner

4

Jungle Love

Sexual Identity

U.S. Landmark Case: Obergefell v. Hodges, 135 S. Ct. 2584 (2015);* (5–4 decision)

Is same-sex marriage a constitutionally protected right? Since the 1970s, this issue has been raised in various parts of the United States. In 1993, Hawaii's Supreme Court suggested that its state prohibition against same-sex marriage might be unconstitutional. This caused both state and federal actions leading to the Federal Defense of Marriage Act (DOMA). In 2004, Massachusetts became the first state to legalize same-sex marriage. Opponents in many other states rushed to enact restrictive state statutes and constitutional changes, generally making it law that marriage can only be between a man and a woman. Yet many other states embraced the same-sex marriage concept and legalized such unions. In 2013, the U.S. Supreme Court decision in *United States v. Windsor*, 133 S. Ct. 2675, struck down the federal prohibition to recognizing same-sex marriage. By 2014, the majority of the U.S. population (70 percent) lived in states that allowed same-sex marriage.

Then, in 2015, in a narrow 5–4 decision, the Supreme Court made marriage equality the law of the land when they held that the Due Process Clause of the Fourteenth Amendment guarantees every citizen the right to marry as a fundamental liberty and that this liberty applies to same-sex couples the same as it applies to opposite-sex couples. With this decision, same-sex marriage is legal across all 50 states and all U.S. territories.

Being Gay

In the introduction you read the marriage story of Jim Obergefell and John Arthur. Their decision to marry led to this U.S. Supreme Court case, which is a consolidation of all same-sex marriage cases from the Federal Sixth Circuit Court. This landmark decision was monumental because it helped gay people feel less like second-class citizens: they now could enjoy the same rights that every other American could. Well, not exactly. Individuals, regardless of their sexual orientation, can marry whomever they want, but discrimination against such individuals is still rampant.

---

*Together with No. 14–562, *Tanco et al. v. Haslam, Governor of Tennessee, et al.*, No. 14–571, *DeBoer, et al. v. Snyder, Governor of Michigan, et al.*, and No. 14–574, *Bourke, et al. v. Beshear, Governor of Kentucky.*

Being anything other than heterosexual is still difficult in the United States and many parts of the world. Many believe that sexual identity, also referred to as sexual orientation, is a choice. This begs the question: If it were a matter of personal choice, so what? Sexual orientation is a person's identity and trying to place people into binary categories is contrary to scientific evidence, as this chapter shows.

It is important to note that writers, authors, and the general public should avoid heterosexual bias in language. Since language is a dynamic entity that changes with time, doing so can be particularly difficult. We get into terminology trouble whenever we label, which is yet another reason why simply using the term "human" for us all makes sense. Yet, we want precision in our words, which is why many advocacy groups as well as writing style guides, offer assistance. For example, using the term "homosexual" tends to be associated with negative stereotypes. The Human Rights Campaign (HRC) provides an excellent guide to help with terminology. Under the website's resource tab, you can find a useful glossary of terms, written to "help give people the words and meanings to help make conversations easier and more comfortable." In this book, the word "homosexual" will be used where appropriate because its use is still widespread and much of the cited research was conducted using the term.

It's often difficult for people to move beyond sexual identity as a binary feature, meaning every person is either male or female. We've been socialized to believe certain things regarding sexual attraction (who we have a romantic interest in), gender expression (the way we "present" to the world through our actions, clothes, and appearance), and masculine and feminine traits (societal pressures and stereotypes related to aggression, athleticism, emotions, caregiving, and others). If people don't conform to our stereotypes or "societal gender norms," many people don't know how to react. Yet, many of us do not express ourselves as totally "masculine" or totally "feminine." It's part of the human condition to be unique. However, cultural pressures certainly can make it difficult for people who do not conform to traditional stereotypes. Macho men can be gentle caregivers like demure women can handle jackhammers. Not all people fit into societal expectations of masculinity or femininity, which in real life cover the spectrum. We can hope that with time, widely held stereotypical images can be recast.

From a historical perspective, the medical establishment has not been kind to gay people. To note, until 1973, the *Diagnostic and Statistical Manual of Mental Disorders* (DSM), a clinical guide published by the American Psychiatric Association, listed "homosexuality" as a disorder. DSM-I (printed in 1952) and DSM-II (printed in 1968) had pathologized homosexuality, but DSM-III (1980) and subsequent editions (the current edition is DSM-5, printed in 2013) removed homosexuality as a disorder, because science had shown it to be a normal state of life and not a clinical condition. As a result, cultural attitudes have changed. So, if there is nothing wrong with being gay, why do moral and legal implications come into play? If a goal of society is to advance sound science, it seems that we should focus more on how to improve the physical and mental health of individuals who identify as something other than heterosexual.

If genes determine skin color, do genes determine sexual orientation? Given what we know about gene expression, it is very likely that complex interactions between multiple genes and the environment play a role in sex, gender, and sexual identity. Thus, like any relationship, it's complicated.

Scientists agree that sex, gender, and sexual identity are inter-related constructs. Here are terms tossed about routinely for sex and sexual identity: male, female, transgender, cisgender, queer, genderqueer, intersex, pansexual, homosexual, heterosexual, gay, straight, lesbian, asexual, questioning, gender fluid, bisexual, curious, and "no label describes me accurately." There are no neatly packaged terms with strict definitions, so confusion abounds.

Approaching this from a human biological perspective, let's begin with some working definitions for our discussion, noting that there are no clear lines of distinction, thus some overlap can exist: "sex" = members of a group based on reproductive function, knowing there can be variation; "sexual identity/sexual orientation" = how one feels about oneself and who they are romantically attracted to; "gender identity" = who you are as a person.

Sex as a label generally falls into two categories, male and female, known as binary sexing. You hear it in phrases such as "Is the baby a boy or a girl?" and "Individuals are the sex they were assigned at birth." At issue in both cases is that the decision-makers who note sex on birth certificates are deciding based on how the external genitalia appear. However, in the natural world, not all people are born either male or female. Babies can be born with genitals that are ambiguous, meaning it is difficult to determine the sex; some of these infants are labeled intersex.

Spending time on labels is important because federal policies are implemented using faulty reasoning or scientific ignorance. To illustrate, the Trump Administration tried to define gender as being binary: one is either male or one is female. Period. They claimed that one can define sex "on a biological basis that is clear, grounded in science, objective, and administrable." It further defines one as being male or female based on "immutable biological traits identifiable by or before birth" and that "the sex listed on a person's birth certificate constitutes proof of a person's sex unless rebutted by reliable genetic evidence." This is blatant denial of science.

Are you aware that within our legal system there is such a thing as the "Gay/Trans Panic Defense"? According to the American Bar Association, this legal defense legitimizes and excuses violent and lethal behavior against LGBTQ people, who comprise 5.6 percent of the total population yet account for nearly 20 percent of hate crime attacks. Gay/Trans Panic Defense is a strategy in which a person claims self-defense against an unwanted "homosexual" advance. People have been brutally assaulted and killed by perpetrators who claim they acted out of violent temporary insanity because of unwanted same-sex sexual advances. This defense is still legal in 42 states. California, Illinois, and Rhode Island, Nevada, Connecticut, Maine, Hawaii, and New York have banned such defenses. The LGBT Bar is leading an effort to ban the "panic" defenses in the other states.

Yes, people, this is real. On October 12, 1998, 21-year-old Matthew Shepard died of injuries sustained from being beaten, pistol-whipped, and left out in the bitter cold in Laramie, Wyoming. His murderers unsuccessfully claimed gay panic for their defense for the brutal murder. The website of the LGBT Bar lists other relevant cases. In July 2018, the Gay and Trans Panic Defense Prohibition Action of 2018 was introduced by Massachusetts Democratic Senator Ed Markey and Massachusetts Democratic Congressman Joe Kennedy. The bill to ban homophobia as a legal defense for anti-gay hate crimes is still pending. This is at a time (2020) when hate crimes have risen 6 percent for gays and 34 percent for the transgender community.

If we simply stayed with the classifying phrase "human" and accepted each other as "human," these other labels would not matter. This does not mean that people should not embrace their personal identity. Everybody should be proud of who they are, be okay with who they are, and accept us all as humans. Why does it matter? Well, from a biological perspective it does not.

Chromosomes

In truth, to know with certainty the sex of humans, we need to look at our chromosomes, and in order to do that, we need a karyotype. Chromosomes contain our DNA, which contains a linear sequence of genes that determine who we are. The non-sex chromosomes are called autosomes or somatic chromosomes, and they are referred to numerically according to a traditional sorting order based on size, shape, and other properties. The 23rd pair of chromosomes comes in two forms, X chromosome and Y chromosome. Humans usually have 46 chromosomes: 22 pairs (44 total) of non-sex chromosomes and one pair (2 total) of sex chromosomes, designated XY (male) and XX (female). Chromosome number varies by species, so your dog has 78 chromosomes and your cat has 38 chromosomes. Look at Figure 4.1 to see a male karyotype that contains 22 pairs of non-sex chromosomes and one pair of sex chromosomes.

Chromosomes are extremely compact and nearly every cell in the body has a complete set. Some chromosomes are longer than others because they contain more DNA, thus chromosomes can be arranged by size. We cannot see chromosomes with our unaided eyes, but when the cell gets ready to divide, the chromosomes wind up like squiggly worms and become tightly packed. At this point, the chromosomes can be viewed using high powered microscopes. Photographs of chromosomes can be taken at this point and used to make karyotypes, arranging them from largest to smallest and numbering them from 1 to 22 plus the sex chromosomes. For example, a karyotype is made by looking at the number, appearance, and arrangement according to size, and the chromosomes are named numerically: chromosome 1, chromosome 2, and so on.

You might be familiar with a karyotype if you've seen the results from amniocentesis or chorionic villus sampling, two procedures performed on pregnant women to test for chromosomal abnormalities in developing fetuses. Humans usually have 46 chromosomes: 22 pairs (44 total) of non-sex chromosomes and 1 pair (2 total) of sex chromosomes, designated XY (male) and XX (female). Using medical shorthand, a female—girl with a vulva, vagina, uterus, uterine tubes, and ovaries—is represented as 46, XX. A male—boy with a penis, testes, and prostate—is 46, XY.

Society provides the gender identity label. For any of us to know "who" we really are, each of us would have to have our sex chromosomes analyzed. With a few exceptions, individuals have two copies of sex chromosomes. This is known as binary nomenclature meaning there are 2 sexes. Be that as it may, so much can happen not only during embryogenesis—the time when we are forming in our mother's womb—but as the individual becomes a person.

An embryo is an organism during the very early stages of development and embryogenesis is the time marked by the development of an embryo. It happens across animal species, and in humans, embryogenesis takes about 8 weeks from the time of conception (fertilization). After 8 weeks the developing organism is called a fetus. (Chapter 5 explores human reproduction in greater detail.) This sounds rather simple, but it is not. Because in our natural world, there are not just 2 sexes.

Karyotyping is not reserved solely for prenatal testing. It also helps us know about ourselves and can help us understand why sex cannot always be determined at birth simply by looking at a newborn's genitals. About 1 in 2,000–5,000 live births display atypical genitalia, meaning that a person is born intersex. Intersexuality results when the person has both male and female characteristics, has non-distinguishable genitalia, or has atypical genitalia. In an attempt to categorize people, biological characteristics don't line up neatly. To illustrate, a person can have XY

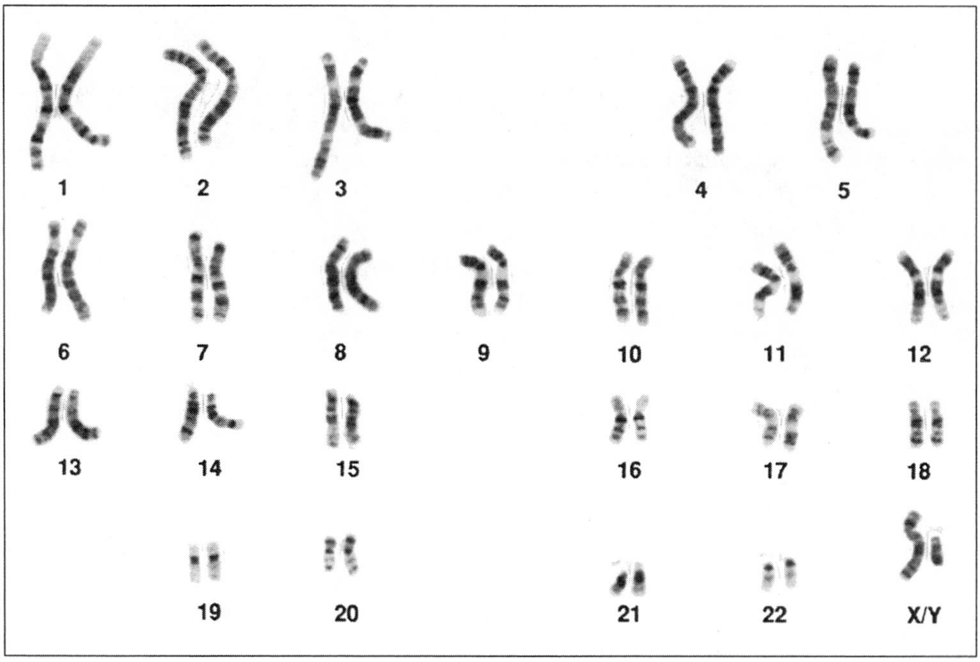

Figure 4.1: Human Male Karyotype (National Cancer Institute).

chromosomes yet be born with female-appearing genitalia. Intersex people are born with any of several variations in chromosomes, reproductive organs (both internal and external), and hormones that do not fit binary sexing, thus "intermediate" between the sexes. An older, inaccurate term for intersex people is *hermaphrodite*. The term *hermaphrodite* is born out of Greek mythology and implies that a single person could be both totally male and totally female, which is anatomically and physiologically impossible.

Intersex characteristics are not always present at birth and some people do not know unless they have been genetically tested. For example, an intersex person may have ambiguous testes and ovaries, but may be unaware. Ambiguous genitalia can be surgically modified to resemble male or female genitals.

Intersex isn't the same as being transgender. Gender is a complicated issue because so many hormonal, physical, psychological, and sociological factors are involved. Gender cannot fit into a nice tidy box, either. According to the Intersex Society of North America, intersexuality is not a discreet category. Some people may be born with mosaic genetics, which means that some cells have XX chromosomes (46, XX) and some cells have XY chromosomes (46, XY). Some people may also go through life not even knowing they are intersexual. Think about that for a moment. Have you ever been genetically tested?

Demonstrating that people cannot be neatly partitioned, is the intersex case of South African distance runner and 2016 Olympic gold medalist Caster Semenya, who competes and identifies as a woman. She is a female athlete with hyperandrogenism, a condition in which her body naturally produces increased amounts of testosterone. In May 2019, the Court for Arbitration in Sport in Switzerland ruled that she must take drugs to suppress her natural testosterone level to compete as a woman. Semenya is aged 29 and holds two Olympic gold medals.

There are also other examples in which people have too few, too many, or a mixture of sex chromosomes: 47 XXY male (Klinefelter syndrome), 47 XXX (Triple X syndrome), 47 XYY, and 45 XO (Turner syndrome). The O in Turner syndrome indicates a loss of the 2nd sex chromosome, so it is written as XO. Knowing our sex can help explain some physical characteristics and reproductive capability, but it cannot totally explain our sexual identity.

Several sentences back, XY meant the person was male. But a person can have XY chromosomes and have female-appearing genitalia because their cells lack receptors for the sex hormone testosterone. If the cells lack receptors for testosterone, the cells don't "know" to develop male-appearing genitalia. This can also happen another way: the person's biology is such that they lack 5-alpha reductase, which is an enzyme that converts testosterone into dihydrotestosterone (DHT), which is the active form of testosterone and is formed from testosterone in body tissues. If a person lacks this enzyme, they will appear female at birth but later in life when they go through puberty, they will develop male sexual characteristics.

And now, for a terrible part of our history: Babies who were born with ambiguous (inexact) genitalia underwent "normalizing" surgeries. This still happens today. Examples include physicians cutting off small penises and testes and surgically creating vaginas. Thus, a baby was "assigned" a sex. As the child grew, more surgeries were necessary, life-long medications were given, and the person was rendered infertile. There are other possible genetic considerations as well. In 2017, Germany became the first country to recognize a third gender of babies as X. Thus, babies may be listed on their birth certificates as something other than just male or female.

There are many non-sex chromosomal variations that also affect sexual development. A familiar example is Down syndrome, or trisomy 21. Individuals with trisomy 21 have an extra copy of chromosome 21. According to the Centers for Disease Control and Prevention (CDC), about 1 in every 700 babies born in the United States has Down syndrome. In real numbers, about 6,000 babies with Down syndrome are born in the U.S. each year. The extra genetic information can come from either the mom or the dad. It is usually fatal to have one too many or one too few of the non-sex chromosomes. And fetuses with extra or too few chromosomes usually spontaneously abort (miscarry). (This topic is covered in Chapter 5). However, trisomy 21 is a special case and the person lives. Women with Down syndrome are able to birth children, but many men are infertile. Although these women maintain pregnancy, there is a 35–50 percent chance that their baby will have Down syndrome. The cause of infertility in males with Down syndrome is thought to be linked to poor sperm development.

Sexual Behavior

Genes do interact to determine who we are and by extension our sexual behaviors. Sexual orientation is innate, meaning we are born with an inherent romantic, emotional, or sexual attraction to another person. Finding a specific "gay gene" per se has been elusive because most genes code for proteins—they do not encode for behaviors. Plus, multiple genes generally code for multiple proteins, not just one. Furthermore, genes interact with environmental factors to influence genetic expression, and the environment is not static. Moreover, sometimes genes in a cell are "off" and sometimes they are "on," so when a particular environmental factor has the ability to exert influence, the gene may not be receptive.

Whether this topic is a source of fascination or an impetus of hatred, it is important to note that homosexuality is normal. And humans are not the only species to be homosexual.

Homosexuality in the non-human animal kingdom does exist and has been observed in both males and females among 1,500 species of mammals and birds, including gay lions in the West Midland Safari Park in England, gay gorillas in the Rotterdam Zoo in the Netherlands, and gay penguins in Australia at Sydney's Sea Life Aquarium.

The gay penguins in Australia are making news because they've hatched a chick. How? When a heterosexual penguin couple left one out of two eggs exposed, aquarium workers gave the exposed egg to Sphen and Magic, who had already built a nest and sat on the egg constantly until it hatched. They are still doting parents. From a biological perspective, homosexuality in the animal kingdom may seem paradoxical since these animals cannot reproduce. However, it is normal behavior in the animal kingdom as a whole.

Another interesting phenomenon occurs as well: gynandromorphy. The term is derived from the word parts *gunandros* meaning *of doubtful sex*, and *morphe*, meaning form. Gynandromorphic individuals, found in some bird and other species, have some male and some female characteristics. In birds, one side can be male and the other side can be female. For example, coloring and corresponding sexual organs on one side of the body typify feather coloring and genitals of males while the other side typifies coloring and genitals of females. Studying gynandromorphs in the natural world is difficult; however, researchers suspect that two sperm fertilize an egg containing two nuclei instead of the normal one nucleus, or perhaps it results when the sex chromosomes fail to separate. The condition is not seen in humans because cellular mechanisms prevent more than one sperm from entering an egg cell.

Per our earlier discussion, it should also be noted that gynandromorphism is different from hermaphroditism, a condition in which an individual has a full complement of both male and female sex organs. Examples of hermaphroditic organisms include snails, earthworms, and some fish.

While the search for a "gay gene" continues, there is some evidence supporting the genetics of homosexuality. For example, fraternal birth order may be predictive of male gayness, also termed androphilia. The greater the number of biological older brothers, the increased likelihood of androphilia. This has also been termed the maternal immune hypothesis and suggests that there is progressive immunization of the mothers to male-specific antigens that are involved with fetal male brain masculinization. This means that with each subsequent pregnancy, the mother is exposed to more of these male antigens, thus increasing the odds with each male birth that the next baby boy will be gay.

Using the term "preference" is a tad tricky because preference suggests people have a voluntary choice when deciding about having sexual encounters with another person. While this may ring true for individuals who identify as heterosexual, for people who identify as something other than heterosexual, it assumes that a person's sexuality is a matter of choice. Thus, when then Supreme Court nominee Amy Coney Barrett used the term "sexual preference" during her confirmation hearing in October 2020, many in the LGBTQ community were rightly outraged.

Finding a so-called gay gene has been elusive, but the quest continues. For example, in a study involving nearly 500,000 people, scientists were unable to find the so-called gay gene. This study also showed that some people who have had sex with people of the same sex do not consider themselves to be gay. Sexual identity and sexual behavior is quite fluid across the entire human spectrum. Perhaps our sexuality is genetically influenced, but not genetically determined. Like other human behaviors, such as attraction, personality, and addiction to name a few, sexual focus is complex.

Given that our purpose is to procreate (remember, from a pure biological perspective, we're here to carry on the species), shouldn't the "gay gene" disappear from the gene pool? If gay people can't reproduce because sex with the same sex cannot produce offspring, why are gay people still present in humanity? The reason is because people who have sex with people of the same sex have also had sex with people of the opposite sex. Bisexual people also have sex with both sexes. And, these sexual encounters have produced viable offspring, who are also capable of reproducing. This has been happening for millennia.

Sexual activity itself is complicated. How we humans experience sex is as varied as the frequency with which one engages in the sexual act. The reasons are plentiful as sex is emotional, behavioral, biological, psychological, and sociological. What constitutes sex is also varied depending on the person, culture, and what sex act will cause arousal. But why discuss it here? Because there's one dark area in our current climate that demands some light. That area is prison rape. And a common misconception is that prisons turn heterosexual people into homosexuals.

Non-consensual sex, or rape, is a common occurrence in our detention centers. Prison rape is the forcing of another person to have sexual intercourse with the offender against their will. Sexual victimization by another inmate or facility staff is a problem. According to a United States Department of Justice Bureau Statistics report, *Sexual Victimization in Prisons and Jails Reported by Inmates, 2011–2012*, sexual assault affects about 4.0 percent of state and federal inmates and 3.2 percent of jail inmates. The Human Rights Watch estimated that at least 140,000 inmates had been raped while incarcerated. Prison rape is almost always same-sex rape. Male prison rapists almost always identify as heterosexuals and the victims are generally heterosexuals as well. It's beyond the scope of this book, but the professional literature is rich with findings that there is more behind prison rape than just the sexual act. Writing about forced sex is difficult, and rape within the prison system oftentimes goes unreported. It's worth noting here that sex is complex, convoluted, and perplexing.

Homosexuality

Sadly, the Republican Party has not caught up with the science. The Republican National Conference, referred to generally as the RNC, took center stage in August 2020. Highlights of the first night showed that the party continues catering to the evangelical Christian right—despite seeing one of its biggest public figures, Jerry Fallwell, Jr., former president of Liberty University and major Trump supporter—resign from his position because of his involvement in a twisted sex scandal. The RNC's platform continues touting a strict definition of marriage as one of solely between a man and a woman. And, quite disturbingly, the party also endorses conversion therapy for LGBTQ+ Americans.

Homosexuality occurs cross-culturally, is worldwide, and it remains a source of public preoccupation. What we do know is that conversion therapy—the practice of trying to make a person something they are not—does not work and is cruel. Conversion therapy is a dangerous practice that has been discredited and based on a false premise that being gay or transgender is some sort of disorder that can be cured. Heterosexual people cannot be made into gay individuals, nor can gay people become straight. A person simply cannot be converted to something they are not.

As you read that last sentence you may be thinking my choice of the pronoun, "they" is incorrect and should be "he" or "she." Pronoun usage is something that has changed with the times. In an effort to eliminate bias, women, regardless of their marital standing are now referred to as "Ms." and individuals choosing not to be identified by binary pronouns of him, her, he, or she are now using "they." It takes some getting used to, but language is dynamic and with time, we'll grow accustomed to it.

Back to conversion therapy. For a harrowing account about this practice, read Garrard Conley's memoir *Boy Erased* or watch the movie of the same name. Conley, the son of an Arkansas Baptist pastor, was outed to his parents when he was 19 years old. He then had to choose between attending a church-supported conversion therapy program or risk losing his family and friends. He also felt he risked losing God. This was a God he had worshipped his entire life. He entered a 12-step program that was supposed to convert him to a straight man. Garrard now lives in New York City with his husband. Gay conversion therapy is still legal in 29 states for minors. Conversion therapy for minors is illegal in these 21 states: California, Colorado, Connecticut, Delaware, Hawaii, Illinois, Maine, Maryland, Massachusetts, Nevada, New Hampshire, New Jersey, New Mexico, New York, North Carolina (partial ban; prohibits use of state funds), Oregon, Rhode Island, Utah, Vermont, Virginia, Washington, and Washington, D.C.

Mainstream mental health professionals along with the American Psychological Association condemn any practice aimed at trying to convert a homosexual person to a heterosexual one. The psychological scarring can be profound and lifelong. In addition to conversion therapy, some methods have included institutionalization, castration, electroconvulsive therapy, induced nausea and vomiting while showing homoerotic images, using shame, or other means to "teach" the person that being lesbian, gay, bisexual, or transgender is a defect or disorder. Science has shown it to be neither a defect nor a disorder.

In the United States, there are many groups that endorse conversion therapy. One such group is the National Association for Research and Therapy of Homosexuality (NARTH). The name sounds credible, but the group is not. Through hypnosis, cognitive behavior therapies, sex therapies, and psychotropic medicines, NARTH attempts to convert its patients. While many of the techniques, such as cognitive behavior therapy and using medications for disorders are credible practices, they are not treatments for LGBTQ people for there is no disorder to treat.

Another horrific story involves Canadian David Reimer (1965–2004). David was born male but raised as a female after his penis was injured beyond repair during his botched infant circumcision. His parents chose gender reassignment surgery. Around age 9, Reimer realized that he was not female. He never felt like a girl and his gender identity was clearly male. Proof that we don't choose who we are, and our genders are not learned. At age 15, he transitioned to male. He went public with his story, hoping to discourage similar medical and psychological practices, but suffered from depression and a troubled marriage. He ultimately died by suicide at age 39.

In June 2019, the Vatican office, which is responsible for overseeing Catholic educational institutions around the world, blasted gender theory. The Congregation for Catholic Education stated that the idea that a person's gender identity falls along a spectrum is "nothing more than a confused concept of freedom in the realm of feelings and wants."

Another significant event occurred in June 2019 when Alabama passed a law requiring sex offenders to undergo chemical castration. Whereas physical castration removes the testicles, chemical castration uses hormones (typically medroxyprogesterone acetate or leuprolide acetate) to decrease the circulating levels of testosterone. The reduced circulating level of this hormone is supposed to decrease so-called pathological sexual behavior. It does reduce sexual urges. It also is used in the treatment of prostate cancer, but in this case, it is called androgen deprivation therapy. Androgens, like testosterone, are hormones influencing growth and reproduction. According to the law, men convicted of a sex crime against a child under age 13 must begin chemical castration and continue it until a court rules he can stop. Critics state that the law violates the Eighth Amendment, which prohibits cruel and unusual punishment. Critics also cite the hypocrisy of the Alabama law stating that in this same state where sex offenders are castrated, women who get pregnant from rape cannot get abortions.

Ten states allow chemical castration for some sex crimes. These states are Alabama, California, Florida, Georgia, Iowa, Louisiana, Montana, Oregon, Texas and Wisconsin. Oftentimes, sex offenders choose this option because it wins or speeds up parole. Chemical castration is generally reversible.

So why are we bothered by homosexuality? I don't know. Several surveys have tried to get at the root of this issue, and it seems that anti-gay people believe there is some secret "gay agenda." Anti-gays also think that if gay people are treated as equals, then somehow they, the anti-gays, lose their own equality. However, this is still not true and before June 2020, gay people in the United States could be fired simply for being gay. In many states it was legal to fire somebody just for being gay or transgender. So what happened? On June 15, 2020, the Supreme Court ruled in *Bostock v. Clayton County, Georgia*, that it is illegal to discriminate against LGBTQ employees in the workplace. This landmark case came close to the 5-year anniversary of *Obergefell v. Hodges* (June 26, 2015), the case that opened this chapter. Both cases involve discrimination against people by reason of sex and sexual identity. Why were there no protections for gay people in the workplace? Do gay or transgender people threaten heterosexual survival? Of course not. Many who oppose homosexuality do so behind the guise of religion—as in, it's a sin to be gay. If a heterosexual person believes it is a sin to be LGBTQ, why work so hard to oppress these people? After all, the so-called sin would be owned by the LGBTQ person, not the heterosexual one.

Another assault on transgender people came in September 2020 when Republican senator Kelly Loeffler introduced a bill that would ban transgender girls from competing in school sports with their gender. The Protection of Women and Girls in Sports Act, co-sponsored in the Senate by Senators Marsha Blackburn (R–TN), Tom Cotton (R–AR), James Lankford (R–OK), and Mike Lee (R–UT), states that "sex shall be recognized based solely on a person's reproductive biology and genetics at birth" in girls' and women's sports. In her tweet on September 22, 2020, @Senator Loeffler stated, "Proud to lead this effort in defense of women & girls. Title IX established a fair chance for girls of all ages to compete—Sports should be no exception. @SenMikeLee @MarshaBlackburn @Senator Lankford @SenTomCotton & I are ensuring a level playing field." The bill does not mention transgender boys. If the bill passes, how sex will be determined remains to be seen.

Beginning on April 12, 2019, transgender people were not allowed to join the U.S. military. This is not the first time there was such a ban. In 1953, President Dwight D. Eisenhower issued Executive Order 10450, placing a ban on gay men and lesbians from working in any agency of the federal government. In 2016, President Barack Obama repealed the military transgender ban. On July 26, 2017, President Donald Trump announced via Twitter that transgender people would no longer be allowed to serve in any capacity in the U.S. military. Keep in mind that according to the National Center for Transgender Equality, there are an estimated 134,000 transgender American veterans and over 15,000 transgender people currently serving in the military. Also keep in mind that President Trump never served a single day in the U.S. military. Transgender people have been serving and dying for this country for many years. When a coffin arrives draped with an American flag, do we know the person's sexual identity, political affiliation, religious bent, or anything else? Would you give up your life for this country? Why do we make such a big deal about oppressing people? Why do we care which restroom somebody uses? In March 2021, President Biden signed an executive order reversing the transgender military ban.

In April 2019, the Texas Senate approved Senate Bill 17, allowing counselors, attorneys, doctors, and other state-licensed professionals to deny services to LGBTQ people based on religious beliefs. Proponents of the bill said it would allow state-licensed professionals the ability to fight to keep their licenses if they were threatened because of faith-based actions. In essence, it gives people the ability to openly discriminate against the LGBTQ community. These are the same church people who say LGBTQ persons are an abomination to their Christian beliefs. Reconcile that message against passages from Matthew (22:34–40) and Mark (12:30–31) in the Christian Bible that state to "love thy neighbor as thyself." The Christian Bible's John 3:18 tells people to love not in word or tongue, but in deed and truth. Basically, actions speak louder than words. Furthermore, the last seven commandments of the Big Ten can be summed up with "love your neighbor as yourself." Does this mean that these state-licensed practitioners are full of self-loathing?

South Carolina's Comprehensive Health Education Act bans sex education teachers from mentioning any relationships other than heterosexual ones. Teachers are allowed to discuss sex education relative to homosexuals if the discussion revolves around sexually transmitted diseases. This is an example of state-sanctioned discrimination against LGBTQ people. Moreover, the law states that any teacher who allows "a discussion of alternate sexual lifestyles" including "homosexual relationships except in the context of instruction concerning sexually transmitted diseases" can be fired.

We're nearing the end of 2021, and dozens of states are considering legislation related to LGBTQ discrimination. These bills include comprehensive nondiscrimination bills, local control blocking bills, anti-transgender bills, and religious refusal bills. Freedom for All Americans is a bipartisan group that works to secure full non-discrimination protection for LGBTQ people in the United States. Their website also tracks relevant bills.

Let's just face it. Gay people—and all other sexual types—are human beings. Period. The landmark Supreme Court case, *Obergefell v. Hodges*, ruled on June 26, 2015, that same-sex couples have the fundamental right to marry. Not allowing people who love each other to marry is discriminatory. If we subscribe to the ideal that every person is created equal, then same-sex marriages should be the rule of the land. However, there has been much opposition to same-sex marriage. Ironically, it comes from Christians, who firmly believe that marriage should be between a man and a woman. Yet, Christians are supposed to love and accept all people, so this seems a bit hypocritical. If every one of us simply took charge of our own lives, paid attention to our personal business, and loved each other, we certainly wouldn't have such issues.

According to the Southern Poverty Law Center (SPLC), an organization committed to social justice issues, fighting hate and bigotry, and teaching tolerance, there are 1,020 hate groups currently operating in the United States. That's an astounding number. The "Hate Map" developed by the SPLC shows there were 457 groups in 1999; they continued to increase until 2011, then started to decrease over the next three years. They are now increasing year-by-year. On February 20, 2019, the SPLC released their latest data showing that the number of hate groups in the United States hit a 20-year high. Many of the hate groups were driven underground but made a dramatic resurgence as the 2016 election approached and have been growing year by year with President Trump's campaign and presidency. Whatever happened to being enlightened? Every state has a hate group. Americans hate each other because of religion, skin color, and sexual identity. The increase in the number of hate groups is driven partially by this statistic: demographic projections show that around the year 2040 white people will no longer be the majority group. After President Obama took office in 2009, the rise in hate groups increased. They then decreased in 2014 but have steadily increased since. The FBI has gathered statistics on hate crimes every year since 1992. You can find the full reports on their website at www.fbi. gov. In 2020, Southern Poverty Law Center (SPLC) tracked 838 hate groups across the United States.

Looking at current issues and figures related to the LGBTQ community, marriage equality marks a shift in public opinion. According to Robert Salem, clinical professor of law at the University of Toledo, two-thirds of Americans support same-sex marriage, yet it is a movement of contradictions. For example, LGBTQ people are still excluded from nondiscrimination protections on the federal and state levels, particularly if it violates a perceived religious conviction. While more youth are coming out as gay, bi, or trans, queer youth are dying by suicide more often than their straight peers. Moreover, 40 percent of homeless youth, are LGBTQ. Transgender women of color are subject to a greater risk of violence and abuse than are white, heterosexual males.

Sadly, there has been a flurry of new discriminatory laws being enacted on the state and federal levels. The Trump administration also rolled back a number of laws protecting LGBTQ people. Fifty-four state bills have been proposed banning trans people from using restrooms matching their gender identity; 13 states have pending bills excluding trans youth from taking part in school athletics; some states now have laws allowing healthcare professionals to refuse treating LGBTQ people if doing so violates their religious beliefs; eight states allow doctors to refuse trans-related healthcare for youth. A recent law in Tennessee allows adoption agencies and foster care facilities to deny services to people who want to adopt children if the applicants are LGBTQ. This means that a gay couple could be married in Tennessee, but they cannot adopt children in Tennessee.

These so-called religious freedom laws supersede the civil rights of LGBTQ people. The same strategy applied here is also being applied to reproductive rights, discussed in Chapter 5. Thirty percent of Americans believe that denying services to LGBTQ people is okay if providing the service violates their religious beliefs. This number was only 16 percent prior to 2014. Today, 20 percent of Americans think it is okay for businesses to deny services to Muslims and Jews if it violates the religious beliefs of the business owners, and 15 percent of Americans think it is okay to deny services to African Americans. Yes, we have religious freedom—a revered right—in this country, yet religious exemptions to deny services to minorities is dangerous and threatens equally-revered civil rights.

While marriage equality has been a boon, there has been considerable backlash. In 2016, there were more than 200 anti–LGBTQ state bills introduced in the United States, and it hasn't subsided since then. Equality for all doesn't apply to everyone yet. Women, older individuals, immigrants, people of color, and the LGBTQ community still face challenges. Professor Salem affirms that as a society we need to turn our attention to the notion that discrimination is okay if it is based on religious conviction. Every person deserves protection and respect.

Surprisingly, when I wrote a blog about this topic in June 2020, this more than any other blog up to that point received the most public comments from haters balanced by the most thankful personal messages from supporters and people not living a binary life. I deleted the hurtful comments because the world is already full of too much negativity. I was touched by the educators who contacted me and asked for references and materials to be used in their classrooms. Others were thankful that I was drawing on science to make the case that we're all people, deserving of basic civil liberties and understanding. To me, these exchanges proved that we need scientific literacy more than ever.

This chapter shows that biological factors make each of us unique, contributing to life within a big wide wonderful world of human possibilities. We are all born into environments we could not and did not choose, with the genes over which we had no say. As a species, our sex and sexual identity fall along a spectrum. And while we can't pinpoint why that is, it is a part of the natural world. No differences among us harm any of us—they simply add to the rich experience of our shared humanity.

Take-Home Message

Logic and science should not be abandoned. Every person deserves to be treated with human dignity and should not be discriminated against. Pseudoscience should not prevail when there is credible science at our fingertips. In 2021, there is no excuse for ignorance.

> *Love is all you need.*
> —Paul McCartney (1942–);
> English rock star

5

Birds and Bees

Human Reproduction

U.S. Landmark Case: Roe v. Wade, 410 U.S. 113 (1973), (7–2 decision)

In its 1973 ruling in the case of Roe v. Wade, the Supreme Court of the United States found that the constitutional right to privacy granted by the Fourteenth Amendment extended to a woman's decision to terminate her pregnancy. This landmark decision declared that a woman has the right to an abortion during the first two trimesters of pregnancy. The decision was not unanimous but was a resounding 7–2 in favor of Roe. This ruling struck down the laws of 46 states against abortion, including the plaintiff's home state of Texas.

Due to resignations, the case was originally heard in 1971 by only seven justices although two new justices had been appointed by President Nixon. The seven-member Supreme Court decision declared the various states' anti-abortion statutes would be struck down as "unconstitutionally vague." Instead of issuing a decision with this basis for the majority decision, Chief Justice Earl Warren chose to have the full court of nine justices rehear the case, which it did in October of 1972. This time the court ruled that the Texas law as well as other such laws violate the due process clause, which protects against state action that violates a person's right to privacy including a woman's qualified right to terminate her pregnancy. The Supreme Court clearly pointed out that this is not an absolute right, but that the state does have "legitimate interests in protecting both the pregnant woman's health and the potentiality of human life." The Supreme Court further stated that the interests of the state increase in importance as the pregnancy progresses. It found that an abortion in the first 12 weeks of pregnancy does not endanger the woman. Another point of the ruling is that state's interests emerge in the second trimester and that states may require that all abortions be performed by licensed physicians under medically safe conditions.

During the third trimester, the interest of the state in "the potentiality of human life," referring to the fetus, greatly increases the power of the state to regulate and/or prohibit abortions except when necessary to save the life or health of the woman. At the time of the third trimester the fetus is determined to be viable and capable of living outside the womb and being entitled to protection from the state.

The Court did not accept arguments that the fetus be regarded as a "person" within the meaning of the Due Process Clause of the 14th Amendment. This Amendment states that no state shall deprive any person of life, liberty or property without due process of law. The Court stated, "There is no medical or scientific proof that life is present from conception. We need not resolve the difficult question of when life begins, when those trained in the respective fields of medicine, philosophy and theology are unable to arrive at any consensus. The judiciary at this point in the development of man's knowledge is not in a position to speculate as to the answer."

In March 1970, when the case was filed, Roe was six months pregnant and chose to remain anonymous. After disclosing her identity as Norma McCorvey of Dallas, Texas, she stated that her decision to move forward with the case was done to help other women. In an interesting turn of events, while McCorvey became an active public advocate for women's reproductive rights, by August 1995 she had switched sides, and become a pro-life advocate. But, there's another twist to this story: In an end-of-life documentary, *AKA Jane Roe* (released in 2020), she revealed that she never truly supported the pro-life movement. A death-bed confession filmed in 2017 shortly before she passed away, Norma McCorvey stated that she "took their money and they put me out in front of the camera and told me what to say, and that's what I'd say." It was all an act, and in her own words, she said she was good at acting. Throughout her life, she gave birth three times and never had an abortion.

Introduction

It's likely that if you had to name one Supreme Court landmark case, Roe v. Wade is the one you'd probably cite. A Google search of the case will provide nearly 11 million hits, none of which is about alternate ways to cross the English Channel. The decision of the court protects pregnant women, yet any Right to Life rally shows it is about much more than this. It is tied to sex, sex education, religion, oppression, and a host of other legal and non-legal issues. Moreover, controversy and misinformation abound. On one side of the debate (and there are many sides), people think that sex education in schools causes teenagers to have more sex, abortion is murder and should be illegal, and abortion clinics should be closed. The scientific view is that sex education teaches students about the anatomy and physiology of their bodies, abortion issues require full understanding of the events of pregnancy, and abortion clinics provide women with access to non-abortion–related health care. Perhaps after reading this chapter, you will come away with a greater comprehension surrounding the issues and realize just how complex matters are.

Having sex is a totally natural human act. Attitudes about sexual intercourse, however, vary from culture to culture. As a topic, sex is inherently interesting as indicated by the volume of information surrounding the subject. Libraries are filled with books on sex. University sex courses are filled with students. No section of a sex course has ever been cancelled due to low enrollment. In my courses, when we reach the textbook chapter on sex and the reproductive systems, this generates more interest, questions, and discussion than any other topic in anatomy and physiology. It's also revealing for me as an instructor to learn what my students really *do not know* about sex and their reproductive systems. Keep in mind that these are college students. The volume of information is counterbalanced with the volume of misinformation, underscoring the necessity of sex education.

Reproduction sounds relatively simple, but in nature there are several types of reproduction. Two important types are asexual reproduction and sexual reproduction. Each of these is worthy of a brief definition so we can make sense of the natural world and our place within in. Our first type is asexual reproduction, and it occurs *without* the union of male and female sex cells. Breaking the word apart, the word part *a* means *without*; therefore, *asexual* means reproducing without fusing sex cells. Organisms that reproduce asexually include bacteria, algae, and fungi. These organisms are also the oldest form of life; they likely emerged 3.8 to 4 billion years ago, and are still with us today. In fact, bacteria play important roles in our bodies by producing vitamins and breaking down food; like plants, algae generate oxygen we need to breathe; and fungi can be good to eat and can be used to make antibiotics.

Sexual reproduction occurs by the union of a male sex cell (sperm) and a female sex cell (oocyte); these sex cells are also called *gametes*. During sexual reproduction, male and female gametes fuse to form a *zygote*.

The topic of reproduction is broad, and as such, there are many controversial issues revolving around it like sex education, contraception, circumcision, and abortion. Nature, however, has no qualms about these matters. Animals either mate and procreate or they don't. If we view this topic from a strict biological perspective, reproduction is simply a means by which humans produce offspring to propagate the species. In short, sexual reproduction allows us to make more humans.

False concepts and idealized notions surround the subject of human reproduction. Thus, it's tough finding a beginning point. One place is to begin by learning the science behind reproductive biology. For instance, we can identify the key reproductive structures along with their functions. From there, we can discuss the process of reproduction and human development while delving into assorted topics.

This book isn't intended for use in an anatomy class, but it will be helpful to learn the structures of the human reproductive system and how they function, especially since the government and others make attempts to control it. Moreover, it's good to know how the body works because the degree of misinformation is extreme, especially where it concerns women. To illustrate this point, in 2016, states enacted 60 new restrictions on women's reproductive care. Contrast this with male's reproductive care in which there are no state restrictions. Oftentimes women's health is synonymous with women's reproductive health. This is important to note because women's *overall health* tends to be tied to their reproductive care. One doesn't hear of men's health being synonymous with men's reproductive health. A woman's health can be in jeopardy because of her reproductive system; but women's health is not solely about reproduction. There are other body systems and the number one killer for both men and women is cardiovascular disease. Cardiovascular disease is a condition that involves the heart and blood vessels, neither of which is part of the reproductive system. Regardless, cardiovascular disease does tend to unfold differently among humans.

In addition to health issues, women also get unfair treatment when it comes to consumer goods. On average, women's personal care products, such as shampoo, razors, shaving cream, and deodorant, cost 13 percent more than these same men's products. Haircuts and clothes also cost more for women. In 35 states, menstrual necessities such as tampons and feminine pads are subject to sales tax because they qualify as "luxury goods." It's gender-based price discrimination. On April 3, 2019, U.S. House of Representatives member Jackie Speier introduced H.R. 2048: Pink Tax Repeal Act, to end gender-based discrimination in the pricing of goods and services. It was referred to the Subcommittee on Consumer Protection and Commerce on April 4, 2019. This bill was introduced in the 116th Congress which ended January 3, 2021. Introduced bills that receive no votes such as this one "die" at the end of that Congress and are cleared from the books. On June 11, 2021, Congresswoman Jackie Speier reintroduced the Act.

Sex Education

Accurate sex education is important for advancing understanding of basic reproductive biology. According to the Guttmacher Institute, an excellent source of information not just on sexually transmitted infection (STI) rates but also abortion, contraception, HIV, pregnancy, and teen reproductive health, the total estimated cost of nine million new cases of STDs that occurred among 15–24-year-olds in 2000 was $6.5 billion. It has only increased since then. The Centers for Disease Control and Prevention (CDC) reported that at the end of 2015, an estimated 1.1 million people aged 13 and older had HIV in the United States. The estimated number of new sexually transmitted infections annually in the United States is 20 million, the total number of infections is 110 million, and the total medical costs are $16 billion. Current data from the CDC show that young people (ages 15–24) account for 50 percent of all new STIs, yet this group represents 25 percent of the sexually active population. Keep in mind that these diseases are largely preventable given appropriate sex education, adequate access to condoms and dental dams (used for oral sex), and scientific information.

As scientific discovery expands, so does knowledge, urging us to no longer remain ignorant. Title X Family Planning Program, also known as Public Law 91–572 or "Population Research and Voluntary Family Planning Programs" is a federal grant that provides people with comprehensive family planning and preventive health services. It was enacted in 1970 while President Nixon was in office. This program, serving more than four million people in the United States, allows access to contraception and health care.

The Trump administration's funding for Title X clinics emphasized abstinence until marriage. Reviews of sex education programs and policies show that abstinence-only-until-marriage programs are ineffective and harmful. Extensive evidence shows that abstinence-only approaches do more harm than good. The abstinence-only programs withhold important scientific information and health knowledge while providing medically inaccurate advice. It is morally reprehensible to advance misinformation.

Reproductive biology is complicated, which is why it's important to understand as much as possible. No person, parent, or legislator can make informed decisions about reproduction and reproductive choices without a basic understanding of how it works.

When studying biology, an underlying principle is that this branch of science deals with the study of life. In fact, the word *biology* comes from two separate word parts: *bio*, meaning "life" and *logos* meaning "study." The word alone evokes emotion and lends itself to some big questions. What is life? When does it begin? Add the word "reproductive" to biology and the questions expand. How much control should humans have over its creation or termination? What is body autonomy? Some of these questions can be answered with science, others are left to the realm of reason or faith.

There are many dictionary definitions for life, but the scientific definition is precise: life includes the capacity for organization, growth and development, homeostasis, reproduction, and continual change preceding death.

Organization is the first criterion. Exploring a bit deeper, this means a hierarchy whereby atoms make up molecules, which make up cells, which make up tissues, which make up organs, which make organ systems, which assemble into organisms. Organisms are us. We humans are multicellular organisms, made up of trillions of interdependent cells. Our body functions through the interrelationships among organ systems, which ensure the functioning of the entire organism.

Growth and development are tied to energy use by our cells. This is also linked to homeostasis. Chemical reactions occur within our cells that enable life to be sustained and for the organisms to achieve homeostasis, which is the tendency toward internal balance. Not New Age, finding your inner-self balance, but rather the physiological state that your body cells strive to maintain. It is a state of equilibrium in the body with respect to various functions and the chemical composition of the fluids and tissues. Think of it as the body's internal thermostat and gauge system that ensures things like oxygen, blood sugar, and pH remain within set ranges. Every cell requires specific conditions in which to thrive, regardless of what's happening outside the body.

Growth and development ensure physical maturation for ultimate reproduction. Putting aside mental acuity and spiritual well-being and other aspects that make us human, the biological perspective is to ensure the species survives. Organisms must be able to grow (get bigger), develop (move from one stage to another), and reproduce. Reproduction in this sense means that cells must be able to divide and pass on DNA from one generation to the next. Guaranteeing success across millennia involves evolution. For our purposes here, evolution refers to a species' ability to adapt and change its genetic makeup over time.

Scientists worldwide agree that these are the components—and they must all be present—that make up life. Life is really the state of existence characterized by such functions that enable the species to carry on. If you put a cell on the table it will not survive, and thus doesn't satisfy the conditions required for life. You can do this experiment at home: scrape the inside of your cheek with a toothpick and place the scraping on a paper towel. This scraping of crud is made up of cells from your cheek mixed in with some saliva. What do you think will happen to these cells? Are they likely to transform into new cells that would resemble your cheek? No. You just have a damp paper towel.

We just discussed the characteristics of life, but in order for life to be maintained, several factors, known as survival needs, must also be met. These survival needs are very basic and include nutrients, oxygen, water, normal body temperature, and appropriate atmospheric pressure. Nutrients are obtained in the food we eat. Food contains the chemical substances (carbohydrates, fats, proteins, vitamins, minerals, and water) used for energy, or for participants in chemical reactions, and for cell building. Chemical reactions, known as oxidative reactions, require oxygen in order to release energy from food. We get oxygen from the air we breathe. These oxygen-requiring chemical reactions make ATP—the cell's energy. So, without oxygen, our cells would die within a few minutes.

Continuing with the necessary components for life, we need water. Water is the most abundant molecule in the body; our bodies are made up of 50 to 60 percent water. The percentage is variable because the more fat you have, the less water you have, thus accounting for the 10 percentage point variation. That means that there is an inverse relationship between fat and water. Why do we need so much water? It's necessary for chemical reactions and for providing the fluid for body secretions (like saliva) and excretions (like urine). We lose water through evaporation when we exhale, through sweat when we are too warm, and in urine and feces as we get rid of wastes.

In order for chemical reactions to occur, the body must maintain a temperature of approximately 98.6° F (37° C). If body temperature gets too low or too high, death can occur. Most body heat is generated through skeletal muscle contractions. This is why you shiver when you are cold: your muscles automatically begin contracting to generate heat. But your body will also do a cost-benefit analysis. If you are generating too much energy trying to keep yourself warm and shivering isn't working, then you'll stop shivering and the body will go into preserve mode. Blood will be shunted from your extremities (you can live without a toe or two) to your core, warming vital organs (those you can't live without like your heart and lungs). If you don't warm up, then death ensues. Keep in mind, we're still talking about all the requirements for life.

Lastly, life can only exist under the appropriate atmospheric pressure. Atmospheric pressure is the force that air exerts on the body. Gas exchange, namely oxygen for carbon dioxide in the lungs, can only happen at the right atmospheric pressure. At high altitudes, like on top of Mount Everest, the atmospheric pressure is low, and the air is thin, so gas exchange is inadequate to support cellular function—which is why you need oxygen tanks to breathe if you intend on spending much time on the mountain top. The cruising altitude of passenger jets is about 30,000 to 40,000 feet. At this altitude, we can't breathe, so airplanes are pressurized to keep our body cells oxygenated. Lots to think about here as we consider everything that is needed to sustain life. While food, clothing, and shelter may be basic needs, things like consistent body temperature and atmospheric pressure are essential.

Vasectomy, Circumcision and the Hymen

Vasectomy and circumcision are other controversial topics related to the body. A vasectomy is the cutting and removal of a little segment of each vas deferens, the tube through which sperm travel to reach the exterior of the male body, to produce sterility. It's like taking a drinking straw, snipping out a section, and then sewing the ends closed. The removed portion goes into a medical waste bin while the severed ends are cauterized, thereby interrupting the passageway for sperm to travel. There are two vasa deferentia, one in each testicle. The sperm are still produced, but they can't be shot out through the penis because the conduit to the exterior has been snipped. The male's body reabsorbs the sperm, and everyone lives happily ever after. This is a form of contraception and is forbidden by some religions.

While vasectomy is forbidden by some religions, it is not legislated like the female reproductive system. Within the man's testicles, sperm can survive about 74 days. For this reason, a man must ejaculate up to 20 times or wait approximately three months after a vasectomy for the vas deferens to be clear of sperm. That means that a man is still capable of producing offspring immediately after a vasectomy. Normally, about three months post-vasectomy, the man brings an ejaculate sample to his physician's office to be tested for sperm. If none are present, he's given two thumbs up on being completely sterile.

Did you know that it is possible to go sperm fishing? Well, the scientific procedure is known as *sperm retrieval* and it can be done posthumously. Posthumous sperm retrieval (PST) is also known as postmortem sperm procurement. Why would anyone do this? One scenario might be the case of a heterosexual couple who wanted to have children, but the man died before they had the chance. So, the sperm was procured from the man's reproductive tract and posthumous conception was achieved. There have been a small number of cases in which sperm have been retrieved (best window of opportunity is within 36 hours after death) from the male; and using invitro fertilization techniques, have resulted in successful pregnancies.

Unlike embryos and fetuses that can't live outside the body, sperm *are* viable. How long can sperm live *outside* the male's testicles? That depends on where it is deposited. Sperm deposited in the female uterus can live about 5 days. This explains why women who are on the tail end of their menstrual cycle can get pregnant. If a female ovulates shortly after she's finished her period, the sperm can still be alive and able to fertilize an available oocyte. The uterine and ovarian cycles are fairly predictable and follow some basic timelines, but nature always has the last say.

Sperm exposed to air die pretty quickly, but environment plays a role in determining viability. As the saying goes, location, location, location. As we've seen, it's about 72 hours in the vas deferens and 5 days in the uterus. They can live up to 72 hours if the sperm is washed and incubated for later use via intrauterine insemination or in vitro fertilization. Frozen sperm can last for years in controlled environments, which is usually done to preserve sperm for later use. For example, if a man is about to undergo chemotherapy, he may preserve sperm in advance of treatment.

Sperm can also survive around the female external genitalia. There's a lesson here, ladies: if you need to produce a urine sample shortly after having unprotected sex, be sure that you supply a clean-catch urine sample. Say what? This is a method of collecting urine so that germs from the urethra, penis, and vagina, as well as semen from a recent sexual encounter, are not getting into the urine sample. How is this done? The easiest way is to sit on the toilet, urinate a little to "wash" the area. Then, spread the labia and while using a clean wipe, wipe the vulva from front to back. Use another wipe and wipe from front to back again making sure to clean the urethra (opening for urine to the exterior) and vagina. Then, urinate into the cup and cap it.

There's a really good reason for wiping from front to back *at all times*, and it's something you should do *every time* you wipe yourself with toilet paper after urinating. The anus (opening from the colon to the exterior) is chock full of bacteria. If you wipe back to front, you will inoculate your external genitalia with bacteria, which can lead to urinary tract infections. If you wipe from back to front with cleansing wipes before giving a urine sample, the urine will contain bacteria from your butt. Urine should be sterile—nothing but water and metabolic wastes and a few other things, but no bacteria.

Here's a true story to illustrate my point. Once while teaching a night microbiology lab, my students were using their own urine to perform a Sedi-Stain procedure. This procedure involves collecting urine, centrifuging a test tube of the urine, adding a drop of the commercially produced Sedi-Stain to the tube, and transferring a drop of the sediment onto a microscope slide, placing a coverslip over the sample, and then analyzing the sample under a microscope to look for blood cells, urinary casts (cylindrical structures produced by the kidneys indicative of certain diseases), uric acid crystals, and other elements in the urine. While peering into the microscope and viewing her carefully made Sedi-Stain urine slide, one of my students kept seeing this "darting structure" on the slide. She called me over to take a look. With a keenly trained eye, I knew in an instant what I was looking at. A living, swimming sperm was front and center. This was one of those rare teaching opportunities that professors can only dream of. As students lined up one by one to see this, the blushing student made sure that we all knew that it was okay, because after all, the guy she had sex with right before class was her husband. Duly noted, and thank you, Stellar Student.

What about circumcision? The word comes from the Latin term *circumcidio*, meaning *to cut around*. In males, circumcision is the surgical removal of the foreskin, which is that flap of skin covering the penis tip. It's routinely done in U.S. hospitals on newborns. Sometimes it is done as a religious ritual. Other reasons for removing a perfectly healthy flap of skin include personal hygiene or preventive health care. From the personal hygiene and preventive health perspectives, boys can be taught to pull back the skin when washing to ensure cleanliness and to reduce the risk of creating warm, moist areas for microbes to live. Jewish, Islamic, African, and Australian cultures engage in circumcision as part of religious or cultural heritage. Yet, the practice can be viewed as unnecessary, disfiguring, and done without consent since newborns have no say. Stop and think about this for a moment: A new, viable, thriving baby arrives into this world and within his first few days of life, an unnecessary surgical procedure is performed. Where are the laws preventing this? Especially when the American Academy of Pediatrics (AAP) does not recommend routine circumcision for all male newborns.

Female circumcision also exists, but in many more forms. It refers to genital cutting, ranging from the removal of the little hood of skin around the clitoris called the clitoral prepuce to removal of the clitoris, labia minora, and labia majora. There are no known medical reasons or health benefits for female circumcision. In fact, female circumcision is also known as female genital mutilation (FGM). The practice is grounded in gender inequality. It is a violation of female human rights, can cause severe bleeding, problems urinating, complication in childbirth, and increased risk of newborn deaths. According to the World Health Organization, 200 million girls and women alive today have been cut, with concentrations in Africa, Asia, and the Middle East.

Another topic of curiosity centers around the female hymen. The hymen is a thin, membranous fold that partially closes the opening of the vagina. Its presence purportedly denotes virginity. Except, of course, it is no more a sign of virginity than having a tail makes a mammal a dog. Some females are born with a hymen, some are born without one. Breaking the hymen supposedly indicates that a woman is no longer a virgin because it was broken by vaginal penetration with a penis. As a membrane, the hymen stretches—it can stretch from bicycling, exercise, or tampon use. The point is that it's not indicative of virginity. And an intact hymen doesn't prevent infection with a sexually transmitted disease.

Rape

Rape. What a horrific crime. Rape is the forcing of another person to have sexual intercourse with the offender. This is not consensual sex. This is someone forcing another person against their will. Rape is usually committed by a man. In the United States, one in five women and one in 71 men will be raped at some time in their lives. That's 20 percent of women and 1.4 percent of men. An estimated 63 percent of sexual assaults are not reported to the police, making it the most under-reported crime. In one three-year longitudinal survey, the incidence of national rape-related pregnancy was 5 percent among victims of reproductive age (12–45). The majority of rape-related pregnancy cases in adolescents resulted from an assault by a known, often related perpetrator. Nearly half (47.1 percent) received no medical treatment related to the rape. Of rapes that resulted in pregnancy, 32.4 percent of the victims didn't realize they were pregnant until the second trimester. Only 32.2 percent of the victims chose to keep the infant; 50 percent underwent abortion; 5.9 percent placed the infant for adoption; and 11.8 percent experienced a spontaneous abortion.

Going back to the previous paragraph, 1 in 71 men are also the victims of rape. In the previous chapter, prison rape was discussed. However, other contemporary rape and sexual abuse scandals are at the fore: These involve the Catholic Church and the Boy Scouts of America.

While the topic is contemporary, the history is anything but. For decades, an alarming number of priests in the Catholic Church have sexually abused children and nuns while weaponizing the Catholic faith. First publicized in 1985, when a Louisiana priest pleaded guilty to 11 counts of molesting boys, books on other cases of sex abuse were published in the 1990s, and a series of articles by the *Boston Globe* blew the scandals out of the water in 2002. With respect to the Boy Scouts, in 2019, more than 800 men across the country have filed lawsuits alleging sexual abuse. Today we are learning of the allegations, cover-ups, crises, and scandals surrounding the abuses. Many documents are being revealed and written. These crimes are atrocities. Tragically, some grown men, unable to come to terms with being sexually abused by men they trusted as children, have committed suicide. The list of accusations is exhausting. All of this is beyond the scope of this book except to point out hypocrisy at the highest levels.

Rape causes many unwanted pregnancies and it is often closely linked with family and domestic violence. Sexual victimization is a horrific crime that stays with a person forever. Even if victims receive physiological and psychological treatment, it remains a lifelong event. Unless it has happened, you can never fully experience what it is like. Imagine if a female were forced to have a child that was the result of rape? It might be easy to say that the person can give the baby up for adoption. But, it's not. Adoption isn't an easy choice as a woman must go through the entire pregnancy plus a lifetime of wondering what happens to the child.

Getting pregnant from a relative is abhorrent. Sex with or without consent with a relative is incest. Children resulting from incestuous relations is called inbreeding. Incest is not biologically normal and non-human animals have evolved mechanisms to avoid inbreeding. From a scientific perspective, inbreeding increases the chances of offspring having deleterious traits and as such, decreases the genetic fitness of the population.

Incest has a considerable "yuck" factor and is taboo cross-culturally. Aside from the yuck factor, there are genetic consequences. Having offspring with close relatives increases the chance by approximately 42 percent that the child will be born with a serious birth defect. Even if the child is part of the 58 percent who are born without a birth defect, imagine what that person may feel or how they will make peace with knowing that they are a product of incest? Uncle John is also Daddy John. Grandpa Dick is also Daddy Dick.

Sexually Transmitted Diseases (STDs)

Bacteria are unicellular organisms that live *in* us, *on* us, and *around* us. These microorganisms (microbes) can be beneficial or harmful. Millions of beneficial bacteria populate our gut and help with food digestion, vitamin production, and overall health. Others can be harmful and cause disease. You most likely have taken an antibiotic at some point in your life to get rid of a bacterial infection. If a sore throat cropped up or a wound got infected, an antibiotic (*anti* = against + *bio* = life) might have been prescribed to get rid of the culprit.

Unlike the digestive tract, the reproductive tract is nearly microbe free. The exception is the genus *Lactobacillus*, which is found naturally in the vagina to help keep it healthy. We also find *Lactobacillus* in dairy products, grains, water, sewage, beer, wine, fruits, and in sourdough—just to name a few. Lactobacilli form part of the vagina's normal flora and produce lactic acid and hydrogen peroxide that inhibit the growth of yeast and microorganisms. If other bacteria are introduced, for example via semen, which can carry microorganisms, infections like sexually transmitted diseases—the dreaded STDs—can result. STDs are also called sexually transmitted infections (STIs). In addition to sexual contact, reproductive tract infections (RTIs) can occur when normal bacteria from the body surface or digestive tract make their way into places where they don't belong. They can also be introduced via surgical or medical procedures.

Bacteria and viruses make up the microbial world and can be helpful, harmful, or merely passengers. We are just beginning to understand the many roles each plays in our body, but it will take some time to fully realize our relationships. It is estimated that there are about 39 trillion bacterial cells living among our 30 trillion human cells. Given these numbers, it seems as though we are little more than a big, walking microbe colony!

How does one acquire an STD/STI? As contagious diseases, they are acquired during sexual contact with a person who is already infected with the disease. Sexual contact includes vaginal, anal, and oral sex. They were once known as venereal diseases, but that term has fallen out of favor. It's not such a bad term if you know that the derivation of the word comes from the Latin word *venereus* meaning "sexual lover."

Sexually transmitted infections are very common, and a person may not even know they have one because some don't have any symptoms. Common bacterial STIs includes chlamydia, gonorrhea, and syphilis. Viral STIs include hepatitis B, human immunodeficiency virus (HIV), herpes simplex virus, and human papillomavirus (HPV). A common protozoal infection is trichomoniasis. Viral STIs generally have more serious consequences than bacterial STIs, in part because there is no cure. For example, hepatitis B can cause liver disease, HIV depresses the immune system and can lead to AIDS, and HPV can cause cancer. While it is possible to live with hepatitis B and HIV/AIDS, there is no cure.

The most common STI in the United States is HPV. There are more than 100 varieties of HPV, which cause genital warts, cervical cancer, and other cancers. While treatment can help, there is no cure. Vaccines given to both males and females are available to prevent some types of HPV infection. According to the CDC, HPV vaccination is recommended for everyone through age 26 years, with vaccination beginning at age 11–12 years. If the vaccine is given at ages 11–12, then only two doses are needed; vaccinating later may require three doses. Adults ages 27–45 who have not been vaccinated may consider being vaccinated after speaking with healthcare professionals regarding their personal circumstances.

Gonorrhea—sometimes called "the drip" or "the clap"—is also quite common, affecting primarily teenagers and people in their 20s. The disease has been called "the drip," probably because it can involve discharge from the penis or vagina. The genesis of "the clap" is less clear. There are several possible explanations for why gonorrhea was called the clap. One makes sense from a word derivation perspective: The French word *clapier* means brothel, and this was certainly a place where STDs could be passed around. Another explanation is that the term comes from a once-prescribed treatment for gonorrhea of clapping the man's penis against a hard surface or clapping the penis with the hands to kill the "offender" by force. Either way, penis pounding sounds painful.

Syphilis causes sores known as chancres on the genitals. You can pronounce *chancre* as SHAN-ker or KANK-er; both are acceptable. Gonorrhea and syphilis infections can be cured with medications; however, if left untreated, each can have long-lasting effects. For example, untreated syphilis can lead to brain damage, paralysis, and blindness. The damage may not show up for years and untreated syphilis can also lead to death. Early treatment with antibiotics can cure the disease; late treatment can cure the disease and stop further damage. Syphilis also left an indelible mark in American history; refer back to Chapter 3 for a recounting of the Tuskegee Study, a study in unabated racism.

Overview of Reproductive Structures

Let's begin with the anatomy (structure) and physiology (function) of the reproductive system. This complex system consists of various organs that differ between the sexes. In males, some of these organs produce sex cells or the organs are used to impregnate females. In females, these organs also produce sex cells, and if pregnancy occurs, they house a developing embryo and fetus, and nurture an infant after birth. For our discussion, males are XY and females are XX. Two terms causing confusion are embryo and fetus. An embryo is an organism in the early stages of development from fertilization to the end of the 8th week. A fetus is the developmental stage that begins after the embryonic stage and ends at birth. The female reproductive system is more complex than the male. The reason is because the male reproductive system's job is to produce and deliver sex cells, while the female reproductive system does this, plus nourishes and harbors a developing individual and can ultimately give birth to and further nourish an infant.

Take a look at these quick reference tables and accompanying illustrations, which give brief descriptions and visual representations of key male and female reproductive structures and organs of pregnancy. These structures will be discussed throughout this chapter.

Male Reproductive System Quick Reference Table and Figure 5.1

Male Reproductive System Structures	Functions
Penis	• Contains erectile tissue enabling erections • Organ of sexual intercourse • Deposits sperm into female vagina • Contains sensory nerve endings for pleasurable sensations
Glans penis	• Head of the penis that is rich with sensory nerve endings
Foreskin (prepuce) not shown in the figure	• Fold of skin that covers the glans penis in an uncircumcised male; this skin is removed when a male is circumcised
Urethra	• Structure common to both the reproductive and urinary systems • Transports semen to body exterior • Transports urine to body exterior
Scrotum	• Pouch of skin surrounding the testes (testicles)
Testis (testicle)	• Gonad • Paired structures that produce sperm and hormones
Epididymis	• Paired duct where sperm mature
Ductus deferens (vas deferens)	• Paired duct that transports sperm from the epididymis in each testis (testicle) to the urethra
Semen not shown in the figure	• Thick, yellow-white fluid that contains sperm, and a mixture of secretions from the testes, seminal glands, prostate, and bulbo-urethral glands
Sperm not shown in the figure	• Male sex cell (gamete) that contains genetic information • Structurally has a head and tail (flagellum) and is capable of "swimming" by whipping its tail

Male Reproductive System Structures	Functions
Seminal gland (seminal vesicle)	• Paired structures that secrete fluid that makes up semen • Opens into the ductus (vas) deferens
Prostate (prostate gland)	• Secretes fluid and enzymes
Bulbo-urethral gland (Cowper gland) not shown in the figure	• Paired structures that secrete fluid that lubricates penis tip

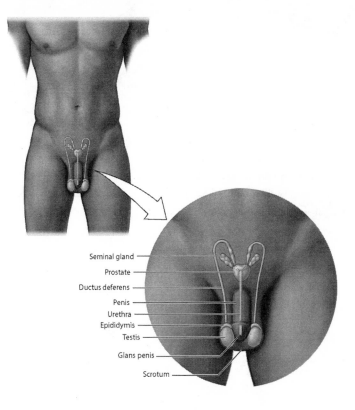

Seminal gland
Prostate
Ductus deferens
Penis
Urethra
Epididymis
Testis
Glans penis
Scrotum

Figure 5.1: Male Reproductive System.

Female Reproductive System Quick Reference Table and Figure 5.2

Female Reproductive System Structures	Functions
Mammary gland	• Structure within each breast that secretes milk
Mons pubis	• Rounded pad of fatty tissue overlying the joint of the pubic bones in the front of the body that is covered with pubic hair
Vulva not shown in the figure	• Collective term for external female genitals
Clitoris not shown in figure	• Small, sensitive structure containing erectile tissue at anterior end of vulva
Vagina	• Muscular tube leading to the body exterior • Organ of sexual intercourse • Canal for menstrual fluids • Birth canal for infant

Female Reproductive System Structures	Functions
Labia—not shown in the figure	• Outer and inner folds of the vulva on each side of the vagina
Ovary	• Paired structures that produce oocytes (egg cells) and hormones
Oocyte not shown in the figure	• Female sex cell (gamete) that contains genetic information • Commonly called an egg cell
Uterus	• Organ that houses developing embryo and fetus • Site of exchange between maternal and embryonic/fetal bloodstreams • Without pregnancy, its lining is shed every month as menstruation
Uterine tube (fallopian tube)	• Paired structures that deliver oocyte or fertilized egg to uterus • Site of fertilization

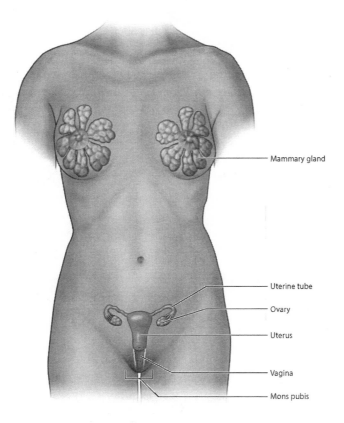

Figure 5.2: Female Reproductive System.

Structures of Pregnancy

Structure	Description
Embryo	• Fertilized oocyte from day 1 until end of week 8
Fetus	• Stage of development beginning at week 9 and ending at birth

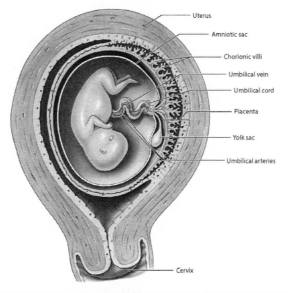

Organ	Description and Function
Placenta	• Organ of pregnancy • Allows for metabolic interchange between the embryo or fetus and mother • Tissue derived from both the embryo and the mother
Amniotic sac	• Fluid-filled bag within uterus
Amniotic fluid	• Fluid within the amniotic sac that surrounds fetus and protects it from mechanical injury
Chorionic villi	• Capillaries carrying blood of the embryonic circulation
Yolk sac	• Membranous sac attached to the early embryo
Umbilical cord	• Connecting stalk between the embryo or fetus and the placenta • Nourishment route • Contains 1 umbilical vein and 2 umbilical arteries • After birth it will become the umbilicus (navel)
Umbilical vein	• Vessel that supplies fetus with oxygenated, nutrient-rich blood from the placenta
Umbilical arteries	• Vessels that transport fetal wastes back to the placenta

Figure 5.3: Organs of Pregnancy.

The human body is made up of trillions of cells, and these cells are either non-sex (somatic) cells or sex (germ) cells. (These terms were also used in the previous chapter, so they should sound familiar, but a recap doesn't hurt.) There are many more somatic cells than sex cells, but our focus here will be on the sex cells—sperm and oocyte. Recall that these sex cells are called *germ cells* or *gametes*. Anatomy vocabulary is full of synonyms and there is no way around it. Sex cells contain half the number of chromosomes as somatic cells and are able to unite with other sex cells from the opposite sex to form a new individual. See the Cell Reference Chart Below.

Cell Reference Chart

Cell Name	Alternate Name	Chromosome Number
Non-sex cell	• somatic cell	2 sets of chromosomes designated diploid (2n)
Sex cell	• germ cell • gamete • Females: oocyte • Males: sperm	1 set of chromosomes designated haploid (n)

Refer back to Figure 4.1 Human karyotype

Oocytes and Sperm

In females, the paired ovaries produce oocytes. Those two o's together are not a typo and are pronounced oh-oh. It sounds a bit like a Santa imitation, and the entire three-syllable term is pronounced oh-oh-sights. It's a fun word to pronounce. The word part *-oo* means *egg* and *-cyte* means *cell*, which gives you a clue about the term: oocytes are egg cells. In everyday language, you'll hear them called eggs. Oocytes are the only cells in the human body that you can see without using a microscope. For perspective, one oocyte is about the size of a grain of sand. Oocytes can be fertilized with male sperm, forming a fertilized oocyte known as a *zygote*. If fertilization does not occur, the released oocyte remains as is and is expelled from the body in menstrual blood.

Sperm are interesting cells because they are the only cells in the body that have a tail. The anatomical term for a tail is a *flagellum*. This flagellum allows the sperm to motor about. Part of keeping sperm viable once they are outside the male body is maintaining them at an appropriate pH. Semen is the fluid surrounding sperm and it has an alkaline pH range of 7.1–8.0, which is perfect for sperm maintenance. Remember from high school biology that a pH of 7 is neutral, values less than 7 are acidic, and values greater than 7 are alkaline. To provide a point of reference, pure water has a pH of 7, vinegar is about 2.5 and ammonia is about 11.5.

Why is pH so important? pH is important because sperm wouldn't survive the inhospitable environment of the vagina if they weren't carried in an alkaline fluid. The vagina is slightly acidic (pH = 3.8–4.5) so that bacteria and other cells—like sperm—cannot survive. Most bacteria thrive in neutral or near-neutral pH, so an acidic pH kills off most bacteria. The alkaline semen protects sperm; once ejaculated into the vagina they wouldn't have a chance of survival were it not for the alkaline coating. The ejaculate is basically a huddle of sperm. Those in the center are protected by those on the outside. The average ejaculate is between 2 and 5 mL (about a teaspoon) and contains about 100 million to 500 sperm! Yet, only about 50 sperm ever reach an awaiting oocyte, and of that amount, only one can fertilize it. Only one.

Since sperm make up only about 1 percent of the total semen ejaculate, what makes up the rest? The bulk consists of carbohydrate (sugar) and protein. Carbohydrates and protein contain calories. So, no surprise, a "typical" ejaculation has 5 to 7 calories. The sugar is fructose, and this fructose is needed because sperm need an energy source to whip the tail and swim. Sperm are cells, so their mitochondria (cellular powerhouses) rely on fructose to make energy in the form of ATP. Besides carbohydrate and protein, other substances in semen are water, vitamin C (ascorbic acid), phosphate, bicarbonate buffers, and enzymes. Water makes the ejaculate a fluid and vitamin C acts as an antioxidant to improve sperm quality (some studies show that taking oral vitamin C supplements helps male infertility). The amount of the ejaculate is determined by age (younger guys ejaculate more than older guys), time since last ejaculation (lesser amounts are ejaculated after each subsequent ejaculation), and the man's personal physiology.

Quantity and quality both matter when it comes to sperm. In order to successfully fertilize an awaiting oocyte, there have to be lots of sperm. Lots. They also have to be robust and have the perfect shape (morphology). These are reasons why so many are produced and expelled. After ejaculation, the female vagina considers the sperm unwelcome visitors. Those sperm that do survive the vagina enter the uterus, where they flounder around looking for the entrances to the two uterine tubes. In most cases, only one uterine tube will have a waiting oocyte. So, half will travel to a dead end. The buffers surrounding the sperm neutralize the vagina's acidity, while enzymes erode the outer lining of the oocyte. Once the outer lining has been eroded by the enzyme *hyaluronidase*, one sperm slips in. That's it. The oocyte's outer membrane seals and prevents other sperm from entering.

The fertilized oocyte, now called a zygote, is capable of developing into a new individual. Each sperm and oocyte contains 23 chromosomes; when they unite, the species number of chromosomes (46 in humans) is maintained. As we discussed in Chapter 4, sometimes there are more than 46 chromosomes.

One method of birth control uses spermicides, chemicals that kill sperm. Spermicides are available over-the-counter in creams, foams, gels, and suppositories and are inserted into the female vagina before sexual intercourse. They work because they contain nonoxynol-9 (N-9), which attacks the membrane surround the sperm known as the *acrosome*. Once the membrane is dissolved, the sperm head detaches from the tail, causing sperm immobilization (if they can't swim, they can't reach the oocyte) and eventual death. Used alone, spermicides are about 71 percent effective.

You might be familiar with N-9 in other products, like sports creams, shaving creams, and cosmetics. Why would a spermicide ingredient be found in common household products? The reason is because N-9 is an organic compound that is an excellent surfactant, a substance that reduces surface tension in a liquid and makes it slippery. Surfactants are compatible with both body oils and water, so they are used in products we put on our skin.

Reproductive Biology

Now, put aside what you *believe to be true* and ask yourself what you *really know* about human reproduction and how the characteristics of life fit into your knowledge base. Take some time to think about this. While reproduction happens on a daily basis, and we humans are quite successful at it, could you explain from start to finish what actually happens in order for a pregnancy to occur and be maintained? Do you know what every step involves? Can you accurately describe what is happening? Where does your information come from? Before explaining human reproduction, revisit the definition of life from the preceding paragraphs. Clearly, life is an interplay among variables, and this is evident most dramatically when we consider the variables at play in the process of human reproduction. Consider one other thing: males cause 100 percent of wanted and unwanted pregnancies. If the conditions are right, ejaculations may result in pregnancy.

Recall from the beginning of the chapter that the ovaries release oocytes. In reality, female ovaries release *secondary oocytes*. A secondary oocyte is a primary oocyte that has already undergone the first stage of meiosis (a type of cell division). A primary oocyte is diploid and contains 46 chromosomes while a secondary oocyte has undergone two stages of meiosis to become haploid with 23 chromosomes. Biology is quite complex, but we'll keep it simple and just call them oocytes. Once fertilized by a sperm, an oocyte is called an *ovum* (the scientific term for egg) or a zygote (as previously noted). However, the oocyte does not have to be fertilized. If fertilization does not occur, the oocyte exits the female body in menstrual fluid—the blood during menstruation.

If you are like many people, you are probably scratching your head thinking, "I sort of recall learning about oocytes in biology." Or, maybe you never heard of the word. Or, most likely you've always heard of eggs. Eggs are easier to remember than oocytes because *eggs* is a one-syllable word and *oocytes* is a three-syllable word. But the distinction between eggs and oocytes is important because this marks the beginning of the pro-life/pro-choice debate.

In human biology, we don't use the term "egg." We use the term "ovum" or "zygote." Reptiles and birds lay eggs, with nice protective shells, membranes, albumin, and yolk for the nourishment of the embryo. An egg is an imprecise term in human biology because it can be variously applied to many stages beginning with the primary oocyte to the stage in which an ovum is implanted into the uterine wall as a blastocyst, which is yet another stage of fertilized cell structure. Regardless of the term, in order for a secondary oocyte to become fertilized, it must meet up with a sperm in the uterine tube. This is important to note: Fertilization occurs only in the uterine tube.

At this point, a brief summary is in order: each month, if a female is of reproductive age (somewhere between age 12 and 50 depending on individual physiology), she will release one (sometimes more) oocyte from her ovary (usually only one ovary) each month. That oocyte "pops" out of the ovary and crosses a very tiny gap to enter the uterine tube. Fringe-like structures of the uterine tube at the end closest to the ovary begin beating and sweep the oocyte into the uterine tube. These fringe-like structures are called *fimbriae*, and they flutter under the influence of hormones. Once inside the uterine tube, if it meets up with a sperm, fertilization can occur. If fertilization occurs, the cell is now a *zygote*. It can also be called an *early embryo* or *pre-embryo*. This zygote then spends another few days traveling down the uterine tube. During this time, the fertilized cell is not increasing in size, but its interior cells are dividing. As you can see, it's much more complicated than sperm + egg = little tiny human. That's not what happens.

Scientific literacy is significant here. To begin, everything we know about reproduction at the cellular level stems from science. When we view the stages of development through the lens of a microscope, it becomes clear that we can no longer view development solely through the lens of those opposing abortion.

Biologically, there are many stages of development, and developmental biology is a branch of science that deals with the developmental process from zygote to full-grown adult. Developmental biologists tend to focus on early development, notably the embryo, because during the embryonic stage, so much is happening.

If a working definition of *conception* is the point in time when an oocyte is fertilized by a sperm, then the zygote represents a soon-to-be human (similarly called an unborn human). By extension, if a person believes that life begins at conception, then frozen embryos also represent unborn humans. So many questions surround frozen embryos. For example, if the embryos are frozen and then implanted in the uterus at a later day, is the subsequent baby older than gestational age? If something happens to those frozen embryos before they can be implanted, who is liable? What about frozen oocytes? Bioethicists and the population at large ponder such questions among many others. And rightly so. In 2018, a woman from Ohio had five of her embryos frozen and stored at University Hospitals in Cleveland before she underwent cancer treatment. Unfortunately, due to rising temperatures in a cryogenic freezer, her five embryos were among 4,000 frozen oocytes and embryos that were rendered non-viable. While we may have differing views on when life begins, it's easy to determine where liability starts.

We arrive at who we are through developmental stages, beginning with hatching the so-called egg, intermediating with the birth of a fetus, and ending with senescence and death. *Prenatal* is the term biologists and obstetricians use for the stages of development that occur before birth. The prenatal stages are divided into the pre-embryonic stage, embryonic stage, and fetal stage, which are further subdivided into terms you probably already know: first trimester, second trimester, and third trimester. Much is happening at the cellular level throughout these stages and trimesters.

Stages are marked by cell divisions, each with distinct names and characteristics. Zygote formation is the very earliest stage and lasts about 0–30 hours. The single-celled zygote contains chromosomes from both the male and the female and undergoes a series of mitotic cellular divisions known as *cleavage*. For the next 30–72 hours, the zygote makes its way down the uterine tube, all the while undergoing cellular divisions. During cleavage, the zygote forms a 2-celled blastomere, then a 4-celled blastomere, and then an 8-celled blastomere. Each of these cell divisions takes place within the space of the original zygote. This means the cell is dividing but not increasing in size. The 8-celled blastomere then divides again, forming a spherical 16-celled mass of cells called a morula. At the morula stage, the cells are about 3–4 days old. The stage following the morula phase is the blastocyst, which lasts about 4–16 days. The morula is the last stage before blastocyst implantation on the uterine wall.

Learning the terminology can be exhausting, nevertheless, these stages show that reproductive biology is complex. At this point, we have a developing ball of cells—the blastocyst—that can implant on the wall of the uterus. The outer and inner structures of the blastocyst are important, too. The blastocyst has an outer mass of trophoblast cells and an inner mass of embryoblast cells. The trophoblasts erode the lining on the uterine wall allowing for implantation and building the placenta. Once the blastocyst has implanted, the woman is officially pregnant. The inner cells form the embryonic disc. Note that the trophoblast cells form the placenta, but do not form the embryo itself. At this point you should start to see that both bodies (female's and embryo's) are required to keep the process moving along. During these early stages of development from fertilization until the end of day 16, the developing cell mass is called a pre-embryo. The embryonic stage is the developmental phase from day 16 through week 8. The fetal stage begins at week 9.

Many events are happening simultaneously between days 9 and 15. The most important events include amniotic cavity formation (around day 9), yolk sac formation (around day 10), and gastrulation (formation of germ layers) by day 15. During this time, known as the *phylotypic stage*, the developing organism looks indistinguishable among other organisms of various species. What does this mean? If you were to look at these cells through a high-powered microscope, you wouldn't be able to tell if you were looking at a human, giraffe, dog, or cat. Gastrulation is a big day in the life of a cell because this marks the formation of three layers of cells called the germ layers: ectoderm, mesoderm, and endoderm. Each of these germ layers will eventually form all the body systems over the next two trimesters.

One structure worthy of noting is the primitive streak. This is a faint embryological structure that appears on day 15 and arises from the mesoderm and endoderm. It provides visual evidence of the cephalocaudal axis. Cephalocaudal is one of those million dollar terms. Breaking the term apart, you get *cephalo* meaning *head* and *caudal* meaning *tail*. Why bring up this term? Because around day 16, heart progenitor cells migrate through the primitive streak to an area in the mesoderm called the primary heart field. These cells will form parts of the heart while secondary heart field cells will form other parts of the heart. Yet, it is still not a heart. It's the foundation of the circulatory system.

By the end of the first trimester, the fetus is getting bigger, human-like features are appearing, and fetal movements will soon appear. Organ systems begin forming in the first trimester and become functional in the second and third trimesters. Many other structures, such as extra-embryonic membranes, an umbilical cord, and the placenta also form to support the developing cell mass. Around week 24 of development, the developing fetus has a 50 percent chance of survival outside the womb with considerable medical intervention in a neonatal intensive care unit.

The average length of human gestation is 40 weeks or about 280 days. It is important to note that this is an average length. Some research has shown that this length of time can vary naturally by as much as five weeks. The amount of time in the womb required to fully develop a human baby place both anatomical and physiological stresses on the mother. For example, the nutrient needs during pregnancy are higher than at any time during adulthood, with energy requirements increasing 300 calories per day, protein needs increasing 10 g/day, and carbohydrate needs increasing 250 g/day! Vitamin and mineral requirements increase, too. Females gain about 25–35 pounds during pregnancy, and that weight is not aligned along the long axis of the body. Think of having a pot belly and the strain that places in trying to move or the strain it creates by pulling on the spine, causing swayback. Just moving utilizes considerable additional energy. As pregnancy progresses, the uterus and fetus become so large that they eventually push the female's abdominal organs out of their normal positions, which makes late stages of pregnancy quite uncomfortable.

From a physiological perspective, maternal breathing rate increases, mammary glands of the breasts develop and begin producing milk, filtration rate in the maternal kidneys increases nearly 50 percent, urine volume increases, maternal blood volume increases by 50 percent, and nutritional requirements of the female increase 30 percent above normal. Although these changes are normal, they can carry significant medical risk. Moreover, medical risks associated with pregnancy increase with age, and for women over age 35, the chances of dying from pregnancy-related complications are almost twice those of being killed in a car accident. While pregnancy is a natural event, the imposed risks make it potentially dangerous. According to the CDC, 700 women die each year in the United States from pregnancy-related or delivery-related complications.

Several paragraphs have been devoted to the events of pregnancy, but how do women know if they are pregnant? Missing a period is often the first sign. Fatigue is also a classic sign. Another sign is the presence of a specific hormone. If a woman is pregnant, trophoblastic cells secrete a hormone called human chorionic gonadotropin (hCG) that can be detected in urine and blood. If a woman suspects that she is pregnant, healthcare professionals usually recommend that she wait until the first day of her missed period before taking a urine pregnancy test. While urine pregnancy tests are relatively inexpensive and can be done in the privacy of your own home, blood pregnancy tests cost more and must be done in a clinical environment. However, blood tests can detect pregnancy earlier than a urine test, at about 7–12 days after possible fertilization. The earliest possible day for an accurate reading of a urine test is 14 days after possible fertilization.

The first trimester is the most dangerous because timing is everything. The groundwork for organ development is happening and complex processes are happening quickly. If the developing embryo-to-fetus is to occur successfully, the timing has to be spot on.

To gain a bit more perspective, here's a chart outlining gestational stages of an embryo and fetus in the uterus. The relative sizes and weights at each gestational stage are given in the table.

Gestational Age, Size and Weight Table

Gestational Age	Size	Weight
1 month	0.2 inches	0.0007 ounces
2 months	0.63 inches	0.04 ounces
3 months	2.13 inches	0.5 ounces
4 months	4.6 inches	3.53 ounces
5 months	6.46 inches	10.58 ounces
6 months	11.81 inches	1.32 pounds
7 months	14.8 inches	2.22 pounds
8 months	16.69 inches	3.75 pounds
9 months	18.66 inches	7.05 pounds

Healthcare providers and anatomy educators take specific courses in embryology. Embryology is a broad field that includes molecular, cellular, and structural levels. Researchers study them all to better understand the nuances and science of each. Thick academic textbooks are devoted to just this topic. It takes approximately nine months of precision throughout the developmental stages to transform a single cell into an infant. In many ways, embryology itself is in its infancy as we're just now learning and expanding our understanding of the role prenatal factors have on our personality, behavior, and adult health.

Problems with Pregnancy

Pregnancy isn't an easy process and so much can go wrong at any time throughout the event. For example, if implantation of the zygote occurs outside the uterus, most often within the uterine tube, an *ectopic pregnancy* results. The word *ectopic* comes from the Greek word, *ektopos* that means *out of place*. The uterine tubes cannot sustain a pregnancy, and the embryo cannot be relocated to the uterus, via surgery or otherwise. Women experiencing an ectopic pregnancy have severe abdominal pain and spotty bleeding from the vagina early within the first trimester. The developing embryo can rupture causing massive bleeding to the extent that this event can be life threatening to the mother. Once an ectopic pregnancy is detected, surgical intervention or laparoscopic procedures to remove the pregnancy are necessary to save the life of the mother. To demonstrate a point regarding the importance of understanding science and showing how science infiltrates public policy, in April 2019, Ohio Republican State Representative John Becker sponsored HB 182 that defies science. The bill prohibits health insurers and public employee benefit plans from covering abortions except when necessary to "re-implant the fertilized ovum" or to save a pregnant person's life. It makes no allowance for pregnancies that result from rape or incest. He falsely claimed that contraceptives cause abortion and that ectopic pregnancies can be fixed by surgically placing the embryo into the woman's uterus.

There is, however, sound science surrounding a uterus transplant from a deceased donor. In June 2019, a baby girl born at the Cleveland Clinic made history: her mother had received a uterus transplant from a deceased donor. The mother had uterine factor infertility (UFI), which means there are abnormalities with the uterus, or the uterus is absent. Uterine transplantation is in its infancy, and at the Cleveland Clinic, three transplants have been successful. The embryos must be cultivated in vitro (in a test tube), but this success story shows that implantation is possible after a complex surgical procedure and immune suppression.

To provide another perspective, in fertile heterosexual couples who have sexual intercourse regularly without contraception, 90 percent will conceive a child within one year. But even in healthy women, pregnancy is still risky. According to the CDC, hypertensive disorders affect many women and the rate has increased substantially from 1993 to 2014. These disorders include gestational hypertension, preeclampsia, and eclampsia.

Gestational hypertension is high blood pressure that develops just as a result of being pregnant. Preeclampsia is high blood pressure with signs of kidney, liver, and other organ problems; it is one reason women die giving birth. Preeclampsia can cause abruptio placentae, a condition in which the placenta separates from the uterine wall. If this occurs at a stage when the fetus is viable (able to live outside the uterus), a cesarean section (C-section) may be performed. Eclampsia is preeclampsia with seizures. In addition to ectopic pregnancy, other disorders of pregnancy include gestational diabetes, hyperemesis gravidarum, placenta previa, deep vein thrombosis, pulmonary embolism, and postpartum hemorrhage. The point? Pregnancy is no walk in the park.

A deeper examination of these conditions reveals that each can place the female at considerable risks. Remember that an ectopic pregnancy occurs when the blastocyst implants outside the uterus, usually within the uterine tube (tubal pregnancy). This is a medical emergency and requires abortion, otherwise, the mother would die from internal hemorrhage. Not all cell clusters implant within the uterine tube; sometimes the blastocyst implants in the abdominopelvic cavity. Again, there is a huge risk of rupturing with internal bleeding and abortion is medically necessary. Gestational diabetes, characterized by the inability to metabolize carbohydrates sufficiently, insulin insensitivity, hyperglycemia, and glucosuria (sugar in the urine), occurs in 10–15 percent of all pregnancies. Most cases resolve after pregnancy, but as many as 50 percent of women will develop diabetes mellitus later in life. Excessive vomiting, nausea, weight loss, and dehydration in pregnancy is known as hyperemesis gravidarum. This is not your typical morning sickness. Signs and symptoms can get better throughout the term, but in some women, it lasts the duration of the pregnancy and must be closely monitored.

Placenta previa is a condition in which the placenta is implanted too low in the uterus and covers the cervix, preventing a vaginal birth. This condition is closely monitored, but birth of the fetus must be by cesarean section. A deep vein thrombosis (DVT) is a blood clot that develops in the lower leg, thigh, or pelvis. It's a problem because it can break loose and travel to the lungs, causing a blockage and condition known as pulmonary embolism. All are life threatening situations.

Another factor is bleeding. Postpartum hemorrhage is excessive bleeding following delivery. Surgical repair of the tissue, ligation, packing with gauze, or hysterectomy are the treatments. However, the female can die before treatment takes place. Yes, if after pregnancy and delivery the bleeding is so severe, a total surgical removal of the uterus may be necessary to stop the bleeding and save the woman's life.

After birthing a child, there are still other factors to consider. Some won't be obvious until later in life. During childbirth, vaginal tears and anal sphincter tears are possible. Such trauma to the female anatomy is common during normal vaginal childbirth. Avulsion of levator ani muscles in the pubic region associated with a vaginal tear after normal vaginal delivery is associated with prolapse of organs later in life. Anal incontinence risks increase if the anal sphincter is torn during childbirth. Basically, the uterus can slip down into the vagina with organ prolapse and soiling oneself or having no control of passing gas or feces later in life can occur. Vaginal delivery is also associated with a significant long-term increase in stress urinary incontinence (involuntarily urinating when laughing, sneezing, running, or jumping). These increased signs occur regardless of female age or number of deliveries.

Another problem with pregnancy has to do with a woman just being pregnant. In June 2019, 27-year-old Marshae Jones was charged with homicide after losing her pregnancy when another woman shot her in the abdomen. You read that correctly. This Alabama woman was five months pregnant and was involved in a fight in which another woman fired a gun into Jones' abdomen, causing Jones to miscarry. Marshae Jones was being blamed for causing the death of her unborn fetus, even though another person, Ebony Jemison, shot her. In July 2019, Jefferson County Alabama District Attorney Lynneice Washington dismissed the case and no further legal action will take place against Jones.

After birthing a child, the female's body is still involved. Nursing a suckling infant is still the female role. And, women do this by choice. Sleep-deprived women with cracked nipples who are recovering from birthing a baby (and may have stitched, sore vaginas) wake up every hour to attend to their newborn. This is not penance. But women should have a say as to whether they want to do this.

Is it possible for men to be surrogate mothers? In the 1994 film *Junior*, Arnold Schwarzenegger is a scientist who becomes the first pregnant man. What if a zygote were surgically implanted into the male abdomen close to the liver where there is a rich blood supply? Could the male body support a pregnancy to full term? It is still not known because there are a host of other factors to consider. How would the male body respond to the stress of pregnancy? Would the cardiovascular system adapt? Would the appropriate hormones and amounts be made by the endocrine system? That said, there are cases of transgender men giving birth if the men kept their female reproductive organs after transitioning. Now, would these men be forbidden to have an abortion? Have lawmakers even considered this? Men still wouldn't be able to breastfeed. That option is reserved for females.

Breastfeeding

Breastfeeding, the nursing of an infant with milk produced by the female, is legal in all 50 states. Good to know, right? Furthermore, the federal Break Time for Nursing Mothers law requires employers covered by the Fair Labor Standards Act (FLSA) to provide basic accommodations for breastfeeding women at work. What are these accommodations? There must be both time and a private space that is not a bathroom for breastfeeding women to pump breast milk. While some may frown on this as being unfair to other women without children or to men who do not breastfeed, consider that we were all infants at one time. In a healthy society, breastfeeding is good. Breastfed babies are healthier, which means that medical costs are lower and time off work to care for sick children is lessened. It has been estimated that health care costs for newborns is decreased by two-thirds when women are able to participate in work-based maternity and lactation programs.

When did this Break Time for Nursing Mothers law take effect? As an amendment to the Fair Labor Standards Act (FLSA) of 1938, one would think that it took effect decades ago.

However, it took effect on March 23, 2010, as part of the Affordable Care Act. Put this into perspective: Women in the Senate got their own restroom adjacent to the Senate floor in 1993. Women in the House of Representatives got their own restroom in 2011.

Breast milk is tailor-made for an infant and supplies the baby with nutrients that cannot be synthesized, made in a lab, or given in formula. During pregnancy, the mammary glands of the breasts start preparing for the birth of the fetus, and immediately following birth, female breasts begin secreting human breast milk. This is known as lactation, a term derived from the Latin word *lac* meaning *milk* and *lactatio* meaning *suckle*. The most abundant proteins in human breast milk are casein, alpha-lactalbumin, lactoferrin, secretory immunoglobulin A, lysozyme, and serum albumin. On the other hand, cow's milk proteins are alpha-lactalbumin, beta-lactalbumin, and casein.

Time for a short diversion about cow's milk. Stop and think about cows. They are mammals. We are mammals. In order for breasts to produce milk, hormones of pregnancy signal to cells in breast tissue to begin producing milk. Without these hormones, milk is not produced. Suckling encourages and continues hormone production and thereby milk secretion. This is called a positive feedback loop: suckling causes more milk to be produced → more milk in turn enables suckling → which causes more milk. One event amplifies the other. Even women who have not been pregnant or given birth can stimulate milk production in the breasts. Women who have not borne children can use breast pumps to prime their breasts for subsequent breast feeding. This is often done for women who adopt infants and want to breast feed.

If you've never thought about how cow milk production occurs, it might be good to know that the same positive feedback that occurs for humans is also true for cows. Cows must either be pregnant, have suckling calves, or be given hormones for milk production to occur. The common hormone used—and approved by the Food and Drug Administration (FDA)—is bovine somatotropin (bST). Research has shown that this hormone increases milk production, is safe for the treated animal and for people consuming animal products, and it does not harm the environment. The trade name for the most commonly used bST is Posilac. Note the letters "lac" in the name. In reality, how does this work? Dairy farmers can buy Posilac over-the-counter. About two months after the cow has a calf, the cow is injected with Posilac. Treated cows are given the synthetic hormone for about eight months of the year after having a calf. Typically, cows lactate for about eight months. The protocol is for the milk producer to stop milking the cow after ten months. This is followed by a two-month period of non-milking to allow the mammary gland to rest before the female cow has another calf and starts the process all over again. This is typical for milk-producing cows in the U.S. What is not typical is for mammals to drink milk after infancy or for mammals (like humans) to drink the milk of other mammals (like cows). Just something to think about.

Returning to human breastfeeding. Breastfeeding is recommended for the first 6 to 12 months of life. Infant supplements are not necessary if the woman's diet is well balanced and the infant is nursing successfully. The chief protein in cow's milk is beta-lactalbumin with its beta-casein fraction being the main major human allergen in cow's milk. Casein is also the main constituent of cheese. Breast milk also offers immunological protection. For example, breast milk protects infants from antibiotic-resistant bacteria. The first milk produced by females is called *colostrum* or *foremilk*. Colostrum is produced for the first few days after giving birth, and it differs from the milk produced later in that it contains more lactalbumin, lactoprotein, and antibodies. The antibodies confer immunity to the newborn. Research also shows that breastfed babies have a reduced risk of sudden infant death syndrome (SIDS), have fewer infections, and are less likely to become obese. Breast milk also encourages beneficial gut bacteria in the infant, leading to a healthier microbiome.

Breastfeeding also confers benefits to the woman. Breastfeeding decreases the risk of breast cancer and ovarian cancer. Additionally, women can save $1,000 or more per year when they don't have to buy formula.

Separation of Church and State

The concept of separation of church and state is to establish non-religious, secular states in which rules are made for the welfare of that state and those rules are not based on religious beliefs. With 50 states plus Washington, D.C., that means the entire country would fall under the principle that laws aren't made based on religion. Right? Well…

Time for a short history lesson. This phrase about separation of church and state is not in the U.S. Constitution. The original phrase, stated by Thomas Jefferson in a January 1, 1802, letter to the Danbury Baptists, was intended to assuage the fears of the Connecticut people that the government would not interfere with their business. Although quite similar to "separation of church and state," Jefferson used "wall of separation between the church and the state." Okay, so we're splitting hairs. The purpose was to keep the state (government) out of the church's business, not to keep the church out of government business.

Here's what the constitution does state: "Congress shall make no law respecting an establishment of religion or prohibiting the free exercise thereof." In summary, both the "establishment of religion" and "free exercise of religion" affect how government cannot interfere with religion. Scholars of the Constitution and people who understand the context in which it was framed know that our common understanding is the opposite of its original intent. The framers of the U.S. Constitution did not want a state-denominational religion, but they did favor state doctrinal religion. And what was the default state doctrinal religion? A religion that believes in God—theism. Examples of theism within our government abound: Presidents are sworn into office reciting the phrase, "So help me God." The Supreme Court building has carvings of Moses and the Ten Commandments. The Declaration of Independence mentions God four times. Each session of the U.S. Senate opens with a prayer given by the Chaplain. The first Chaplain of the U.S. Senate was Episcopal Bishop of Maryland Samuel Provoost, who began service in 1789. There is also a Chaplain of the United State House of Representatives, who begins each day's proceedings with a prayer. The list goes on, and these are just a smattering of examples.

Theism is worldwide and finding an exact number of the world's religions is difficult. According to the Pew Research Center, in 2015, there were 7.3 billion people on Earth. Of that 7.3 billion, most, but not a majority were Christian (2.3 billion; 31.2 percent); however, 1.8 billion (24.1 percent) were Muslims, followed by 1.2 billion (16 percent) who were unaffiliated. Here's the breakdown for the remainder:

Hindus—1.1 billion (15.1 percent)
Buddhists—0.5 billion (6.9 percent)
Folk religions—0.4 billion (5.7 percent)
Other religions—0.1 billion (0.8 percent)
Jewish—0.014 billion (0.2 percent)

Does religion enter into healthcare practices? Ask obstetrician-gynecologists who work in Catholic-owned hospitals. Catholic-owned hospitals, which comprise at least 10 of the 25 largest hospital systems in the United States, impose restrictions on reproductive health, including contraceptive services and abortion. Catholic doctrine will permit abortions to save the life of a woman, but ethics committees at Catholic hospitals determine how much health risk constitutes a threat to a woman's life. Recall the explanations of reproductive biological concepts covered so far. This means that care to a pregnant woman can be delayed, even while her own health is deteriorating. Some policies prevent physicians from carrying out an abortion if the fetal heartbeat is present. Several states have passed "heartbeat bills" aimed at banning abortion as soon as a fetal heartbeat is detected, and they make no provision for incest or rape. This is a time before many women even know they are pregnant. Here's a real case for consideration: According to police reports, an 11-year-old in Ohio was raped multiple times by a 26-year-old, resulting in pregnancy. In April 2019, Ohio passed a strict abortion law preventing women— much less a child—from having an abortion. A child, whose body is still developing, would have to carry and deliver a rapist's child.

If we review human development, the heart really isn't even formed at this stage and any electrical activity detected is coming from the fetal pole, a thickening on the margin of the yolk sac. This so-called heartbeat is politically charged terminology to make people think there is a child with a beating heart. With such restrictive policies preventing physicians from intervening, a woman could be dying before the eyes of her physician. According to a study published in the *American Journal of Public Health*, interpretation of Catholic doctrine has interfered with medical judgment. Some patients had to be transferred to the nearest hospital—which could be 90 miles away—for treatment. Treatment has been delayed in other cases; and in some cases, doctors go ahead and do what they think is medically necessary to save the woman's life. These hospitals accept public funds and are subsidized by public funds. Public funds refer to American tax dollars. If a hospital accepts public funds, then the public should be able to use the facilities without judgment.

Perhaps you are thinking, "Don't go to a Catholic hospital if you disagree with their values." Sounds reasonable. Yet, 15.2 percent of our country's hospitals are Catholic-owned and accept public funds. This means that such a hospital may be the only one in the region for pregnant women. Physicians take a Hippocratic oath, named for the Greek physician regarded as the father of medicine, Hippocrates (460–377 BCE). There are several versions of the oath, which makes sense because the original one swears to gods and goddesses such as Apollo, Asclepius, Hygieia, and Panaceia. At its core, the oath states the obligations and proper conduct for physicians, which includes saving women whose pregnancy threatens their lives.

The Tension Between Science and Faith: Choice

Americans want individual freedom. As the pandemic revealed, Americans went to great lengths to protest wearing face masks, even when doing so would protect their fellow citizens. However, when freedom contradicts personal or religious beliefs, many are willing to forego individual choice, at least for other people, in order to impose their beliefs. Freedom of choice is a clear example. Choice is selecting or making a decision. Women can choose to have a baby or women can choose to have an abortion. Regardless of how you think or feel about this issue, keep in mind that without the female body's role in pregnancy, there is no pregnancy.

There is a dizzying array of state bans on abortion throughout pregnancy, with laws aimed at various definitions including last menstrual period (LM), postfertilization, post-implantation, physical health, and viability. At best, it is a patchwork of laws and restrictions, none aimed at preserving access to healthcare for women.

As of September 2021, ten states have passed early abortion bans. The laws vary, but Alabama's is most extreme, followed by Texas. Alabama has banned abortion except if a woman's health is at risk. Ohio, Missouri, Kentucky, Louisiana, Mississippi, and Georgia have passed abortion bans at six to eight weeks. Some of these laws have been permanently or temporarily enjoined by court order and have not taken effect. Utah and Arkansas have passed abortion bans at 18–22 weeks. Each is intended to challenge the 1973 Roe v. Wade Supreme Court ruling. Anti-abortion laws are based on either ignorance or faith. The information is there, but many people, including lawmakers, refuse to educate themselves. A classic example of this occurred in August 2012 by U.S. Republican representative Todd Akin of Missouri. When asked if he believed abortion is justified in cases of rape, he replied that rape does not result in pregnancy. He went on to say that "It seems to be, first of all, from what I understand from doctors, it's really rare. If it's a legitimate rape, the female body has ways to try to shut the whole thing down." Wow. Legitimate rape? This sounds harsh, but in order to make informed decisions, let's agree that reproductive biology is complicated. Shouldn't people making decisions for somebody else be well-versed in the facts? As more states challenge Roe v. Wade, it's likely the conservative-leaning Supreme Court will be revisiting this landmark case in 2021.

The basic arguments on the two sides of the pro-life versus pro-choice debate boil down to questions about (1) when life begins and (2) whether a woman's autonomy supersedes the interests of an embryo or fetus. People on the pro-life side, often with strong religious underpinnings, tend to insist that terminating a pregnancy destroys a life—life, the thinking goes, begins at conception, and abortion ends that life and is therefore equivalent to killing a human.

On the pro-choice side, people tend to argue that pregnancy, especially before the point of fetal viability, falls under the purview of women's health, and as such, is entirely the realm of the woman and her doctor. No third party should be involved in its decision making, and certainly not third parties who have no experience with pregnancy, birth, lactation, or other women's health conditions.

It seems that pro-lifers are really just pro-birthers, for once the child is born, the woman and child are left behind. Consider these points: pro-life legislators are more likely than pro-choice legislators to vote against affordable contraceptives, sex education, paid maternity leave, and access to affordable childcare. Research has shown that each of these decreases abortion rates. And, as discussed in Chapter 9, pro-life politicians are more likely to support the death penalty, defense spending, and wars—all of which aim to kill living people in other ways. If we want all babies to be born, we can't deprive them as babies, children, and adults of life-promoting services, aid, education, and assistance. Pregnant women understand the consequences, which is why each abortive decision is personal.

Beginning to write these pages in fall 2020, it was often difficult to stay focused, because the political rhetoric was as high as many have ever experienced. Within my inner scientific circle, we tried hard to use our platforms to diffuse angry, inaccurate voices. Single-issue voting—and there were many single issues to debate during this troubled time—oftentimes was at the core and moved a friend to write this poem.

My Birth

When I was a beggar, you ignored me.
When I was a slave, you persecuted me.
When I was a refugee, you shunned me.
When I was an orphan, you relinquished me.
When I was a prostitute, you abused me.
When I was an LGBTQ, you renounced me.
When I was a girl/woman, you confined me
When I was a native/gypsy, you ridiculed me.
When I was white/elite, you disdained me.
When I was black/brown, you despised me.
When I was in the womb you fought for me.
After that, how was it okay to abandon me?
How come *my birth*, became a criterion
for you to then just forget about me?
—Dr. Anjali Dogra Gray

The issue is thorny to say the least. However, thorough consideration reveals that some basic assertions fall apart on closer inspection. One of the big issues with the idea that abortion should be illegal, for example, is that outlawing it doesn't make it go away, it just drives it underground. According to the World Health Organization (WHO) during 2015–2016, about 22 million unsafe abortions were performed worldwide. Twenty-two million women is almost the equivalent of the entire population of the state of New York, which has about 19.6 million people. The remaining 2.4 million people is nearly the population of New Mexico, which has roughly two million people. Of these 22 million women, an estimated 47,000 died and another five million have a disability as a result of the unsafe abortion. Deaths and disabilities can be prevented through myriad measures including education, family planning, provision for safe and legal abortions, and healthcare for complications of abortion. We have a worldwide health issue requiring global compassion and cooperation. We don't want to ignore science, criminalize abortion, or punish women for being the sole sex whose bodies carry a pregnancy.

What is the medical definition of *abortion*? According to *Stedman's Medical Dictionary*, an abortion is the expulsion from the uterus of an embryo or fetus before viability (20 weeks' gestation [18 weeks after fertilization] or fetal weight less than 500 g or 1.1 pounds). Viability means that a fetus is capable of living on its own outside the mother's body. If you'll recall, at around week 24 of gestation, the fetus had a 50 percent chance of survival outside the uterus with intense medical intervention.

In 1973, the United States Supreme Court's decision on Roe v. Wade protected a woman's right to an abortion because it was protecting a woman's right to privacy. To criminalize abortion is to violate a woman's right to privacy. Health matters are only the business of the person involved along with her healthcare provider. Even in the day of social media, we still like to keep some things to ourselves. Shouldn't we each have autonomy over our own bodies? To protect our personal health privacy, laws were passed to prevent others from knowing your business. Enacted in 1996, the Health Insurance Portability and Accountability Act (HIPAA) is a set of federal regulations that protect the privacy and security of health information. Compliance is mandatory in the United States.

Abortions can also occur due to natural causes. These are known as spontaneous abortions or miscarriages. It means that a pregnancy ends before birth when an undeveloped pre-embryo,

embryo, or fetus separates from the uterus. Most occur early in the pregnancy, though they can occur any time before 7 months of development. Ten to 15 percent of pregnancies result in spontaneous abortions. Spontaneous abortions are usually because of chromosomal abnormalities in the embryo or fetus or fetal deformities that are incompatible with survival. If the miscarriage occurs after 7 months, it may be caused by the woman's physiology. That is, gestational diabetes, hypertension, infectious disease, or drug abuse can cause a miscarriage. Sometimes, the cause of miscarriage is just not known. According to the CDC, current pregnancy rates and abortions are both the lowest since they started tracking. Moreover, nearly as many women miscarried (17 percent) as had abortions (18 percent).

Closely related to a spontaneous abortion is a stillbirth. A stillbirth is the birth of a fetus that has died in the uterus. Sadly, these are fetuses after about the 24th week, which is near the end of the second trimester. Many of the same causes of spontaneous abortion also cause stillbirths. Premature births are those born after the stage of viability, but before 37 weeks' gestation. When a stillbirth occurs, it is often difficult to give grieving parents much time to spend with the fetus/baby because decomposition begins so quickly. A device called the Flexmort Cuddle Cot is a soft cooling pad that can be used to keep the body cold and slow deterioration, giving parents and other family members time with the baby.

Since 1969, the CDC has documented the number and characteristics of women obtaining legal abortions. In 2014, the abortion rate was 2 percent less than in 2013. The majority of women who had abortions were in their twenties and the vast majority of the procedures (91.5 percent) were performed at less than or at 13 weeks' gestation. At 13 weeks, the fetus is about 2.5 inches and about the size of a lemon.

Decisions regarding abortion are extremely personal, thus only the person involved can make the decision. However, the decision should be based on the best medical information, not hearsay or vivid, fabricated images on the Internet.

Depending upon the stage of pregnancy, there are two main ways of ending a pregnancy: medicine (abortion pill; nonsurgical) or medical abortion (in-clinic procedure). The two types of medical abortions are therapeutic (ending a pregnancy because the mother's life is in danger) and elective (ending a pregnancy by choice).

The abortion pill can be taken within the first 10 weeks of pregnancy. The pregnancy "time clock" begins with the number of days since the first day of a woman's last period. The abortion pill is actually two pills, one containing mifepristone and the other misoprostol. The first pill blocks progesterone, a necessary hormone for pregnancy and the second pill causes uterine cramping and bleeding, which empties the uterus. An in-clinic abortion involves the use of gentle suction to remove the pregnancy. Medical procedures are used through week 24. Third trimester abortions are rare because they are high risk and can cause serious complications. Third trimesters abortions are also often due to fetal abnormality and are the result of extremely difficult decisions by the family. Abortions are safe and 1 in 4 (25 percent) women in the United States have an abortion by the time they are 45 years old.

There are legal restrictions on abortion, and these vary from state to state. Regardless of where you live, very few abortions are done after 16 weeks, and abortions are rarely done after 24 weeks of pregnancy. Abortions are safe in this country and they do not prevent a woman from becoming pregnant later.

What we do know is that restrictive laws or abortion bans result in unsafe abortions. There is no evidence to suggest that restrictive laws lower abortion rates. Research has shown that unsafe abortions are prevented by ensuring that girls and women are able to prevent unintended pregnancies, that they have access to comprehensive sex education, that effective and affordable contraception is available, and that girls and women have access to safe, legal abortion and post-abortion care.

Lawmakers are currently making it very difficult for women to have safe abortions or receive affordable contraception. In 2020, the Republican senate was scheduled to vote on a bill that would ban abortions after 20 weeks. Would our federal and state policies legislating abortion be different if women were in charge of making them? Consider that the 116th Congress was not representative of the diverse population of the United States, although the 116th House of Representatives had more women of color than ever before. As of January 3, 2021, the 117th Congress of the U.S. convened in Washington, D.C. This Congress is the most diverse in history with 127 women serving in the House of Representatives, making up 27% of that chamber's total. There are 24 women in the Senate, which is one fewer than the prior Congress. Additionally, 124 lawmakers identify as Black, Hispanic, Asian/Pacific Islanders or Native American, and collectively they constitute 23% of Congress. In spite of this increase in diversity, Congress is still less diverse than the nation as a whole.

We also have laws in place that curtail federal spending on medical research that uses tissue from aborted fetuses. In June 2019, the Trump administration said it would stop funding research using fetal tissue from abortions. But legislation goes back several decades. In 1977— three years after Roe v. Wade—Congress passed the Hyde Amendment. The amendment is legislation that bars the use of federal funds to pay for abortions except to save a woman's life or if the pregnancy is the result of rape or incest. The Hyde Amendment affects funding under Medicaid and restricts funding under the Indian Health Service, Medicare, the Children's Health Insurance Program, the military's TRICARE program, federal prisons, the Peace Corps, and the Federal Employees Health Benefits Program. Medicaid is a program jointly funded by the federal and state governments, so states can choose to cover the service. Women living in 35 states and Washington, D.C., and covered by Medicaid have extremely limited coverage for abortion because of the Hyde Amendment. Women who are poor, are part of a minority group, or who are in poorer health rely on Medicaid. According to the Guttmacher Institute, women who get abortions are disproportionately low-income, young, and racial/ethnic minorities.

According to the latest Census Bureau numbers, there were roughly 169.36 million females versus 162.02 million males in the U.S. However, women live longer than men and women outnumber men by a ratio of 2-to-1 in the 85 and older population. Thus, it begs the question: Why would we allow a male-dominated Congress to make decisions and pass laws that affect women exclusively? It's baffling. It's like asking One Billion Rising if men are allowed to have vasectomies. (Just in case you're not familiar with this phenomenal organization, One Billion Rising is a global movement founded by Eve Ensler in 2012 to end rape and sexual violence against women. She is also known for her play *The Vagina Monologues*.)

Lopsided politics prevent the passage of the Equal Rights Amendment (ERA), which began its career in 1923. The proposed amendment, first written by Crystal Eastman, proposes to end legal distinctions between men and women in matters ranging from divorce and property to employment. As an amendment to the Constitution, it had to be ratified by 38 states by March 22, 1979.

A staunch opponent who likely single-handedly prevented the passage of the Equal Rights Amendment was Republican conservative, Phyllis Schlafly (1924–2016). During the 1970s, she organized the Stop Taking Our Privileges (STOP) Campaign arguing that passage of the ERA would take away dependent wife benefits under Social Security, disallow separate restrooms for men and women, and make women eligible for the military draft. As you might imagine, she opposed abortion, same-sex marriage, and linked LGBTQ issues with the ERA. Schlafly also published several conservative books with dubious claims. Ironically, while opposing women's independence, Schlafly took advantage of equal opportunities in education having earned a law degree, toured the country speaking on her platform issues, served as an editor, and remained politically active throughout her life.

The amendment still hasn't passed, despite women's movements, female empowerment, and bipartisan support between then and now, most recently in January 2020 when Virginia's General Assembly passed a ratification resolution for the ERA. Moreover, before 1979 and the 2020 Virginia vote, five states, Idaho, Kentucky, Nebraska, South Dakota, and Tennessee, voted to rescind their ratifications. Two issues are currently at the fore: (1) Virginia's 2020 vote was past the deadline and (2) the legality of states rescinding their constitution ratification remains a legal conundrum.

If this amendment is to gain any traction, a new resolution must be made. To that end, House Speaker Nancy Pelosi brought forth a formal intention for legislative consideration. In essence, the amendment approval process is starting fresh. In addition to creating equal footing for the sexes, advocates for the ERA link women's rights with control over their bodies, including access to abortion.

Legal and safe abortion is always on the political docket. The official Republican platform asserts that "the inherit dignity and sanctity of all human life and affirms that the unborn child has a fundamental individual right to life which cannot be infringed." There are no exceptions. Rape? Not excluded. Incest? Not excluded. Mother's life? Not excluded. Where is the sanctity of human life here?

Aside from politics and the anatomy and physiology of childbearing, we must also consider the psychological impact on the woman. To illustrate, many women are fully aware of their ovulation cycle, although some cycles are not as regular. When a woman misses her period, she might think she is pregnant if she had unprotected sex, or she might think she's just late for any number of reasons. If she is pregnant, however, the gamut of emotions begins. If a woman chooses to terminate the pregnancy, only a few options are available. One such option is Plan B. Plan B (morning-after pill) is an over-the-counter pill that must be taken within 48 hours of unprotected sex. At this time, the woman has no idea whether conception has occurred because it's too soon to tell. Anywhere between 6 and 14 days after fertilization, she can take a home pregnancy test, in which she urinates over a strip that detects the particular hormone of pregnancy, hCG. Note the range, indicating again that not everything can be exact and precise. Some home pregnancy tests claim to detect the hormone within four days. But, it's much more complicated than this. If a woman knows what stage of her ovarian/menstrual cycle she's in, perhaps four days works for her. Most doctors recommend waiting until the first day after a woman misses her period. Again, menstrual cycles are not precise either and can be quite variable. Note that while it takes two to tango, only the woman has any responsibility at all these stages.

If a woman chooses abortion, and she lives in a state with a "heartbeat" bill, the waters can be quite muddy while the bill's cruelty can be crystal clear. Here's a classic scenario: A few days after a positive pregnancy test, a woman checks in with her obstetrician-gynecologist—commonly called an ob/gyn—for an ultrasound. If the embryo hasn't implanted yet or has implanted somewhere else (as in an ectopic pregnancy), the ultrasound might not be able to confirm the pregnancy. In such a case, the woman has to come back a few days later and try the ultrasound procedure again. Many ob/gyns want a visual ultrasound confirmation of the pregnancy. Now, if you're counting the days, it's very likely that the magical six-week timeframe has passed. Some states also require a 72-hour waiting period after deciding on pregnancy termination.

On June 29, 2020, in a 5–4 vote, the Supreme Court ruled in *June Medical Services v. Russo* that physicians who perform clinical abortions did not have to have admitting privileges at a nearby hospital. Having such a law would burden women seeking pre-viability abortions. The law was intended to shut down abortion clinics in Louisiana and had no benefits for patient health.

A woman and her doctor are the only ones in the position to decide what is best. There is no other medical procedure within our entire healthcare setting in which legal codes dictate what goes on between a patient and their medical provider. Okay, maybe a few, but they apply to both men and women for the common good of society, like, hey, you can't sell your kidney to support your cocaine addiction. There certainly are no laws that legislate whether a man can have a vasectomy or be circumcised. Legislation on the reproductive system only applies to women.

Planned Parenthood, Obstetrics and Gynecology

For many people, Planned Parenthood is often synonymous with abortion. It should not be. It is a nonprofit organization over a hundred years old that provides sexual health care in the United States and globally. In addition to abortion services and abortion referral (not all Planned Parenthoods provide abortions), they also provide birth control and information on birth control; this includes birth control implant, IUD, birth control shot, birth control vaginal ring, birth control patch, birth control pill, condom, internal condom, diaphragm, birth control sponge, cervical cap, spermicide, fertility awareness methods; withdrawal (pullout method), breastfeeding as birth control, outercourse and abstinence, sterilization (tubal ligation), and vasectomy; emergency contraception (morning-after pill); general health care; information on sex and relationships (sexual consent, masturbation, sexual and reproductive anatomy); HIV services; LGBT services (including hormone therapy for transgender patients); men's health services (addressing cancer, STDs, fertility, sexual dysfunction, birth control, and routine checkups); patient education (including up-to-date information on sexual and reproductive health); pregnancy testing and services (ensuring women get the care they need to stay healthy); STD testing, treatment and vaccines; and women's services (breast exams, pelvic exams, cancer screenings, and pregnancy-related services). The nonjudgmental services they provide is lengthy. Sex is a normal part of life. Denying people healthy sex lives is not normal.

Unlike Planned Parenthood, where women can receive quality information, a crisis pregnancy center (CPC), also called a pregnancy resource center (PRC), operates using different tactics. Their goal is to counsel women against abortion. They are fake health centers that sound like the real deal. In fact, state tax dollars often go toward funding them. Using phony ads to trick pregnant women, they claim to offer free ultrasounds and pregnancy support. These facilities look like bona fide medical centers, but they are predatory organizations that shame women, provide misleading and outright false information, and push women past the deadline for legal abortion. CPCs do not provide abortions; they are unregulated, anti-choice groups. Furthermore, studies have shown that they provide inaccurate and misleading information about condoms, sexually transmitted infections, and methods to prevent the transmission of sexually transmitted infections. This is harmful to society.

If the United States is a country in which freedom of religion can be practiced, then all religions should have equal footing. Jewish law, for instance, does not consider the fetus to be a being with a soul until it is born. Stated another way, it does not have personhood and therefore abortions should be safe and legal.

The abortion issue isn't relegated solely to Planned Parenthood. Abortion is an issue that infiltrates medicine as practiced by medical professionals on the frontline of women's health. To that end, there are areas of medicine specialized for the medical care of women. These medical professionals at the forefront of female reproductive health are obstetricians, gynecologists, and midwives.

Obstetrics is the specialty of medicine concerned with the care of women during pregnancy, during labor (parturition), and for a short period of time after childbirth. Gynecology is the medical specialty concerned with diseases of the female reproductive tract as well as the endocrinology and reproductive anatomy and physiology of females. Midwives are medical professionals with specialized training in obstetrics and childcare who can practice midwifery (assisting women in childbirth).

Medical professionals often belong to professional societies whose role is the advancement of a particular medical specialty based on science and evidence-based practices. The American College of Obstetricians and Gynecologists (ACOG) is the premier professional organization for physicians practicing obstetrics and gynecology. This professional society is dedicated to improving women's health. According to their website, the American College of Nurse-Midwives (ACNM) is the professional association that represents certified nurse-midwives (CNMs) and certified midwives (CMs) in the United States. With roots dating to 1929, ACNM sets the standard for excellence in midwifery education and practice in the United States and strengthens the capacity of midwives in developing countries. Its members are primary care providers for women throughout the lifespan, with a special emphasis on pregnancy, childbirth, and gynecologic and reproductive health. ACNM reviews research, administers and promotes continuing education programs, and works with organizations, state and federal agencies, and members of Congress to advance the well-being of women and infants through the practice of midwifery. The Reproductive Health Access Project (RHAP) is a national nonprofit organization that works directly with primary care providers, helping them integrate abortion, contraception, and miscarriage care into their practices so that everyone can receive this essential health care from their own primary care clinicians.

Each of these organizations and its medical professionals understands all the nuances of abortion. So, when the president of the United States used inflammatory language and rhetoric in his 2019 State of the Union Address, stating how a New York law would "allow a baby to be ripped from the mother's womb moments before birth" and said that "these are living, feeling, beautiful babies who will never get the chance to share their love and dreams with the world," he greatly exaggerated what really happens and continued to perpetuate myths, misconceptions, misunderstandings, falsehoods, and fear. In early February 2019, the ACOG released a fact sheet about abortions and stated that "politicians should never interfere in the patient-physician relationship." Would you go to your senator and ask his or her opinion on whether you should have your appendix removed? Of course, you wouldn't. Politicians are not medical experts.

Legislating Reproductive Health

Lots of the language used in politics, such as "later-term abortion" isn't even used by the medical establishment. It is inaccurate and purposefully confusing language used for political reasons. Going back to the stages of development, like much in life, it is not clear cut. For example, obstetricians can use 24 weeks as the threshold for viability of a fetus. Remember that viability refers to the age at which a fetus can survive outside the uterus. But there are cases in which a 23-week fetus may survive and cases where 25-week fetuses will not survive. There are cases in which the woman's life is at stake if the pregnancy were to continue. There are cases where both cannot survive. There are cases where everything is going just perfect in the pregnancy and then at say week 26 or week 27, a lethal anomaly in the fetus is discovered. Remember that not everything develops at once and some things cannot be diagnosed until quite late in the pregnancy. There are also anomalies that are completely incompatible with life outside the uterus. Imagine the devastation of the woman whose pregnancy was clicking along perfectly fine and then suddenly it is not. If politicians have their way, a woman would have to walk around with a virtually dead fetus because laws would prevent an abortion. Every single situation requires individual assessments between the woman and her healthcare provider. Every. Single. Time.

During the 2019–2020 political cycle and Supreme Court nominee Amy Coney Barrett hearings (she was approved and became a Justice), abortion rights were front and center once again. Inflammatory language surrounding "late-term abortions" were at the fore. But, in 2019, Democrat Pete Buttigieg gave the best response that captures how most American women feel. In a May 19, 2019, town hall hosted by Fox News, newscaster Chris Wallace asked Buttigieg about abortion. Here's the exchange:

> **WALLACE:** Do you believe, at any point in pregnancy, that there should be any limit on a woman's right to an abortion?
>
> **BUTTIGIEG:** I think the dialogue has gotten so caught up on when you draw the line that we've gotten away from the fundamental question of who gets to draw the line. And, I trust women to draw the line.
>
> **WALLACE:** You would be okay with a woman well into the third trimester to obtain an abortion?
>
> **BUTTIGIEG:** These hypotheticals are set up to provoke a strong emotional reaction.
>
> **WALLACE:** These aren't hypotheticals—there are 6,000 women a year who get an abortion in the third trimester.
>
> **BUTTIGIEG:** That's right, representing less than one percent of cases a year. So, let's put ourselves in the shoes of a woman in that situation. If it's that late in your pregnancy, that means almost by definition you've been expecting to carry it to term. We're talking about women who have perhaps chosen the name, women who have purchased the crib, families that then get the most devastating medical news of their lifetime, something about the health or the life of the mother that forces them to make an impossible, unthinkable choice. That decision is not going to be made any better, medically or morally, because the government is dictating how that decision should be made.

The current political climate is rife with misinformation, proposals aimed at oppressing women, and tactics suppressing personal freedoms. In November 2018, U.S. Secretary of Education Betsy DeVos proposed reshaping federal Title IX policy to protect accused perpetrators of sexual assault instead of protecting survivors. One in five women experience attempted sexual assault or actual sexual assault while in college. New rules from the Trump administration let employers and universities refuse to provide health insurance coverage for birth control. This is ridiculous, especially since some birth control is used medically for other reasons, such as regulating menstrual cycles, lessening painful menstruation, taming hormonal acne, reducing uterine cancer risk, reducing ovarian cyst risk, relieving symptoms of premenstrual syndrome (PMS), relieving symptoms of premenstrual dysphoric disorder (PMDD), managing endometriosis, lessening menstrual migraine pain, reducing anemia risk, reducing bleeding associated with fibroids, treating polycystic ovary syndrome (PCOS), treating amenorrhea, treating primary ovarian insufficiency (POI), and treating heavy menstruation. Proposed health insurance rules could make it impossible for women to get coverage for abortion.

Adding salt to the wound, there is pregnancy discrimination. Many states have no protection for working pregnant women who can be forced out of their jobs if they cannot fulfill their duties while pregnant. Think about women who may have lifting restrictions while pregnant. Without accommodations, they could be sent home without pay and without a job to come back to. H.R.2694, known as the Pregnant Workers Fairness Act, was introduced in May 2019. It would require employers to provide reasonable accommodations for pregnant workers. In May 2021, the House passed H.R. 1065, which prohibits discrimination against and making reasonable accommodations for qualified employees affected by pregnancy, childbirth, or related medical conditions. These work adjustments are similar to accommodations provided under the Americans with Disabilities Act (ADA).

Currently, 27 states have laws that protect pregnant women against workplace discrimination and require employers to apply the same benefits, terms, and conditions to pregnancy, birth, recovery or associated conditions as are applied to other temporary disabilities. States without such laws or that have pregnancy discrimination laws but with restrictions (i.e., number of employees) include Alabama, Alaska, Arizona, Arkansas, Colorado, Florida, Georgia, Indiana, Massachusetts, Mississippi, Missouri, Nebraska, New York, North Carolina, North Dakota, Pennsylvania, South Dakota, Tennessee, Utah, Vermont, Virginia, and Wyoming.

Note that Alabama, Arkansas, Georgia, Iowa, Kentucky, Louisiana, Mississippi, Missouri, Ohio, Texas and Utah have already passed anti-abortion bills, while Florida, South Carolina and West Virginia are working to pass anti-abortion bills. In states where there are no protections for pregnant working women, there are also laws forcing them to remain pregnant. Maine, New York, and Vermont have added abortion protections.

The Department of Health and Human Services encourages health care providers to use personal beliefs to deny patients basic health care. The administration continually tries to block patients from care at Planned Parenthood health centers, even though 2.4 million people use their services, including for cancer screenings, wellness exams, STD tests, birth control, and education. The administration illegally slashed funding for sex education programs and shifted toward abstinence-only programs. This last item is particularly disturbing because it is draconian and study after study after study over many years have shown abstinence-only education is ineffective. These programs do not reduce teen pregnancies or sexually-transmitted diseases and they do not delay intercourse.

Education about something as important and normal as sex is important. Abstinence-only-until-marriage (AOUM) or sexual risk avoidance are scientifically and ethically problematic. Findings of a 2017 study in the *Journal of Adolescent Health* demonstrate that "AOUM programs threaten fundamental human rights to health, information, and life. Young people need access to accurate and comprehensive sexual health information to protect their health and lives."

TAKE-HOME MESSAGE

The biology of reproduction and human development is highly complex, but it needs to be understood to form an informed position on policy decisions around these issues. Legislation written without a full understanding is often nonsensical at best and, at worst, can be extremely harmful. Public policies formulated around personal beliefs rather than scientific evidence can lead to higher rates of unintended pregnancy, abortion, sexually transmitted infections, and other adverse public health effects.

> *Real knowledge is to know the extent of one's ignorance.*
> —Confucius (551–479 BCE),
> Chinese philosopher

6

Staying Alive

Vaccines and the Immune System

U.S. Landmark Case: Bruesewitz v. Wyeth, 562 U.S. 223 (2011); (6–2 decision)

Hannah Bruesewitz was scheduled to receive a series of five doses of Tri-Immunol DTP (diphtheria, tetanus, pertussis) vaccine made by the pharmaceutical company Wyeth. Shortly after receiving the third dose of vaccine, Hannah began having seizures and suffered brain damage. Her condition was termed residual seizure disorder. Hannah's parents filed suit in the "Vaccine Court," which was created within the United States Court of Federal Claims under the National Childhood Vaccine Injury Act of 1986. The Court's purpose was to settle the numerous claims of vaccine injury that were then flooding the court system. The government felt that such injury cases would cause pharmaceutical companies to stop developing vaccines out of fear of unending trials and debilitating damage awards by sympathetic juries. Two stipulations of the Vaccine Court were that the petitioner had to show an injury occurred immediately after vaccination and the injury had to be listed in the Vaccine Table of side effects maintained by the Vaccine Court. Residual seizure disorder was not on the list of side effects and compensation was denied. The Bruesewitz Family then sued under Pennsylvania tort law. The case then came before the U.S. Supreme Court. The Court sided with the pharmaceutical maker, stating that allowing Vaccine Court appeals would set a precedent that could be potentially devastating to vaccine makers to the extent that no future vaccines would be produced.

Preventing Disease

Outside of hand washing and fluorinated water, vaccines have been one of the greatest boons to public health. As I write, we're in the midst of the COVID-19 pandemic and pharmaceutical companies have raced to find a vaccine. A pandemic refers to a disease outbreak that stretches over a whole country or the entire world. The word comes from the Greek word *pandemos*, which is derived from the word parts *pan* meaning *all* and *demos* meaning *people*. Compare this with the word, *epidemic*, which means a widespread occurrence of an infectious disease in a community at a particular time.

Vaccines protect against numerous diseases and have prevented millions of illnesses. In short, we want pharmaceutical companies to continue working with researchers to develop effective vaccines without fear of constant litigation. The opening landmark case speaks to the issue of vaccine development without fear of perpetual lawsuits. Yet, as a society, we also want the sound practices of science to be upheld so that vaccines do not pose a risk to public health. To date, vaccines have proven to be safe and effective, but many people believe they are bad, can cause disease, and are responsible for autism spectrum disorder (ASD). From a scientific and human health perspective, vaccines do prevent disease, and they do not cause autism.

The immune system is one of the most complicated, elaborate, and individualized body systems. Its purpose is to protect us. As such, it is an intricate complex that relies on interrelated cellular, molecular, and genetic components to provide a defense against foreign organisms like bacteria and viruses, and against rogue cells that cause cancer. It also wards off aberrant native cells. When the immune system fails, disease results.

But what is disease? In broad strokes, disease is a disorder that causes an interruption of normal body functions. It can affect specific organs, organ systems, or the entire body. Disease is not the direct result of some sort of physical injury. Diseases are caused by many factors, including microorganisms, changes in a protein-coding gene, personal habits such as smoking, poor nutrition, and weakened immune systems among others. Finding underlying causes is complicated and sometimes the answer to what causes disease is just not known. Regardless of its cause or effects, a person with a disease just doesn't feel good.

Many terms are used to mean disease. Some of these terms include disorder, illness, sickness, ailment, malady, pathology, and homeostatic imbalance. Despite which term is used, they all refer to the same thing and mean that something is wrong with the body. To sum up, it's something nobody wants. Modern medicine has taken great strides toward preventing disease. We know that eating well and exercising do much toward thwarting disease. Preventing disease is also a good place to start, which means that vaccination is a means for preventive treatment.

Vaccines Do Not Cause Autism

One singular beneficial thing to human health and staving off disease has been vaccines. Vaccines, also called *vaccinations* and *immunizations*, are medical preparations that actually *protect* us against diseases. This protection is known as *immunity*. Vaccines work because the substances in them stimulate antibody production, and these antibodies provide immunity against specific diseases. In the body, when normal, native antibodies encounter something that is foreign or has a foreign protein, the antibodies attack the foreigner to get rid of it. Vaccines are usually prepared using harmless causative agents of a particular disease. These harmless agents are antigen substitutes for the real thing. When these antigen substitutes are injected into the body, the body automatically begins producing antibodies to attack this foreign agent. If the body ever encounters the real antigen for which the substitute was created, we already have a pre-formed antibody ready to go, or we can make one quickly, to prevent the disease.

We don't have vaccines for every illness, but we do have vaccines against some major horrible infectious agents, including viruses, bacteria, fungi, and parasites. We know them as germs. Over the years, vaccines have gotten a bad rap because people who know little or nothing about science or who disregard science, have been quite vocal about vaccines causing autism spectrum disorder (ASD). In many circles, just the term *autism* is used. A mountain of scientific evidence shows that vaccines do not cause ASD. Autism spectrum disorder is a developmental disorder of variable severity that affects communication and behavior. "Spectrum" is used because there is great variability in the manifestation of the disorder. It occurs across all ethnicities and among all socioeconomic groups. While no parent wishes ASD upon their child, vaccines do not cause it.

The first COVID-19 vaccines rolled out quicker than expected, using new technology that enabled a first-of-its-kind vaccine that uses messenger RNA (mRNA). Shortly following the new Pfizer/BioNTech and Moderna mRNA vaccines, Johnson & Johnson brought a third vaccine using "standard" DNA technology to the market. All three have proven safe and effective. Yet, even with the scientific evidence of the current COVID-19 vaccines proving their safety and efficacy, anti-vaccine activists are condemning their use on social media, alternative news outlets and the streets.

So, when and where did this false correlation between vaccines and autism spectrum disorder originate? This myth originated in 1998 in an article published in a prestigious medical journal, *The Lancet*. The lead author was Dr. Andrew Wakefield, who attempted to show that the measles, mumps, and rubella (MMR) vaccine caused autism spectrum disorder. Measles, mumps, and rubella are diseases caused by three different viruses and the MMR vaccine is used for immunization against the respective diseases. The first dose is given to infants 10 to 12 months of age and the second dose is administered at four through six years of age. This published "research" was a shoddy study in that it covered only 12 children, it was not randomized, it was not placebo controlled, and it was not even a scientific study. Why it was even published continues to flummox scientists to this day. It has subsequently been identified as sensationalized and fraudulent. Since then, Wakefield and his colleagues have been found guilty of deliberate fraud, the article has been retracted, and Wakefield has been stripped of his license to practice medicine. However, the cat is out of the bag and this deceptive finding started an anti-vaxxer crusade.

This crusade had been fueled by Donald Trump and Jenny McCarthy. On March 28, 2014, Trump tweeted "Healthy young child goes to doctor, gets pumped with massive shot of many vaccines, doesn't feel good and changes—AUTISM. Many such cases!" While in office, he was silent on encouraging vaccination—something every president since Franklin Roosevelt has gotten on board with.

Jenny McCarthy is an American actress, former nude model, anti-vaccine activist, and author of *Louder Than Words: A Mother's Journey in Healing Autism*. McCarthy's son had been diagnosed with autism spectrum disorder, and like many parents searching for a cause and a cure, she linked vaccines with causing ASD. In 2007, while promoting her book, she was interviewed on the Oprah Winfrey show. Oprah Winfrey is a trusted name in America, so if Oprah has this person on her show, she must be speaking the truth. Right? Wrong. The airing of this broadcast brought the anti-vaccine movement into millions of homes. Later, in 2010, McCarthy and Jerry Kartzinel, M.D., published *Healing and Preventing Autism: A Complete Guide*. Subsequent to that publication, McCarthy made the rounds promoting her cause and her book. Keep in mind that McCarthy has no science background and never attended medical school. In fact, she declared to Oprah that she got her degree from "The University of Google." However, her beliefs on this topic and anti-vaccine advocacy became dogma for a nation of people searching for an autism cause. It also had horrific consequences.

According to the Centers for Disease Control and Prevention (CDC), more than five million American adults have an autism spectrum disorder. Moreover, about 1 in 54 children has ASD. Let's break this down. The CDC is the leading public health institute in the United States. It is staffed with real scientists, epidemiologists (scientists who study disease occurrences and health-related conditions in a population), physicians, microbiologists (scientists who study microorganisms like viruses), and other qualified people whose sole purpose is to protect the public health. Formed in 1946, it focuses on infectious disease, environmental health, and health promotion, among other equally important areas. These people are serious about getting things right. Three and a half million people is a very large number, so it is not surprising that many people are searching for answers about and cures for autism spectrum disorder. However, one should not look to former *Playboy* bunnies for medical advice.

As stated previously, there is no evidence linking autism spectrum disorders with vaccines. In the largest-ever study of over 95,000 children, researchers found no MMR-autism link in vaccinated versus unvaccinated children. These findings, published in *JAMA* (*Journal of the American Medical Association*) in 2015, found no increased risk of autism spectrum disorder with vaccination at ages, 2, 3, 4 and 5 years. The study did find that autism spectrum disorder rates were actually *lower* in vaccinated groups. In another study, also published in *JAMA*, researchers also reported that receiving the MMR vaccine was not associated with increased risk of ASD, even if an older sibling had ASD.

Before the anti-vaccine crusade, the MMR vaccine had nearly eradicated measles, mumps, and rubella in the United States. It is very likely that you are not even familiar with these diseases because vaccines have been so effective. To date, there are vaccines against 27 different diseases available in the United States to prevent disease.

Vaccine History

In order to gain an appreciation for vaccines, let's take a step back in history to a time that pre-dates vaccines. We'll begin in Asia around the 10th century (901–1000), more than a thousand years ago. During this time, smallpox was endemic. Smallpox is a contagious disease caused by a virus, specifically the *Orthopoxvirus*. Although there are slight nuances between the terms *contagious* and *infectious*, there is little to no difference when they are applied to the spread of disease. Language purists will tell you that *contagious* means that a disease is transmitted by physical contact, while *infectious* means the disease is transmitted by microorganisms in the air or water. When you come down with an illness, it doesn't matter the term: you're sick!

Smallpox is marked by chills, high fever, backache, headache, and characteristic ugly bumps. Within two to five days of being exposed to the virus, hundreds of circular bumps erupt on the skin. These bumps can cover the entire body as they transform into fluid-filled blisters that eventually progress to dry scabs that fall off. To get an idea of the grotesqueness, picture a horny toad. Now envision yourself wearing that horny toad's skin. The permanent scars left behind are called pock marks. And they aren't pretty. For millennia, smallpox was a dreaded scourge across the planet, and fatality rates ranged from 20 to 30 percent.

Around the 12th century (1101–1200), Chinese people noticed that children who developed smallpox—and who survived—did not contract this dreadful disease again. Thus, the people came up with this bright idea to deliberately infect young children with the particles of the ground-up smallpox scabs from survivors. This practice was known as *variolation*, a term derived from the scientific name of smallpox, *Variola*. The scab particles were blown into the noses of children or scratched into the skin surface, causing a mild form of the disease. When they recovered, they too were immune to subsequent smallpox infections. This method was also introduced to England and North America in the 1720s. Variolation had risks: infected people could also die from the procedure itself, or the mild form of the disease contracted by the patient could also spread the disease, causing an epidemic. Variolation is now an obsolete practice and has been replaced by safer vaccination. Enter English physician, Dr. Edward Jenner.

Dr. Jenner noticed that dairymaids who were infected with the harmless cowpox were immune to smallpox. Cowpox is a viral disease that affects cow udders. When humans come into contact with the virus, which is closely related to smallpox, they soon develop cowpox lesions on the skin. These lesions look a lot like smallpox lesions, but cowpox is much milder than smallpox. The close resemblance to smallpox inspired Dr. Jenner to experiment with the disease.

In 1796, Dr. Jenner deliberately infected an eight-year-old boy, James Phipps, with cowpox. Then, he exposed the young boy to smallpox. Young Phipps never contracted smallpox, and thus was borne the first vaccination, a term coined by Dr. Jenner and named after the alternate name for cowpox: *vaccinia*. In 1798, Jenner's findings were published.

The story doesn't end here, however. Because vaccination was viewed as a way to prevent disease and maintain public health, governments began mandatory vaccination programs. As you might suspect, resistance to vaccination ensued, because people just did not want to be compelled to do something. (Think about present-day protests against wearing face masks to limit the spread of COVID-19.) There were other reasons for the resistance as well. For example, people believed it to be dangerous (keep in mind there was not much science supporting the practice), and others—especially in India, where those practicing Hinduism consider cows to be sacred—balked at having particles from a cow injected into their bodies. (This is not the case today as researchers at the University of Michigan's Institute for Healthcare Policy & Innovation have shown that Hindu and rural children have the highest immunization rates in India.) Irish playwright, George Bernard Shaw (1856–1950) even stated that vaccines were a "filthy piece of witchcraft" that did more harm than good.

Okay, I have been vaccinated. I'm starting to understand how it works. Besides being blown into my nose, how do those germs get in my body in the first place? Those pathogens (disease-causing germs) get into the body through portals of entry. This is the scientific name for entrance gates. Think of any place in the body with a hole or a mucous membrane. The mouth, nose, respiratory tract, gastrointestinal tract, and cuts in the skin are the major routes. Infectious agents must gain entry in order to live and multiply. In epidemiology, we talk about horizontal transmission and vertical transmission. *Horizontal transmission* refers to all body surfaces, the bloodstream, or arthropod (insect and spider) bites. *Vertical transmission* occurs from female to offspring, before birth (from the oocyte or zygote), during pregnancy (across the placenta), during birth (through the vagina), or after birth (through breast milk). Within an hour of birth, newborns are routinely given antibiotic eye drops to prevent eye infections from bacteria introduced during their passage down the birth canal for a vaginal birth or across membranes during a cesarean section (C-section). It is a prophylactic measure to prevent conjunctivitis. Is it necessary? Maybe not. Is it good medicine? Yes, it is.

Back to smallpox. Smallpox was still a scourge into the 20th century (beginning in 1901).

As worldwide trade expanded, immigration rose, and leisure travel increased, the virus could spread easily. In 1967, the World Health Organization (WHO) launched a global smallpox eradication program. Founded in 1948 as an agency of the United Nations, WHO's mission is to promote health and control communicable diseases. In 1949, the last smallpox case occurred in the United States. The last naturally occurring case of smallpox worldwide was reported in October 1977 when Ali Maow Maalin of Somali contracted the disease. During the 1970s, routine vaccination against smallpox ended in the United States; however, it has resumed for military, health care personnel, and others who might be at risk if smallpox were to be used as a biological weapon or as part of bioterrorism. Forget bombs. Biological weapons can bring you down just as easily.

Our Immune System

We have the history of vaccines, but how do they actually work to prevent diseases? The answer lies in our immune system. Our immune system is as individualized and variable as we are. For example, allergies are an example of a hypersensitive response by the immune system. Some of us have allergies, some of us don't. Immunotherapy is a treatment that uses the body's own immune response to get rid of diseases like cancer. Immunotherapy works in some people, but not in others.

The immune system is our personal defense against many diseases. Immune system structures include lymph, lymph nodes, lymphatic vessels, leukocytes (white blood cells), lymphocytes, lymphatic tissues (tonsils), and lymphatic organs (spleen and thymus). That's a lot of anatomy, so let's take it structure by structure, keeping in mind that these immune system structures function as an integrated whole. Look at the table and Figure 6.1 below for quick references and then resume reading about these structures and their functions.

Structures and Functions of the Immune System

Structure	Function
Lymph	• Fluid that contains white blood cells; bathes tissues and drains through the lymphatic system into the bloodstream
Lymph nodes	• Oval-shaped structures along the length of lymphatic vessels that filter lymph
Lymphatic vessels	• Tubular structures separate from blood vessels that transport lymph
Leukocytes (white blood cells)	• Cells that circulate in the blood and body fluids that counteract foreign substances • Includes granulocytes (neutrophils, eosinophils and basophils) and agranulocytes (lymphocytes and monocytes)
Lymphocytes	• Type of white blood cell (cellular warriors) • T cells and B cells
Tonsils	• Lymphatic tissues in the throat, including the pharyngeal tonsil (adenoid), palatine (arch of mouth cavity) tonsil, and lingual (tongue) tonsil
Spleen	• Organ in the left, upper abdomen that filters blood and forms part of the immune system
Thymus	• Organ in the anterior neck that plays a role in immunity

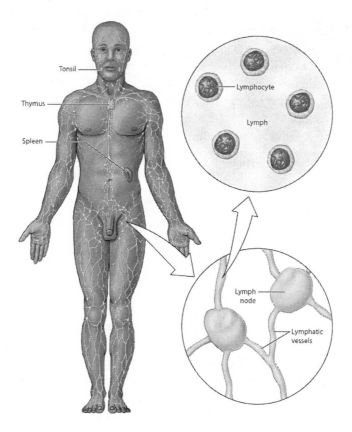

Figure 6.1: Structures of the Lymphatic System.

Lymph nodes are oval-shaped bodies located along the length of lymphatic vessels. The lymphatic vessels are networks of tubes, much like blood vessels, that transport lymph. Lymph is a clear fluid that flows within lymphatic vessels that contains leukocytes (white blood cells), and lymphocytes (a type of white blood cell). These leukocytes and lymphocytes are key to our immunity.

Let's go back to those lymph nodes. You are likely familiar with them already, especially if you've ever had a sore throat or some type of infection. During infections, lymph nodes enlarge. Some of these lymph nodes can be felt at the body surface, specifically, in the neck region, about three inches below each ear. Others that you may be able to feel are in the armpits and groin. During a physical exam, your clinician may have prodded and probed, known technically as palpated, these areas checking for inflammation. Beneath the skin, they feel like hard beans. Their job, though, is quite important: they filter lymph.

To help you understand lymphatic vessels, lymph nodes, and lymph, imagine a string of pearls. The string represents a lymphatic vessel while the pearls represent lymph nodes. Fluid outside our blood vessels is captured in lymphatic vessels where it becomes lymph, which is then filtered through lymph nodes. Ultimately, the cleansed lymph returns to our bloodstream. Lymph contains those all-important lymphocytes, a type of white blood cell. White blood cells are the body's warriors and their job is to attack and rid the body of pathogens, viruses, bacteria, and just about anything that shouldn't be there. Lymphocytes are distributed through the body via lymph and in lymphatic tissues and organs, such as the tonsils, spleen, and thymus.

Tonsils are really just lymphatic structures in your throat that act as sentinels, detecting foreign particles. When you get a sore throat, your tonsils are working overtime to rid the body of invaders. The spleen is a large, very vascular organ lying in the upper part of the abdominal cavity on the left side between your stomach and diaphragm (dome-shaped muscle that is involved with breathing). It contains lymphatic tissue and also serves as a blood filter, identifying and destroying invaders. The thymus is another lymphatic organ that is located in the upper chest, behind your breastbone (sternum). You cannot feel the thymus because it resides behind bone, but its job is to train lymphocytes to recognize normal body cells (self cells) from other cells (non-self cells). Thus far, we have components of the immune system that work together to rid the body of stuff that doesn't belong and to distinguish between self and nonself. Our anatomy lesson isn't quite complete, though. We still have to close the loop in terms of how this relates to vaccines. That's where the lymphocytes come in. First, we need to lay some ground work.

Lymphocytes originate in our bone marrow. Bone marrow is the soft, pulpy tissue that fills the center part of bones. If you are a meat eater, you very likely have seen bone marrow if you've ever cut into a T-bone steak. T-bone steaks are so named because of the characteristic T-shape of the bone. This bone is actually part of the animal's lumbar vertebra and its interior contains marrow. The meat is muscle, namely abdominal internal oblique muscles. Our bone marrow is important because not only does it produce red blood cells, it also produces lymphocytes. Break the word apart: *lymph* refers to *spring water* and *cyte* means *cell*. Lymphocytes are of 2 varieties: T cells and B cells. T cells are called T cells because these are lymphocytes that travel to the **t**hymus (a word that begins with the letter "T") and mature there. B cells mature in the bone marrow, and as you can see, **b**one begins with the letter "B." Not everything in science is difficult. Regardless of where they mature, their roles are basically the same: They produce antibodies, which circulate in our bloodstream, namely the fluid part called *plasma*. Antibodies are molecules that attack pathogens (viruses or bacteria) or they direct other cells to attack these pathogens. Now, we're getting close to how vaccines come into play.

When our lymphocytes encounter a pathogen, they build antibodies to attack that pathogen. The scenario looks like this: Our immune system is on surveillance 24/7/365. It doesn't take time off for holidays. No spring break for the immune system either. When it encounters a germ, it attacks it. This attack is known as mounting an immune response. Part of battling this germ involves building antibodies against that germ. Antibodies attack, weaken, and destroy that invading germ. Antibody building takes time—usually a few days—and we usually get sick while we battle that pathogen and build those antibodies.

Our bodies are quite remarkable, though. Once we have successfully battled that pathogen, the body builds memory cells to that specific pathogen so that the next time it encounters that same pathogen, the memory cells "remember" it and can attack that pathogen without us getting sick again. However, this antibody system is very particular, and this protection against certain diseases is called *immunity*. In many instances, immunity lasts your entire life. For example, if you had chickenpox as a child, you generally don't ever get chickenpox again. The chickenpox virus is still in the environment, but your memory cells "remembered it" and fought off the virus without you ever being aware. Not only did it "remember" the evildoer, it mounted a greater response. Thus, for subsequent infections, the time for combat is much quicker and the numbers of antibodies much greater than an initial contact. However, your pre-built antibodies to chickenpox won't do a thing against some *other* virus.

Let's use two sham diseases, dog disease and cat disease, as examples to clarify this point.

If you come down with dog disease, a condition that causes uncontrolled barking, you will bark continuously until your body mounts an immune response and you quit barking. After dog disease has been successfully eradicated, your lymphocytes build memory cells to dog disease. Several weeks later, cat disease, a condition that causes uncontrolled meowing, is sweeping through your neighborhood. If you encounter cat disease, you will come down with cat disease, because the memory cells you have built and stored for future use are for dog disease. However, if you encounter dog disease again, you will be protected because you have memory cells that can quickly transform into antibodies. This is the premise behind vaccines.

A current treatment for COVID-19 is to receive plasma from patients who have recovered from the virus. Plasma is the fluid portion of whole blood, and this is where antibodies are found. Plasma recovered from patients who have survived a bout of COVID-19 can be given to certain hospitalized COVID-19 patients. Such "convalescent plasma" may lessen disease severity, shorten duration, or help people recover from the disease.

Vaccines and Vaccine Development

Vaccines are substances that stimulate the production of antibodies, thereby providing immunity against a particular (specific) pathogen. Vaccines induce an immune response, but they don't produce the disease itself. So, they prime the immune system for a "just in case" encounter. Made of weak or dead germs or messenger RNA (mRNA), vaccines are harmless substances that initiate the production of antibodies without inducing the disease. If you ever encounter a germ for which you've been vaccinated, you have pre-built antibodies from memory cells to thwart the infectious agent. Vaccines give you long-lasting immunity to some pretty serious diseases. Wouldn't you rather get a vaccine instead of a disease? If you are still doubting the validity of vaccines, think about COVID-19 or Ebola.

Ebola is a highly infectious and frequently fatal disease. Its host species is still unknown, but outbreaks wreak havoc and kill nearly every person who contracts the disease; thus, the development of a vaccine was imperative. The development of that vaccine was the quickest vaccine to be developed before COVID-19. Today, the Ebola vaccine is stockpiled with 500,000 doses kept ready for deployment at the first sign of an outbreak. In the Democratic Republic of the Congo (DRC), for example, over the past four years there have been 4 outbreaks, each outbreak was suppressed by the quick response of vaccinating over 40,000 people. Vaccines work!

Right now we have three vaccines available in the U.S. to fight COVID-19, and more on the way. However, we will not totally suppress this virus until the vast majority of the world is vaccinated. Until then, there will be COVID-19 outbreaks scattered around the globe causing illness and death.

Getting vaccinated is also like "taking one for the team" or "doing it for the common good." People with weakened immune systems, such as very young infants, people with some autoimmune diseases, those with cancer, or people with HIV/AIDS cannot be vaccinated because they may not be able to mount an immune response and build those ever-important antibodies. However, if you and enough people around you are vaccinated for particular diseases, people are protected through *herd immunity*. This is also known as *community immunity*. Think of it as insurance and it is part of living in society.

Infectious diseases survive because they have hosts—you and me. Populations of hosts create "herds" or "communities." For germs to thrive, they need enough hosts. When they have enough hosts, outbreaks occur. Yet, if enough people are vaccinated against particular diseases, it makes it pretty hard for the germs to survive. If a person does get sick, there's less chance others can get sick because it's harder to spread from person to person. Eventually, if enough people are immune to a particular germ, that germ has nowhere to go, it cannot thrive, and the disease can be wiped out. The number of people necessary to achieve herd immunity varies from disease to disease. The goal is to break the chain of infection. Widespread vaccination programs are effective because if enough vaccinated people are in the population, the chain can be broken. Smallpox was wiped out because of vaccination. With the help of the Bill and Melinda Gates Foundation, the goal of eradicating the world of polio may be successful through vaccination.

Epidemiologists determine how contagious a disease is by tracking its spread throughout a population. Each infectious disease is assigned a number, referred to as R_0 (reproduction number). The bigger the number, the more easily it spreads through a population. For example, measles has an R number of 18 (R_{18}), which means one person can infect 18 others. The COVID-19 R number is more challenging to determine because it has lag time—the average time between each infection varies. This R number is then used to determine the minimum percentage of the population who need to be vaccinated to reach the herd immunity threshold. Although various models exist for calculating the herd immunity threshold, they all are very similar.

Models predict that 60–95% of the population will need to be vaccinated for COVID-19 to bring the virus under control. However, with the virus variants due to mutation, these predictions are educated guesses. The variants so far seem to spread much more easily than the original virus, and that would make the percentage of population needed to be vaccinated higher. Additionally, we may find that COVID-19 continues to evolve and produce increasingly more difficult-to-control variants. It is quite possible that we will need annual vaccinations to fight COVID-19, much like annual flu shots.

Here is a chart identifying the disease with the percentage of the nearby population that needs to be vaccinated to achieve herd immunity.

Disease and Percent of Vaccine Coverage Necessary

Diphtheria—85%	Pertussis—92%–94%
Influenza—30%–75%	Polio—50%–93%
Measles—92%–95%	Rubella—83%–85%
Mumps—75%–86%	Smallpox—80%–85%

The polio vaccine has an interesting history. Polio is the short name for *poliomyelitis*, derived from the word parts *polio-* meaning *gray matter* + *myelos* meaning *marrow* + *-itis* meaning *inflammation*. Polio is an inflammatory process involving the gray matter of the spinal cord. As an infectious viral disease, it affects the central nervous system, where gray matter is found, and causes temporary or permanent paralysis. There are various types of polioviruses, belonging to the *Enterovirus* genus. They are spread by direct contact between people, by nasal and oral secretions, and through contact with contaminated feces. It enters through the mouth. In most cases, polio is mild with no outward symptoms; in paralytic polio, the virus enters the digestive system, is then absorbed into the bloodstream, and makes its way to the nervous system. One to two percent of people who contract paralytic polio become paralyzed.

In the 1950s, polio killed 3,000 Americans per year and paralyzed thousands more. Enter the Jonas Salk and Albert Sabin polio vaccines. Jonas Salk (1914–1995) was an American immunologist and the father of the original poliovirus vaccine. This vaccine was made of virus that was propagated in monkey kidney tissue cultures and then inactivated; it is known as the inactivated poliovirus vaccine (IPV). Albert Sabin (1906–1993) was a Polish-American virologist. He developed an orally administered vaccine containing live, attenuated strains of poliovirus; it is known as the oral poliovirus vaccine (OPV). Salk's vaccine was delivered by injection and was largely replaced by Sabin's vaccine, which was administered by mouth. By 1979, the contagious or "wild" form of polio had been eradicated in the United States, but 8–10 people, mostly infants, contracted the disease yearly from the Sabin vaccine itself, yet millions were protected from the devastating disease. Many people who contracted polio spent time in an iron lung. An iron lung is a mechanical respirator in which the entire body, except the head, is enclosed in a metal tank. The tank is sealed at the neck with an airtight gasket and artificial breathing is induced by making the air pressure inside negative, thereby enabling the lungs to inhale and exhale. Fortunately, iron lungs are relics today and reside in museums.

David Salamone, who was born on May 25, 1990, and died 28 years later, was one of those 8–10 people who contracted the disease from the Sabin oral poliovirus vaccine. Unbeknownst to his parents at the time of the vaccine administration, David was born with a genetic immune deficiency called Bruton agammaglobulinemia, which made him susceptible to many infections, including polio. David's father was dogged and determined and became a staunch advocate for Salk's vaccine because it used inactivated virus. He was able to convince administrators at the U.S. Immunization Program at the CDC to change tack. Since January 1, 2000, the only polio vaccine used in the United States has been Salk's injected vaccine, which was developed 50 years prior.

Polio has been around a long time. Egyptian carvings from around 1400 BCE show leg deformities similar to those brought on by polio. During the 1900s, polio reached epidemic proportions. Because of widespread vaccination, it was eradicated from the Western Hemisphere in 1994. It still continues to circulate in other countries, so widespread vaccination is still recommended.

As with any medication, vaccines can cause side effects. These range from mild to serious. Of course, nobody wants to be a sacrificial lamb. For a complete list of vaccines and side effects associated with them, visit the Centers for Disease Control and Prevention's website and search "Possible Side-effects from Vaccines."

Some of you are probably thinking, well, I hear what you're saying, but I have gotten my flu (influenza) shot and I still wound up with the flu. Yes, that is quite possible. Here's why. The influenza virus is a quick-change artist. Its biology is such that the virus has the ability to change to ensure its survival. The flu has various subtypes, strains, and viral properties. Without going into the deep science of viral replication using RNA, influenza predictions involve the worldwide scientific community. Using the best available data, scientists look at antigenic change (sudden change in the molecular structure of the virus) to predict which strains of flu will be circulating and then formulate vaccines based on this information. Recall that antigens are substances that induce an immune response in the body and that specific antibodies are built for specific antigens.

Antigenic shift produces new strains. The production of new strains means that hosts (people) previously exposed to other strains have little to no acquired immunity to the new strain. Antigenic shift causes recombination or genetic re-assortment of viral strains in a given host. This can lead to large-scale epidemics or pandemics.

The 1918 influenza pandemic involved the H1N1 influenza virus and killed an estimated 50 million people. About 675,000 deaths occurred in the United States. At the time, one-fifth of the world's population was infected with the deadly virus. The genes of the virus were of avian (bird) origin, but the genesis of the outbreak is not known. During 1917–1918 it was so severe that life expectancy in the U.S. decreased by 12 years. This flu was unusual: death rates were exceptionally high for people between age 20 and 40. Typically, this age group is the healthiest in the population.

The pandemic occurred during World War I at a time when soldiers were living in close quarters and troop movements spread the disease. It is also known as the "Spanish Flu" but not because the flu originated in Spain. Spain was neutral during the war and reported freely on flu activity. Thus, a name and a misconception were born.

The reason for its virulence (harmfulness) is also not known, although we do know that there were no vaccines to protect people against it. Non-pharmaceutical measures, such as isolation, quarantine, face masking, and good personal hygiene were used, but not applied evenly. The body's number one line of defense against disease is our skin and mucous membranes. So, washing off the germs is always good practice!

In order to maintain herd immunity, typically 95 percent of the population needs to be immunized against a particular infectious disease. Many states require students in public schools to be vaccinated. However, vaccine exemptions do exist. From a child's health perspective, medical exemptions are allowed when the child has a condition that prevents them from receiving a vaccine. However, many states allow nonmedical exemptions for religious or philosophical reasons. To break this down, all 50 states have legislation requiring specified vaccines for children attending school, all 50 states allow for medical exemptions, 45 states and Washington, D.C., allow for religious exemptions, and 15 states allow for exemptions based on philosophical reasons. Moreover, as I write these words, we are in the midst of a coronavirus (COVID-19) outbreak and the nation has plenty of vaccinations.

The coronavirus has been on the scientific radar since 2003, when the world experienced the first coronavirus outbreak known as Severe Acute Respiratory Syndrome (SARS). COVID-19 is not SARS, but it belongs to the same family of viruses. SARS first appeared in China and spread worldwide within a few months. Fortunately, it was quickly contained and today there are fewer than 1,000 cases per year. After that, there was Middle East Respiratory Syndrome (MERS), which was first reported in 2012 in Saudi Arabia; it too is rare. Yet, transmission of infectious diseases are worldwide events because we are a global society.

Because vaccine development is a rigorous and complex process, vaccines generally take years to develop. So, when the President of the United States held a coronavirus press conference on February 26, 2020, and told people that the U.S. is "rapidly developing a vaccine" for COVID-19 and "will essentially have a flu shot for this in a fairly quick manner," he was misleading the American people. The director of the National Institute of Allergy and Infectious Diseases, a person who understands science and vaccine development, stated that a vaccine won't be ready for 12–18 months. Why were they wrong? Because vaccines typically go through stages of development, testing, tweaking, and more testing; it is a rigorous process. Ebola and COVID-19 are two exceptions to this lengthy process mainly due to their extreme deadliness and virulent nature. COVID-19 vaccines did have somewhat of a head start due to research done on earlier corona viruses coupled with the worldwide research community coming together to work on a vaccine. Additionally, the COVID-19 vaccine attained emergency use authorization, which is a fast-track route for priority review with the goal of bringing safe vaccines to market quicker. In August 2021, the FDA fully approved the Pfizer-BioNTech COVID-19 vaccine for ages 16 and over.

The first stage of vaccine development is the Exploratory Stage in which scientists identify the natural or synthetic antigens—such as weakened viruses or bacteria, weakened bacterial toxins, virus-like particles, or other substances derived from pathogens—that might work. Then comes the Pre-Clinical Stage in which one of the antigens is tested in laboratory cell cultures or on animals, typically mice and monkeys. This is done to give scientists an idea as to how it might affect humans. This stage may take one to two years and many "candidate vaccines" never make it beyond this stage and the process must begin again. Then, an Investigational New Drug (IND) application is filed with the U.S. Food and Drug Administration, who has 30 days to approve it. This application provides data, summarizes findings, and describes the purpose of the study. An institutional review board then approves (or disapproves) the next step, which is Phase I Vaccine Trials with humans. From here, it moves on to Phase II Vaccine Trials involving several hundred people, using randomized double-blind placebo (usually saline solution) controlled experiments. Up to this point, the safety of the vaccine is being tested. If all goes well here, it moves on to Phase III Vaccine Trials, using the same protocol but with more people. Additionally, the efficacy is tested: (1) Does the vaccine prevent disease? (2) Does it prevent infection with the pathogen? (3) Does it lead to antibody production or another type of immune response to the pathogen?

The next step is approval and licensure. If Phase III was successful, the vaccine developer submits an application to the Biologics License Application to the FDA. The FDA inspects the factory where the vaccine will be made. If all goes well here, the FDA will license the manufacture of the vaccine and continue monitoring for safety, potency, and purity, all the while maintaining the right to conduct its own testing of the vaccine. The vaccine manufacturer may also continue monitoring for safety, potency, and purity in its own Phase IV Trials; however, this phase is optional, not mandatory.

In May 2020, Moderna gained fast track status for its investigational mRNA-based vaccine, followed by Pfizer-BioNTech in July 2020. The first vaccine administered in the United States was given to ICU nurse Sandra Lindsay on December 14, 2020, in New York. A third vaccine, the Johnson & Johnson vaccine, was approved on February 27, 2021, and quickly put to use.

Another government program administered by the U.S. Department of Health and Human Services centers on vaccine injury and compensation. The National Childhood Vaccine Injury Act of 1986, as amended, created the National Vaccine Injury Compensation Program (VCIP) to compensate people who may have been injured by certain vaccines. You might be familiar with this program if you've ever read the information sheets you are handed after receiving any vaccine. People do have a limited time to file a compensation claim.

In 1990, the CDC and the FDA established the Vaccine Adverse Event Reporting System (VAERS) to "detect possible signals of adverse events associated with vaccines." Flu vaccines are usually given with a fine gauge needle into the deltoid, the muscle in the upper arm. A common injury related to flu vaccinations can occur if the needle is inserted into the wrong area of the upper arm or too deeply, causing shoulder injury related to vaccine administration (SIRVA). Each year, approximately 30,000 adverse events are reported, and 10 to 15 percent of these events are serious, including hospitalization, life-threatening illness, disability, and death. Thus, vaccine development is not taken lightly because the consequences can be horrific.

Vaccine Myths

Vaccine myths continue to persist. In addition to the myth that vaccines cause autism, another myth is that mercury in vaccines acts as a neurotoxin. Neurotoxins are substances that cause destruction of nervous tissue, and mercury does cause neurological issues. In fact, the phrase "mad as a hatter" arose from mercury poisoning. Used by hat makers, mercurous nitrate was a chemical used to cure the felt (hair cut from the pelt). Chronic exposure to mercury vapors by workers caused mental confusion, emotional disturbances, kidney damage, and other signs and symptoms. The medical term for the neuropsychiatric symptoms of mercury poisoning is *erethism*. Today, we know much more about mercury exposure, which is why we need to limit our consumption of fish and shellfish and why there is no mercury in childhood vaccinations.

With regard to mercury and vaccines, we are talking about thimerosal, a compound that contains mercury and is a naturally occurring metal that prevents microbial growth. Thimerosal is used as a preservative in some vaccinations, such as multi-dose vials of flu vaccines, and as an antiseptic used topically (on the skin). There is no evidence that thimerosal in vaccines caused any neurological disturbances, including causing autism. In 1999, as a precautionary measure, the U.S. Public Health Service recommended thimerosal removal to limit mercury exposure in babies. Today, no recommended childhood vaccines contain thimerosal. The evidence never did support the premise that thimerosal in vaccines caused health problems in kids, much less caused autism. If that were true, then we would have seen the number of cases of autism decrease or level off after 1999—neither of which happened. Once again: there is no association between vaccines and autism.

This next myth could also have been placed in Chapter 7: Snake Oil because it speaks to the issue of peddling false science. In the mid 2000s, Dr. Mark Geier, a medical doctor, and his son, David, a guy with a bachelor of arts degree, began chemically castrating autistic children. Say what? As you scratch your head and think about this, consider that parents of autistic children are desperate to search for cures. Using pseudoscience against a medical backdrop, the Geiers claimed that testosterone contributes to autism. They claimed that testosterone binds to the mercury in vaccines, thereby making them more toxic to the brain. Again, playing off the neurotoxin idea. The Geiers used Lupron (leuprolide acetate), a drug that shuts down testosterone production. Back up a minute: Why is there even such a drug? Lupron is a legitimate drug used to treat metastatic prostate cancer. It can also be used to treat precocious puberty, which occurs when puberty begins at an unexpectedly early age, somewhere before the age of 8 in girls and 9 in boys. Without following standard protocol for diagnosing precocious puberty, the Geiers gave daily Lupron injections to kids and misled parents into believing this was an approved treatment for autism. Incidentally, Lupron is also given to chemically castrate sex offenders. This was a money-making venture, billing parents and insurance companies $5,000 or more per month, claiming they were treating precocious puberty. Dr. Geier's Maryland medical license was revoked, and his son was charged with practicing medicine without a license.

Another vaccine myth is that spreading out the timing of vaccines can be safer for kids. During the September 16, 2015, presidential Republican debate on CNN, Donald Trump stated, "I'm in favor of vaccines; do them over a longer period of time, same amount but just in little sections and I think you're going to see a big impact on autism." By April 2019, while the U.S. found itself dealing with a measles outbreak, Trump did an about-face and told CNN reporter Joe Johns, "They have to get the shots. The vaccinations are so important. This is really going around now. They have to get their shots." Spreading out vaccines has not been scientifically proven. Experts, pediatricians, and family doctors conclude that vaccines should not be spread out. The reason is that on a daily basis, a child's immune system is exposed to thousands of foreign environmental antigens each day. Remember that these little foreign particles induce an immune response. The 2020 vaccination schedule exposes a child to about 300 antigens by the age of two. Even if 11 vaccines were given to an infant at one time (**which does not happen**), only 0.1 percent of the child's immune system would be "used." Over the years, the number of recommended vaccines has increased, but since vaccine development has gotten better, the number of antigens in vaccines has decreased. What has not decreased is the number of cases of autism.

Dr. Paul Offit of the Children's Hospital of Philadelphia in Pennsylvania is a world expert on vaccines. He has publicly stated that spacing out vaccinations for children leaves them vulnerable to dangerous disease for a longer period of time. Moreover, one "spaced-out" schedule would require 19 doctor visits over six years, with 12 visits within the first two years. No parent has time for this, especially when it is unnecessary. In 2019, the United States experienced the largest measles outbreak since 1992. This outbreak was catalyzed by vaccine hesitancy. From January 1 to December 31, 2019, there were 1,282 confirmed cases of measles in 31 states. Measles outbreaks are linked to travelers who get measles abroad and bring it back into the United States, where it can be spread because we have pockets of unvaccinated people. The World Health Organization has declared vaccine hesitancy one of the top ten threats to global health. During 2019, there were more than 500,000 confirmed cases of measles reported in more than 180 countries. In 2018, there were 353,236 confirmed cases worldwide. Measles is highly contagious, and the virus can linger in a room for up to 2 hours after an infected person as left the room. It lingers like the smell of a candle after the flame has been extinguished. There's a very simple solution to stopping the spread: vaccination.

Besides not having to experience the misery of having measles, *not getting the measles* has longer-lasting effects related to immunity. Recent studies have shown that in unvaccinated children who contracted the measles, the virus seems to wipe out immunological memory of other disease-causing agents. Measles virus infection diminishes pre-existing antibodies to other pathogens. For children who contracted measles infection, they lost 11 to 73 percent of their antibody repertoire, leaving them vulnerable to infections they already had and for which they already had antibodies. We rely on immune memory to keep us safe from future infections.

Peanut allergy: peanut-free zones have been created in all sorts of places, from school yards and lunch rooms to airplanes and grocery stores. What do peanuts have to do with vaccines? Nothing yet, but researchers are getting closer to developing a vaccine against peanut allergies. Allergies result from an immune system on overdrive. With allergies, a person is hypersensitive to a particular antigen. Recently, the American College of Allergy, Asthma, and Immunology (ACAAI) updated its guidelines for the prevention of peanut allergy. It created separate guidelines for infants at high, moderate, and low risk for developing peanut allergy. Following an informed protocol, parents introduce peanut-containing foods (never, ever whole peanuts!) to prevent peanut allergy. Basically, the infant's immune system becomes primed for peanuts through repeated exposure to ever-increasing amounts of peanut antigens. Peanut allergies can be life threatening, so it's always wise to seek medical advice. However, the ACAAI has a great, information-packed video for you if you go to their website.

Further proof that science crosses numerous aspects of life and that topics within this book could be placed in several chapters is the next topic, which addresses vaccines and aborted fetuses.

If somebody told you that vaccines are made from aborted fetuses or if you heard such news, it probably would give you pause. I've seen interviews with people who refuse to be vaccinated or get their children vaccinated on moral or religious grounds because they believe vaccines are made from the byproducts of abortion. Does a person believe that the pharmaceutical company has a standing order for fetal tissue at the abortion clinic? Now, to shed some science on this.

Two cell lines, WI-38 and MRC-5, are two human cell lines used to make some vaccines. We should begin by defining a cell line. A cell line is a clone of cultured cells derived from an identified parental (primary) cell type. Scientists can make tissue cultures from subcultured cells from a primary cell. One cell can be cloned and cloned and cloned ad infinitum, creating cell lines that can be used for research. Rebecca Skloot wrote about an immortal cell line in her 2010 best-selling book, *The Immortal Life of Henrietta Lacks*, which has been made into an HBO film. Henrietta Lacks was a poor African American woman whose cervical cancer cells were cultured to create the HeLa cell line at the Johns Hopkins University Hospital in Baltimore, Maryland. This was done unbeknownst to her or her family in 1951. Informed consent was not given. Through research and interviews, Skloot was able to piece together the truthful scenario, reveal historical evidence, identify unethical and ethical behavior, and demonstrate this woman's profound impact on science past and present. These infamous HeLa cells are still being used today for medical research. Something to definitely ponder.

Fast forward to WI-38 and MRC-5. The first, WI-38 cell line, was developed from lung tissue obtained from a therapeutically aborted female fetus at 12 weeks gestational age in 1951. Therapeutic abortions end pregnancies when the woman's life is in danger. The timing is merely coincidental with the Henrietta Lacks story. The cells were fibroblasts, early cells that form collagen fibers in connective tissues. While working at the Wistar Institute (WI), Leonard Hayflick developed the WI-38 cell strain. The MRC-5 line was developed from lung tissue obtained from an aborted male fetus at 14 weeks gestational age in 1966. These cells were also fetal lung fibroblasts. This cell line was isolated by J.P. Jacobs and colleagues in England and MRC stands for Medical Research Council. These two cell lines are the only fetal cell lines currently used for manufacturing vaccines. Animal cell lines are now used. Vaccines developed from the WI-38 and MRC-5 cell lines include the following: hepatitis A, rubella, varicella (chickenpox), and zoster (shingles).

For Catholics who are morally opposed to being vaccinated with vaccines developed from aborted fetal lines, the National Catholic Bioethics Centers states this: "One is morally free to use the vaccine regardless of its historical association with abortion. The reason is that the risk to public health, if one chooses not to vaccinate, outweighs the legitimate concern about the origins of the vaccine. This is especially important for parents, who have a moral obligation to protect the life and health of their children and those around them."

Take-Home Message

Vaccines are advantageous to public health and do not cause autism spectrum disorder. Without vaccination programs, preventable diseases can emerge and those most vulnerable among us, such as babies, older people, and those with compromised immune responses, can become needlessly sick and/or die.

Ignorance more frequently begets confidence than does knowledge.
—Charles Darwin (1809–1882);
English naturalist, geologist, and biologist

7

Snake Oil

Complementary and Alternative Medicine

U.S. Landmark Case: Dent v. West Virginia, 129 U.S. 114 (1889); (9–0 decision)

The state of West Virginia passed a law in 1882 that established restrictions on anyone practicing medicine in the state. It basically required anyone wanting to practice as a "doctor" to appear before the state board of health and produce evidence that they met the state's requirements and then be certified as a doctor in West Virginia. The statutory requirements included: proving that he was a graduate of a reputable medical college or that he has practiced medicine in the State of West Virginia continuously for the period of 10 years prior to the 8th day of March 1881, or that he has been found, upon examination by the board, to be qualified to practice medicine in all its departments. (Note that the masculine pronoun "he" was used.)

Frank Dent had been practicing in West Virginia for six years, which was not long enough for him to qualify to be certified. However, he did produce a diploma from the Eclectic College of Medicine of Cincinnati. The West Virginia board of health denied Dent a certificate to practice medicine because they felt his college was not "reputable." He appealed to the U.S. Supreme Court based on the Fourteenth Amendment making the West Virginia statute unconstitutional as it deprived Dent of life, liberty, or property without due process of law—Dent claiming that the right to practice his profession was a deprivation of his vested right.

During that time in America, there were a number of medical professions that were deemed "alternative" medical theories or practices competing with "regular" medicine. Alternative practices included hypnosis, homeopathy, and eclecticism (basically herbal medicine). Eclectic practitioners included a number of non-schooled practitioners with their magic cure-alls, snake oils, liniments, and other fraudulent "medicines." Thus, the West Virginia board of health viewed the Eclectic College of Medicine as not being reputable.

The U.S. Supreme Court found for West Virginia. It stated that while the government cannot arbitrarily deprive someone of their right to pursue and work in any lawful business or profession, subject to such restrictions as the state imposes on all persons of like age, sex, and condition, such an arbitrary deprivation did not occur here. Dent was not granted a certificate to practice medicine for failure to comply with the statute's conditions. Such conditions are allowed so the state can provide for the general welfare of its citizens. Since this 1889 decision, the Supreme Court has upheld the principle that states may regulate the practice of medicine and determine what is and is not lawful in their jurisdiction.

Outside Mainstream Medicine

The Dent v. West Virginia landmark case stretches back more than 130 years and places medical licensing solely on each state. Because states regulate professional licensing of all sorts, criteria vary from state to state and licensure in one state doesn't guarantee licensure in another, although some states do have reciprocity agreements. Some states are stricter than others which makes the area of complementary and alternative medicine ripe for quackery and less-than-honest practices. Add in the fact that it's easy to dupe people, especially those without any medical or scientific background, or for whom mainstream medicine hasn't worked, or those truly desperate for relief for their ailment. Practices outside traditional medicine are so commonplace that people view complementary and alternative medicine as practices that work as well as traditional medicine or make up for medicine's shortcomings. Yet, swallowing the proverbial snake oil isn't always the best solution. Without rigorous testing and scientific validation, complementary and alternative medicine is suspect.

Confusion and turmoil are prevalent in present-day health care treatments. The great expense of new, modern pharmaceutical medications, as well as instrumentation and surgical costs, and the uncertainty of the future of current health care programs and coverages have caused many Americans (nearly half of all U.S. adults) to look outside the conventional medical health system. Many are looking toward other means of complementary and alternative treatment for their health care.

What is complementary and alternative medicine? Both include treatments and practices that lack scientific plausibility or that have not been studied enough to know for sure whether a treatment works. The names sound fancy and make it seem like this stuff is real, but many therapies have been untested or shown to be ineffective. We hear much about herbal remedies used for mood disorders and treatment for a number of medical problems. Acupuncture is used to alleviate pain. Yoga (qigong), tai chi, and biofeedback are used for pain, anxiety, and relaxation. For aching muscles, tension, and stress relief, there is massage therapy. Meditation and hypnosis are touted for tension and stress relief. Even aromatherapy and guided imagery (a relaxing technique in which patients lie in bed with their eyes closed and imagine they are visiting a happy place of their choosing by using their five senses) are also employed. There's naturopathy and Ayurveda. However, many of these ancillary treatments and therapies, for the most part, have not been fully scientifically checked, rigorously tested, and evaluated to date. That's not to say that they don't work; it's to say that we should view them with some skepticism as there is not enough known about them yet. Rigorous testing using the scientific method must occur.

The effectiveness of many practices is still greatly debatable, and there may be both beneficial as well as harmful effects. Induced harmful effects are particularly true with a great variety of herbals, resulting from their excessive and unregulated use. There may be allergic reactions due to toxic impurities and incorrectly mixed herbs as well as interactions with prescription drugs. Yet, very few people dispute their potential significance as the treatments are widely used and the public appears to be "positively convinced" of their beneficial effects, even though they are not abandoning conventional medicine entirely. Belief in healing powers of herbs cannot be discounted because the placebo response—beneficial effect of something that cannot be attributed to the properties of the herb itself—is real.

In 1998, the National Institutes of Health turned its Office of Alternative Medicine (OAM), then a very small enterprise, into a full-strength federal agency and called it the National Center for Complementary and Alternative Medicine (NCCAM). The NCCAM is now known as the National Center for Complementary and Integrative Health (NCCIH) and is part of the 27 institutes comprising the National Institutes of Health (NIH) within the federally-funded Department of Health and Human Services. It has an annual budget of $151.9 million (fiscal year 2020). The NCCIH is focused on research, research training and career development, integration, and outreach. It funds research in five main areas: whole medical systems, mind-body medicine, biologically-based practices, manipulative and body-based practices, and energy therapies. Their purpose is to study what looks promising, what helps and why, what doesn't work, and what is safe.

The formation of the NCCIH spawned new research endeavors at many leading medical schools that developed centers for integrative medicine, and courses are now being offered in complementary and alternative medicine (CAM). In the 1960s and 1970s, CAM was formerly known as "alternative medicine." Then, in the 1990s it got a name change—or marketing makeover—to complementary alternative medicine. Now, it's marketed as "integrative medicine." For this chapter, I'm sticking with complementary and alternative medicine because when I googled "complementary and alternative medicine," I got 123,000,000 results, whereas "integrative medicine" returned 50,400,000 results. Mainstream medicine might call it integrative medicine, but consumers are still calling it complementary and alternative medicine. Change takes time.

Large clinical trials are being designed to assess the merits of popular therapies. For example, there are studies evaluating the use, risks, and benefits of the supplements glucosamine and chondroitin sulfate in the treatment of osteoarthritis; others study the use of acupuncture to ease arthritis pain; some are looking at the use of selenium and vitamin E to prevent prostate cancer; and many are assessing the ability of Ginkgo biloba to preserve mental function as we age. The list is virtually endless.

Nevertheless, there are many skeptics who fear that science-based medicine may suffer. Billions have been spent studying alternative and complementary practices. Scientific journals, including *Science*, and other publications have been extremely critical because they find the studies based not in science but in politics and other non-meritorious agendas. In addition, Medicare and most private health plans are biased against holistic care and will cover only what is conventional. Clinical practice may, in the near future, be revolutionized as more information is made available through intense research, trials, and evaluations, and we may one day experience a new form of integrative medicine in modern health care. In fact, CAM treatments show up in the American Medical Association's directory of billable procedures and a number of small pilot studies are finally being undertaken by insurers and Medicare. As some CAM therapies are currently under investigation, it is wise to point out that physician monitoring is very important and highly recommended.

Weight loss diets, supplements, herbal remedies, and alternative therapies are big business. I get it. However, trouble arises when bastions of science and medicine have departments devoted to practices that are not backed up by science. I'm talking about the Cleveland Clinic in Ohio. If you have a heart problem, this is THE place to be. If you have kidney cancer, you want to be here. It is nationally ranked in many fields as the best. So, it's hard to reconcile its evidence-based, solid science practice with its Center for Integrative Medicine. Their listing of complementary therapies includes acupuncture, acupressure, biofeedback, craniosacral therapy, guided imagery, hypnosis, integrated dry needling, frequency specific microcurrent therapy, myofascial release, osteopathic manipulation, pain management, reiki, relaxation/breathing strategies, therapeutic touch (babies), and yoga. The Cleveland Clinic's Integrative & Lifestyle Medicine also offers Chinese herbal therapy. They also have a Center for Functional Medicine. An entire book can be written on each of these practices. While some can actually make us feel better (who doesn't love a nice massage?), we need science-based medicine and evidence-based practice. Perhaps opening such centers in big-name clinics will allow real science to be done and we'll get real data to evaluate. Integrative medicine makes money, so I suspect that even if many of these treatments are shams, as long as nobody's health is adversely affected, they will persist.

The Placebo Response

Why are some therapies helpful even when there is no scientific evidence to support the treatment? That is, they work even though there's no underlying scientific explanation. The answer lies in the placebo response (also called the placebo effect). A placebo is a harmless pill, medicine, procedure, or touch that is prescribed for its psychological benefit or its suggestive effect. Without medicinal properties, a placebo should have no physiological effect. In fact, before drugs are marketed to the general public, they are tested using placebos. In drug testing, an inert compound that is identical in appearance to a new drug being tested in experimental research is administered to test subjects. The results are then evaluated to determine whether the new drug under study has any effects or if the drug action occurred by suggestion. For the most robust studies, the researcher does not know which drug is the treatment and which drug is the placebo.

Interestingly, placebos can have just as great or greater effect than the actual drug, showing that our mind plays a huge role in making us better. A systematic review of the orthopedic literature found that some sham surgeries were just as effective as actual surgery in reducing pain and disability. Surgeries that failed the test were the following:

> vertebroplasty = stabilization of a vertebra by injecting surgical cement
> intradiscal electrothermal therapy = burning nerve fibers in the inner vertebral disc
> arthroscopic debridement for osteoarthritis = removing or "polishing" rough joint surfaces
> open debridement of common extensor tendons for tennis elbow = scraping elbow tendons

Much of the placebo response is based on our own release of endorphins ("feel good" substances), neurotransmitters (chemical messengers), and hormones (regulatory chemicals) in response to someone trying to make us better or in the belief that a treatment will work. The placebo response is also correlated with how caring and supportive a patient feels their medical practitioner is. Care, compassion, and kindness are always good, regardless of the situation. Practitioners can ethically take advantage of this when prescribing real treatments. In other words, treatments are more effective if you like and trust the person who is taking care of you. Social scientist Daralyn Brody and physician Howard Brody wrote an interesting book, *The Placebo Response: How You Can Release the Body's Inner Pharmacy for Better Health*, that highlights the connection between the mind-body healing process.

There is also mounting evidence that in the United States, the placebo response is getting stronger! Because placebo responses are seen so frequently, it can be difficult to prove just how effective a drug or treatment is. For these reasons, greater study is now focused on the mechanism of the placebo response. Why would something with no additional chemicals or no known physiological effect work to alleviate some problem such as pain?

Acupuncture

Time to break down some of these most popular alternative treatments, beginning with acupuncture. Acupuncture is an ancient Asian system of therapy in which long, fine needles are inserted into discrete areas of the body that are considered linked to symptoms or disease. In addition to being examined as a physical treatment, acupuncture also turns out to be a ritual in which a therapist touches, as well as talks, to the patient receiving treatment. The patient feels she is being cared for and treated humanely and compassionately. Chinese medicine is based on four main beliefs:

1. The human body is a miniature version of the surrounding universe.
2. Two opposing, complementary forces of yin and yang support health. If disease results it is because yin and yang are out of balance.
3. Fire, earth, wood, metal, and water represent all phenomena. These elements also play a role in health and disease.
4. The circulating life force known as qi (pronounced "chee") is the vital energy that maintains human health. Western medicine recognizes other factors as underlying causes for disease. These factors are easily remembered using the mnemonic **VINDICATE**:

V = vascular
I = infection/inflammation
N = neoplasm (cancer)
D = drugs
I = intervention/iatrogenic (caused by medical treatment)
C = congenital (born with it)
A = autoimmune
T = trauma
E = endocrine

Neuroscientist and positron emission tomography (PET) inventor Zang-Hee Cho has used functional magnetic resonance imaging (fMRI) to observe changes in the human brain when acupuncture needles are placed in the foot. He has shown that needling specific points on the foot can modulate activity in four brain areas that are involved in pain perception. He has also demonstrated that an acupuncture point traditionally used by the Chinese to ease eye problems stimulated the brain's visual cortex.

Acupuncture has been the focus of much research in the West, having been tested for everything from ringing in the ears to asthma, with a variety of results. The strongest evidence for its use is for pain and nausea. It appears to boost the level of the body's own opiates or endorphins—the feel-good chemicals—explaining its effect on pain. It also seems to increase the level of brain serotonin, conferring a sense of well-being in the individual. Serotonin is a natural body chemical that affects our mood. The more serotonin in the bloodstream, the better we feel.

People with chronic, moderate, or severe depression may take drugs known as selective serotonin reuptake inhibitors, or SSRIs for short, which work to keep the levels of serotonin high, thereby making a depressed person feel better. The wonders of acupuncture may even lessen one's craving for drugs.

It appears that an acupuncture treatment for chronic pain may be safe, but its efficacy is unclear. Acupuncture for nausea from chemotherapy and for dental pain is probably safe and may be effective.

Homeopathy

The sheer number of alternative practices is mind boggling. One that continues to baffle is homeopathy. Homeopathy is a system of therapy developed by Samuel Hahnemann based on the "law of similia." Huh? It's from the Latin aphorism *similia cimilibus curantur*, which means *likes are cured by likes*. In everyday language, it is a philosophy that a medicinal substance can evoke certain symptoms in healthy people so it may be effective in the treatment of illnesses having similar symptoms if that medicine is given in very small doses. Hahnemann (1755–1843) was a German physician whose practice pre-dates the germ theory of disease (founded by Louis Pasteur between 1860 and 1864). Today, homeopathy uses serial dilutions to create medicines following the "Law of Infinitesimals." This means that the smaller the dose, the larger the effect. For example, drugs are serially diluted in water or alcohol—up to 1:100—and then given. Between the dilutions, the mixture is shaken to release the drug's medicinal properties. Studies have shown that homeopathy is no more effective than a placebo. Alcohol may numb the senses, though.

Naturopathy, Supplements and Herbal Remedies

Naturopathy is a system of therapeutics that rely on natural, nonsurgical, nonmedicinal agents to heal the body. Practitioners of naturopathy are called naturopaths. The underlying principles sound solid: first do no harm, treat the whole person, prevention is key, nature can heal, treat the cause, and physician as teacher. These are the same tenets of medical doctors, too.

If you were not informed, you'd likely think the scope of practice for naturopaths is the same as that for medical doctors—those with MD or DO after their name. MD = medical doctor and DO = doctor of osteopathy. In the United States, the scope of practice for MDs and DOs is virtually the same although the underlying philosophy differs between the schools of thought. Both, however, take the same licensing exams to practice medicine. Naturopaths complete a four-year, graduate-level program at a naturopathic medical school accredited by the Council on Naturopathic Medical Education. Some, but not all, U.S. states and territories require naturopathic licensing. Some states also refer to these practitioners as naturopathic physicians.

Supplements are substances that complete or enhance something else when taken along with that something else. Supplements can also be taken to remedy a dietary deficiency. For example, if a person is deficient in vitamin C, they can take a supplement of vitamin C and boost the level. Herbal remedies rely on medicines made from plants to treat illness or disease. Store shelves are chock full of such supplements and herbs and include St. John's Wort, saw palmetto, chondroitin sulfate, ginkgo biloba, and vitamin E. The herbal St. John's Wort is used for depression and may work but may not be safe. Research has shown that the best treatment for depression is a combination of drug therapy and psychotherapy. St. John's Wort also has adverse interactions with many prescribed drugs. Another herbal, saw palmetto, used for an enlarged prostate, may work, but its safety is uncertain. Chondroitin sulfate used for osteoarthritis and ginkgo biloba used for improving cognitive function in dementia may work, but again, the safety is uncertain.

You see a common thread here: uncertain safety. Much of this has to do with the fact that supplements are not regulated by the Food and Drug Administration (FDA); thus, there is no way to be certain of what's in the supplement or how much of any ingredient is in the supplement. If you want to be sure that any supplement you buy actually contains the ingredients listed, look for USP (United States Pharmacopeia) on the label. Such labeling ensures the quality, potency and purity of the product. It's important to keep in mind that delaying or replacing essential medical treatments (that may save your life) for unknown treatments is not recommended. Moreover, many herbals interact with conventional medications that do work well. This interaction may render a perfectly effective medication ineffective.

In addition to the available, powerful, mainstream medications and treatments, patients want easing of emotional distress, comfort, relaxation, and relief of physical signs and symptoms such as pain and insomnia. Even when there is no credible evidence that something can be cured by untested alternative medicine that differs from established treatment, it may nurture hope and help improve the quality of life. We also repeatedly hear the word "natural" (food and herbs) as a substitute for artificial (medications). However, "natural" does not always signify safe. Nonetheless, complementary therapy may make the treatment more bearable and endurable.

There are questions as to whether or not there should be an integration of complementary and alternative medicine with the many treatments that are available. No alternative or complementary therapy has, as yet, been shown to arrest tumors or lengthen one's life span. Some supplements may actually be harmful if not taken with knowledge and care. For example, high doses of vitamin E and ginkgo biloba have anticoagulant effects that can cause bleeding in patients undergoing surgery. Soy contains plant estrogens that could, under the right circumstances, stimulate the growth of breast or endometrial cancer. Some supplements may have the ability to actually counteract conventional cancer medications. St. John's Wort (often used as a mood booster) has been shown to lower blood levels of a drug used to treat colorectal cancer. Even the long touted and highly recommended antioxidants used to limit adverse effects of chemotherapy and radiation sometimes make these treatments ineffective.

Nevertheless, studies are in progress to evaluate risks and benefits of integrating herbs, vitamins, and other supplements with conventional treatments that relate to interventions based on diet, body manipulations, and mind-body techniques. Some studies have shown that stimulation of a specific acupuncture point near the wrist may alleviate the nausea and vomiting due to cancer chemotherapy. Gentle massage may help improve the arm swelling that oftentimes appears after radical breast cancer surgery that is accompanied by lymph node removal. One easy and cheap treatment that helps many disorders is moderate exercise. Moderate exercise eases both physical and emotional symptoms of stress, and it improves strength, balance, mobility, and range of movement.

Ongoing research at the National Cancer Institute has shown that vitamin E may inhibit cancer-causing nitrates (found in smoked and cured foods) and prevent nitrates from producing potent carcinogenic chemicals in the body and tumors. Although vitamin E may protect against the development of certain forms of cancer (particularly among smokers), the degree of protection is still unknown.

The herbal huang lian, being studied at Memorial Sloan Kettering Cancer Center in New York, contains several main compounds and 30 minor compounds. Unlike many single herbals now being used and tested (Chinese remedies appear to be a blend of many herbs), the mixture appears to kill cancer cells in the laboratory by interfering with the cells' development. It will be interesting to see what these studies ultimately reveal.

Chiropractic

Chiropractic is a system that, in theory, uses the body's recuperative powers and the relationship between the musculoskeletal structures and functions—particularly of the spinal column and nervous system—to restore and maintain health. The word is derived from the word parts *chiro-* meaning *hand* and *prakikos* meaning *efficient*. The practice is focused on spinal manipulation. It was founded in 1890 by Daniel David (D.D.) Palmer and expanded by his son, Bartlett Joshua (B.J.) Palmer in Davenport, Iowa. D.D. Palmer (1845–1913) practiced magnetic healing, the pseudoscience using magnetic fields produced by a magnet to induce healing. B.J. Palmer (1882–1961) was also a magnetic healer and school teacher, developing chiropractic in his spare time.

In some people, spinal manipulation appears to positively treat low back pain. Low back pain is a very common ailment, and many people seek chiropractors to alleviate achy backs. According to one study using the Medical Expenditure Panel Survey (MEPS), between 1997 and 2006 Americans spent $5.9 billion on chiropractic care.

Scientific evidence is mixed with respect to purported chiropractic benefits. A 2010 review of the literature concluded that in addition to low back pain, it may be helpful for treating migraine and neck-related headaches, neck pain, upper-extremity and lower-extremity joint conditions, and whiplash-associated disorders. The review also found no evidence supporting chiropractic treatment for asthma, hypertension, and menstrual pain. The evidence was inconclusive for fibromyalgia, mid-back pain, premenstrual syndrome, sciatica, and temporomandibular joint (TMJ) issues.

Like physicians and naturopaths, chiropractic training is a four-year program with coursework in biomedical sciences, public health, and research methods. Some chiropractors complete a two to three year residency. Licensing varies by state and thus the scope of practice varies by state. Some states allow dispensing and selling of dietary supplements and treating patients using acupuncture and homeopathy.

Safety issues remain a concern for this practice. Side effects post-treatment include temporary headaches, tiredness, or pain in the treated area. Other reported effects include stroke, cauda equina syndrome (pinched nerves in the lower spinal column) and worsening herniated discs. Results of a 2009 study examined 818 cases of vertebrobasilar artery (VBA) stroke. These arteries supply blood to the back of the brain. This study found an association between visiting a health care practitioner and VBA stroke. However, the researchers concluded that there was no evidence that a chiropractic visit put people at greater risk than a primary care physician visit.

It is likely that people suffering the VBA stroke were seeking treatment related to headache and neck pain before their stroke. Of great note, however, is that many chiropractors shun vaccination, and anti-vaccination attitudes abound within the profession.

Cardiovascular Disease Treatment

Cardiovascular function is an area where complementary treatment can be beneficial. Exercise and good nutrition can be used to control cholesterol, relieve angina (chest pain), and unblock clogging arteries. These events play a very important role in the aging process. The most important point being prevention, rather than the use of modern drugs and surgical procedures in advanced heart disease. Both exercise and good nutrition have been shown to improve one's health during most stages of cardiac illness.

The effectiveness of complementary or supplementary approaches are now being thoroughly studied. For example, diets low in saturated fats and trans fats—those high in fruits, vegetables, fish, and whole grains—may reduce the risk of a heart attack or stroke by 20 to 80 percent. Moderate exercise, such as walking briskly for three or more hours per week, has been shown to reduce the risk of heart attack and stroke by 30 to 40 percent. Stress management and smoking cessation, combined with diet and exercise, may not only improve one's health and well-being, but may in some instances, actually interrupt some disease processes. Rigorous lifestyle programs have been demonstrated to slow arterial narrowing and clot formation that may lead to heart attacks.

Supplements containing folic acid (a water-soluble vitamin of the B complex), omega-3 fatty acids (found in fish oils), and diets rich in antioxidants such as vitamin E, A, and beta carotene, may help protect or reduce the incidence of heart disease. Studies are currently underway to compare pill supplements with actual diets containing the ingredients; to date, it appears that real food is better than taking supplements.

Research using controlled studies has found no specific herb to reduce heart attacks or strokes, but some have been shown to lower cholesterol and triglyceride levels. You should be warned that some herbals may actually be dangerous and create risks. For example, using ginkgo biloba could result in excessive bleeding if used along with blood thinners such as Coumadin (warfarin) or heparin. Ephedra (*ma huang*), an herbal stimulant, may lead to dangerously high blood pressure and heartbeat irregularity.

As an antioxidant, vitamin E is considered to be one of the best natural defenses against free-radical damage. A small study in France has demonstrated that combining vitamin E with a cholesterol-lowering drug may prevent the buildup of plaque on arterial walls and may help prevent heart attacks and strokes. These results, however, are still controversial.

Coenzyme Q10 (CoQ10 or ubiquinone) is a strong antioxidant that protects the body from free radicals and aids the mitochondria (powerhouses) of the cell in the complex process of transforming food into ATP, the energy on which the body runs. It has been investigated for use in angina, high blood pressure, cardiomyopathy, congestive heart failure, diabetes, gingivitis, heart attack and mitral valve prolapse. Keep an eye on the latest findings as they become available.

Mind-body therapies such as meditation, prayer, and relaxation therapy, may help improve one's attitude. This may lead to better diets and exercise programs and in general improve one's quality of life. Thus, it serves to help diminish depression and the risk of heart attacks and stroke.

Osteoarthritis Treatment

Conventional osteoarthritis treatment consists of exercise along with a pain-relieving drug. Some popular pain relievers are acetaminophen (Tylenol) and non-steroidal anti-inflammatory drugs (NSAIDs) such as aspirin, ibuprofen (Advil or Motrin), or naproxen (Aleve, Naprosyn). These drugs may cause serious gastrointestinal damage. Prescription drugs like celecoxib (Celebrex) and rofecoxib (Vioxx) are easier on the gastrointestinal tract but may increase the risk of heart attack.

Supplemental remedies now being tested include chondroitin, chondroitin sulfate, glucosamine, and S-adenosylmethionine (SAMe). Chondroitin sulfate is often combined with glucosamine. It may help, but more research is needed. One must be careful because it may elevate blood sugar level (not good in diabetic patients) as well as promote excessive bleeding if one is also taking a blood thinner (warfarin or heparin).

Glucosamine may reduce osteoarthritis pain and improve flexibility, particularly in the knees and hips. You must be careful here, too, since it may interfere with insulin, causing blood sugar levels to rise. Avoid this supplement if you are allergic to iodine or shellfish.

S-adenosylmethionine (SAMe) has been used for many years in the treatment of liver disease, osteoarthritis (particularly with pain in the knees, hips and spine), and depression. It may cause stomach upset and nausea in high doses, and in some people, it has been shown to cause anxiety, restlessness, insomnia, and even mania.

Although more evidence is needed for the beneficial effects of herbal treatments for osteoarthritis, the following have been tried and are still being used with good effects. These treatments include boswellia, turmeric (curcumin), evening-primrose oil, ginger, guggul, horse chestnut seed extract, shark cartilage, stinging nettle, and CBD (cannabidiol).

Acupuncture is being carefully studied for its possible effects on pain in the knees, hips, and spine. Its beneficial effects, if any, are uncertain and as yet not fully researched.

Back Pain Treatment

Conventional back pain treatment usually centers around taking anti-inflammatory or muscle-relaxing drugs—with limited relief—or surgery, if warranted. Thus, many people turn to complementary and alternative therapies, such as chiropractic manipulation, massage therapy, and acupuncture, for relief.

Chiropractic therapy and spinal manipulation are popular alternatives for back pain. However, it is often modestly helpful or not at all. Manipulation of the lower back is considered low-risk.

Massage therapy is regulated, and massage therapists are licensed in 43 states. Preliminary studies suggest that it can help persistent back pain and improve the patient's ability to function reasonably well in daily activities.

Acupuncture for back pain is receiving much publicity. Its effectiveness is still uncertain, and the research is still in its early stages to draw any definitive conclusions. If performed by an expert, it is unlikely to be harmful.

Stretching, strengthening and aerobic exercises, yoga, and stress-reduction techniques are personal ways of treating back pain. Persistence, accompanied by some professional help to avoid undue stress and making the situation even worse, is important. Lack of total movement and inactivity (due to the pain or fear of making things worse) may only increase the problem and make the situation worse.

Depression and Anxiety Treatment

In terms of depression and anxiety, natural products are not always the best or safest treatment for these conditions. Herbs and tonics are not regulated by the FDA, and they also have chemical actions and some side effects not often detailed. One must be very aware of the risks, especially noting that an alternative therapy may interact with other medications. St. John's Wort may change blood levels of prescription antidepressant drugs, undercut the positive effects of some treatments for HIV or cancer, and can also complicate anesthesia during surgery. One needs the advice of a health care professional before undertaking self-help therapy.

TAKE-HOME MESSAGE

If a complementary practice or alternative medicine stands up to scientific scrutiny, then it enters mainstream medical practice. The placebo effect is quite strong and should not be totally discounted because there is a strong mind-body connection. However, complementary and alternative medicine should not replace traditional medicines for which there is scientific validation.

> *The art of pleasing is the art of deception.*
> —Luc De Clapiers (1715–1747); French writer

8

I Want a New Drug
The Good, the Bad and the Unknown

U.S. Landmark Case: Gonzales v. Raich, 545 U.S. 1 (2005); (6–3 decision)

In 1996, the state of California passed the Compassionate Use Act that legalized marijuana use for medical reasons. This was done to offer help for people suffering from otherwise untreatable pain. California's law was in complete conflict with federal law, the Controlled Substances Act (CSA), which is Title II of the Comprehensive Drug Abuse Prevention and Control Act of 1970. This case arose with two marijuana users who grew the plants for personal use. Angel Raich and her doctor claimed that they had tried numerous remedies for Angel's excruciating pain, including medications to which she was allergic, but marijuana was the only substance that gave her any relief. Diane Monson, a co-petitioner, had suffered chronic pain as the result of a car accident some 10 years earlier; she used marijuana to relieve the pain and muscle spasms in her spinal region. Both people used home-grown marijuana and did not buy or sell it whatsoever.

A combined force from the local sheriff's department and agents of the Drug Enforcement Administration (DEA) conducted a raid, and Raich's and Monson's plants were confiscated. Raich and Monson sued claiming that the enforcement of the Federal law violated the Commerce Clause, the Due Process Clause of the Fifth Amendment, the Ninth Amendment (there are other unlisted rights that cannot be violated), the Tenth Amendment (rights not given to the federal government are reserved to the states), and the legal doctrine of medical necessity.

Congress passed the CSA based on the federal government's power to regulate *interstate* commerce. The Ninth Circuit Court of Appeals ruled that the CSA was unconstitutional as it was used here to prohibit *intrastate* medical marijuana use—use that happened only within California—which did not substantially affect interstate commerce.

The U.S. Supreme Court held that it is firmly established that the commerce clause does give Congress the authority to regulate or prohibit the cultivation and use of marijuana in spite of state law to the contrary. The Court reasoned that this was a permissible control of marijuana to prevent the marijuana from being used in other ways. All parties agreed that there was an illicit market in marijuana. The Court reasoned that the demand for illicit marijuana both in California and across state lines was sufficient reason for allowing the DEA to stop marijuana cultivation—even if it was grown for home consumption—because there was no guarantee it would not get into the interstate market.

A Story About Epinephrine (Adrenaline)

The decision rendered in this 2005 Gonzales v. Raich landmark case is quite interesting because as of March 2021, 36 states plus Washington, D.C., and the territories of Guam, Puerto Rico, and the U.S. Virgin Islands have legalized medical marijuana. Sixteen states (including California, the home of the defendant) and Washington, D.C., have legalized marijuana for adult recreational use, and about 70% of adults support legalizing marijuana. It should be noted that this issue and these stats are really in flux with states passing new laws frequently—some making marijuana more accessible, while other states such as Florida, are considering a rollback of their current law. Marijuana legislation is murky, because some states have totally legalized its use, while other have listed medical and decriminalized use, medical use only, decriminalized status, and fully illegal. Like tobacco and alcohol, marijuana is a drug; but regardless of where you live in America, it is perfectly legal to grow your own tobacco and brew your own beer. There is still much controversy surrounding marijuana, partly because much is not known about it and partly because there is a mindset that contends that all drugs are bad, so you should just say no to them. From a scientific perspective, many drugs have therapeutic merit, others require more study, and individual biology influences drug addiction.

Let's look at this from a natural physiological perspective. The number of naturally made chemicals circulating in the body is impressive. Their functions are numerous and range from enhancing mood and regulating blood sugar to affecting heart rate and muscle relaxation. Because these chemicals affect us physiologically and psychologically, in many ways they make us who we are. Thus, understanding them is important to understanding ourselves. This chapter focuses on drugs, those chemicals that mimic natural body chemicals, enhance or hamper their effects, or have totally different actions. Drugs can be legal and illegal. Our discussion begins with an intersection of the two.

Before we begin the intersection story, keep in mind that just as drugs are legal and illegal, drug abuse can also occur with both. For a frame of reference, here is an alphabetical list of the most commonly abused drugs. It's a lengthy chronicle of who's who on the drug scene and includes alcohol, club drugs, cocaine, fentanyl, hallucinogens, heroin, inhalants, marijuana, methylenedioxymethamphetamine (MDMA: Ecstasy/Molly, methamphetamine), opioids, over-the-counter (OTC) medicines, steroids (anabolic), synthetic cannabinoids (K2/Spice), synthetic cathinones (bath salts), and tobacco/nicotine and E-cigarettes. Believe it or not, this isn't even a complete list. There are a host of other drugs, such as betel quid, flakka, caffeine powder, krokodil, and N-bomb that have entered the scene over the last five years.

Some of these names likely sound familiar because they've entered mainstream news. Or, you might have heard the name tossed around at your local pharmacy. For instance, you went to your local Rite Aid, CVS, or Walgreens to buy some allergy medicine and were told you needed to go see the pharmacist. You obliged but felt like you were being sent to the principal's office. The reason you couldn't buy your Claritin-D off the shelf was because of the Combat Methamphetamine Epidemic Act (CMEA) of 2005. Enacted on March 9, 2006, as part of the Patriot Act and signed by President George W. Bush, this federal legislation regulates over-the-counter sales of pseudoephedrine, a common ingredient in cold medicine and used as a decongestant. Biochemically, pseudoephedrine is the isomer of ephedrine, a compound with actions that are similar to our body's naturally-produced epinephrine (also known as adrenaline). An isomer is two compounds with the same formula but a different arrangement of atoms in the molecule, giving it different properties. Ephedrine is also similar in molecular structure to the street drug methamphetamine. Epinephrine is a natural hormone released from the adrenal glands, two triangular-shaped organs capping each kidney.

Because of pseudoephedrine's chemical and molecular similarities to methamphetamine, it can be used to synthesize illegal methamphetamine (meth). Meth is a white powder that can be swallowed, smoked, snorted, or injected. It can be used legally to treat low blood pressure. It is also used illegally as a stimulant. For a cultural reference, in the hit series *Breaking Bad*, Walter White (Bryan Cranston) and Jesse Pinkman (Aaron Paul) cooked up meth through five seasons.

Here's a quick reference for the chemical names:

pseudoephedrine = decongestant
methamphetamine (meth) = illegal drug
ephedrine = compound in pseudoephedrine that is like epinephrine; used to make meth
epinephrine = natural chemical released from adrenal glands

Epinephrine causes vasoconstriction (blood vessel narrowing), vasodilation (blood vessel expansion), and airway dilation. As a decongestant, pseudoephedrine works by constricting blood vessels lining the nasal passages and sinuses. This narrowing decreases blood flow to the nose and sinuses thereby reducing congestion and mucus production and allowing for expansion of our breathing passages. It is used to treat congestion caused by the common cold, hay fever, and allergies. At low doses, ephedrine can treat low blood pressure (hypotension) and slow heart rate (bradycardia); at high doses it causes high blood pressure (hypertension) and rapid heart rate (tachycardia)

If you want to buy pseudoephedrine at your corner pharmacy, you'll have to ask for it by name, provide photo identification, and limit your purchase to a 30-day supply. Specifically, purchases are limited to 3.6 grams of pseudoephedrine base per day. Stores are also required to keep personal information of purchasers for two years. Buyers must also be adults because it can't be sold to minors.

Some situations make the purchase limit seem nonsensical. For example, if you are going on an extended vacation or need to buy some decongestants for yourself and another family member, that other family member will have to be an adult and make their purchase separately. You can't buy any for you and your child, who will likely tag along on the family trip. Just ask Tim Naveau, who in 2006 was arrested and charged with a Class-B misdemeanor for buying several packages of Claritin D to combat his allergies and those of his son. It appeared as though he was stockpiling for illicit activity. In reality, he was stocking up because his son would be away at church camp. Either the son was making meth at camp or the guy was being a good dad and helping his son, a fellow allergy sufferer.

That was the story of a regular consumer, just buying a product. Digging a little deeper about epinephrine and a pharmaceutical company, we come across a little dirt. Among its various actions, epinephrine is also used to treat asthma and allergic reactions by dilating the breathing passages (bronchi) within the lungs. A life-saving synthetic form of epinephrine comes packaged as an auto-injectable EpiPen. You probably have heard about EpiPens because a ruthless pharmaceutical company named Mylan purchased the rights to EpiPen and gradually hiked the price of this product from about $100 for a 2-pack to over $600 for a 2-pack. This increase—500 percent—boosted EpiPen profits to $1.1 billion.

The U.S. House Oversight Committee noticed in September 2016 and reviewed this action in relation to antitrust laws. Antitrust laws do not prohibit price gouging, but they do prohibit the use of unreasonable restraints of trade to facilitate or protect a price increase. Thus, the Oversight Committee asked the Federal Trade Commission to review this price increase. Mind you, Mylan didn't develop this drug. They simply bought the rights to it. The company just wanted to turn a big profit at the expense of consumers who really needed the medication.

In 2011, Primatene Mist, an over-the-counter (OTC) epinephrine inhaler for asthma was taken off the shelves because it contained chlorofluorocarbon (CFC) propellants. Chlorofluorocarbons are chemicals known to deplete the ozone layer and were part of the 1989 Montreal Protocol of Substances that Deplete the Ozone Layer and the Clean Air Act of 1990. Primatene Mist was re-released in 2018 without the CFCs and is used to treat intermittent asthma. It retails for about $30 for a canister that contains 160 doses. As I write this, it is still sold OTC; however, many healthcare providers think it should be a "behind-the-counter" drug like ephedrine. Why? Because when you use Primatene Mist you are *inhaling epinephrine*. While it does indeed relax bronchial smooth muscles, resulting in bronchodilation and easier breathing, it also increases the rate and force of heart contractions and thus it is also used in a clinical setting to treat heart failure. It's also a vasoconstrictor, squeezing blood vessels, so it will increase blood pressure. Your body doesn't know why you're inhaling epinephrine, so its mechanism of action won't change. These are all sympathetic effects—fight or flight responses—which means it'll rev up your metabolism. Which means you'll see Internet sites touting its use for weight loss. Don't use it for weight loss.

Pharmacology

Pharmacology is the branch of science concerned with the sources, uses, appearance, chemistry, effects, mechanism of action, and interactions, among many other features, of drugs. Granted, this field of study is quite comprehensive. Before exploring each aspect, break the word apart into its components, *pharmaco-*, which means *medicine or drugs*, and *-logy*, which means *study of*. Pharmacology dates back to the 1st century BCE with the work of Greek physician and father of pharmacology, Pedanius Dioscorides, who authored *De materia medica*. This book is a compendium of medicinal plants and the medicines that could be derived from them.

Two main areas of pharmacology are pharmacodynamics and pharmacokinetics. The biochemical changes, physiological effects, and the mechanisms of drug action are grouped under the term pharmacodynamics. It's a complicated area that studies the uptake, movement, binding on cell receptors, and interactions of drugs at the cellular and tissue levels. Knowing the pharmacodynamics about a particular drug isn't enough; we also have to know its pharmacokinetics. Pharmacokinetics is the study of the drug's absorption, distribution, biotransformation (drug metabolism or processing), and excretion. Pharmacokinetics is related to the concentration of the drug, its chemical byproducts (what it gets broken down into along the way in the body), and the time required for drug concentrations to change.

As an analogy, think about a drug as a bite of food: food contains fats, proteins, carbohydrates, vitamins, minerals, and water. Each has to be broken down into particles that can be used, those particles are transported in the bloodstream to the cells, and what's not used is excreted. Some food, like saltine crackers, begins breaking down immediately in the mouth, but other food, like whole kernel corn, never gets fully broken down. Drugs also have those same characteristics. Some drugs like nitroglycerin are absorbed rapidly under the tongue while other drugs have to be metabolized to release their potential.

Drug physiology involves many different and interacting aspects from absorption, metabolism and function, to routes of administration and excretion. In order for drugs to do any good, the active ingredient must reach its target cell after the body absorbs it. Drug absorption depends on the drug's ability to get into the bloodstream, which is dependent upon the route of administration. Common routes include orally (by mouth), sublingually (under the tongue), intramuscularly (via injection into a muscle), subcutaneously (just beneath the skin), transdermally (across the skin), intravenously (in a vein by needle), and by suppository. Let's begin our discussion with the oral route.

If the drug is swallowed, the first barrier it encounters is the acidity of the stomach, which can change or break down the drug. The body doesn't know the difference between food or drug, and stomach acid, called hydrochloric acid (HCl), has a good purpose: it destroys bacteria in food, and it helps make an enzyme important for protein digestion. Another stomach factor is food itself. Food in the stomach can alter a drug's absorption. This is why some drugs have to be taken on an empty stomach, others can be taken on a full stomach, or still others should be taken *with* food.

If the swallowed drug survives the stomach, then it moves through the small intestines of the digestive tract where it is continually absorbed into the bloodstream and directed to the liver, which plays a role in how the drug is metabolized (broken down). Your body doesn't know the difference between swallowing food or swallowing a drug; both will be metabolized. When speaking of drugs that are taken orally, first-pass metabolism has to be considered. First-pass metabolism (also called the first-pass effect) is the intestinal and liver breakdown of a drug after it has been absorbed into the bloodstream. So, you swallow the pill, it goes through the digestive tract, and as it is absorbed into the bloodstream, its active substances are removed from the blood before entering the general circulation. Thus, the drug that you initially put in your mouth gets changed to a different form. The challenge in drug development is to make sure the components of a drug can still reach their target tissues with enough potency to do some good. Another challenge is to be sure that *not too much* of the drug reaches its target and causes harm. It's a balancing act: too much is bad and too little is ineffective. Striking a balance is what drug research and development are all about.

Organ systems do not operate in isolation, and the digestive, cardiovascular, and urinary systems are examples of different organ systems working together. The drug is metabolized by the liver (digestive system), transported to tissues via the bloodstream (cardiovascular system) to exert its function, and unused components are excreted in urine via the urinary system or expelled in feces via the digestive system.

A fast-acting route of drug administration is sublingual. Some drugs are designed to dissolve quickly and begin acting immediately. These drugs are placed under the tongue where the active substance diffuses directly into the blood. Drugs can also be administered via needle into a muscle, known as an intramuscular (IM) injection. This may sound odd, but practitioners must take obesity into account when administering an IM drug because the needle must be long enough to pierce through the fat and reach the muscle. Given the obesity epidemic, studies have been done on this topic, because problems arise if the drug is designed to be absorbed from muscle tissue but only makes it to the fat tissue between the skin and muscle. Another route is just beneath the skin (subcutaneously). Subcutaneous drugs are absorbed less rapidly because there are fewer blood vessels just below the skin than found in muscles. Another popular route is *on* the skin, as in transdermal patches or creams. This route is called transdermal because once on the skin the drug has to cross several layers of the skin. You may have applied a topical oint-ment on the skin to relieve pain or itching.

The intravenous (IV) route is the most direct and fastest acting because full-strength drug is injected into the veins, bypassing the digestive tract, and is readily available. Lastly, drugs can also be given in suppository form and inserted into the rectum or vagina. Drugs adminis-tered rectally or vaginally cross the mucous membranes quite easily. "Body packers" are cocaine smugglers who insert bags of drugs into body cavities, namely the mouth, vagina, and rectum. Body packing is used for international drug transport, concealment, or to get drugs into pris-ons. This isn't the smartest thing to do, especially for drugs inserted into the rectum, because cocaine packets aren't always air-tight or they rupture, resulting in sudden death due to acute cocaine toxicity. Because drugs are rapidly absorbed across the membrane, this highly concen-trated form of cocaine results in fatal cocaine intoxication. Even if the bags don't rupture, they could cause bowel obstruction. The cocaine high is characterized by excited delirium, aggres-sion, unexpected strength, and resistance to pain.

Classifying drugs is not always clear cut. Drugs can be classified according to their mech-anism of action (for example, do they act on the nerve receptors or blood vessels?), their effect on a specific body system (do they treat nervous system disorders or skin conditions?), their class (are they beta blockers or antihistamines?), or their targets (do they work against viruses or bacteria?). This is just a small sampling. Keep in mind that semester-long courses are taught on these topics and entire careers are based on pharmacology. For our purposes, we'll learn enough of the necessary science to be well-informed on a very complicated issue.

When thinking about any drug, consider that *anything* you put in your body has the poten-tial to have an adverse effect. Whether it is food or drug, how your body reacts may be totally different from how some other person's body reacts. A classic example is peanut allergies. If you are a peanut allergy sufferer, you know that eating just one peanut may be enough to cause ana-phylactic shock, a life-threating condition. On the other hand, plenty of people enjoy them with great zeal without any bad effects. The point is that we all have the *potential* to physiologically react differently to the same thing. With that in mind, let's go a little bit deeper into the science of pharmacology.

As with peanuts and how you will respond to eating them, drug efficacy depends on the particular drug and individual. Drug dose and the body's response is variable. It can also be dependent on the amount, potency, and duration of drug exposure. Toxins work the same way. Take carbon monoxide, a colorless and odorless poisonous gas. Breathing a little may make you drowsy; but in greater doses, breathing carbon monoxide can kill you. Fuel-burning equipment, like furnaces and automobiles, produce carbon monoxide as a byproduct of combustion. Man-ufacturers add sulfur, a rotten egg odor, to natural gas so that if there is a gas leak in your house, you are warned by the smell. Carbon monoxide detectors with alarms are also a good idea to install because they warn of carbon monoxide in your breathing space. Breathing carbon mon-oxide is deadly because carbon monoxide (CO) has a greater affinity for our red blood cells than does oxygen (O_2) in the air. The hemoglobin in our red blood cells likes oxygen—the O_2 form—because it "knows" what to do with the oxygen. If the hemoglobin attaches to carbon monoxide, it robs our blood cells of the O_2 form, and the result is carbon monoxide poisoning.

How a particular substance within a drug works, known as its mechanism of action, is also another drug feature, and many factors affect drug actions. Some of these seem like common sense and others might be new to you. One factor is a drug's half-life. Half-life is the amount of time it takes for a drug to decrease from full concentration in the blood to one-half concentration in the blood. A drug's half-life determines how long the drug will remain active. The longer the half-life, the longer the drug's activity. When new drugs are developed, its half-life is important to figure out. If you are taking a pain reliever for an ache in your little toe, you want to be sure that once you swallow the drug that enough of it remains intact to have an effect on a distant body part. Figuring out the half-life is important, too, because it also helps in determining dosage amounts and times. If a drug has a short half-life, it has to be administered more often than a drug with a longer half-life.

Some medications have a special coating on the capsule to extend the half-life. The tiny drug particles *within* the capsule can also have a special coating to extend the half-life. In the case of the specially-coated particles, the outer ones dissolve first while the inner ones dissolve later, causing an extended drug release time. To use a football analogy, think of a player carrying the ball (inner particles) surrounded by teammates (outer particles). If tacklers are in the area, the outer teammates will get hit first, but the player in the middle will get to advance a little further on account of the outer protection.

Other factors that affect drug action are age, sex, body weight, circadian rhythm, and disease. Aging affects metabolic rate, so drug dosages are adjusted for developmental stages. Children, young adults, and older adults metabolize drugs at varying rates. Moreover, some drugs cannot be given to very young children because they are not "anatomically equipped" yet. What exactly does this mean? The next time you pick up your bottle of over-the-counter pain medicine, read the label. It will have a warning about giving to children under a particular age. The reason is because the pediatric brain is still developing and only about 90 percent of the brain is developed by age five. An important structure in the brain, known as the blood-brain barrier, is a selective membrane that allows the passage of some substances while restricts the passage of others. It's our gatekeeper for determining what gets to meet our brain and what doesn't. Alcohol can readily cross the blood-brain barrier; this is one reason why you get stupid pretty quickly after drinking too much. But the barrier also protects our brain from some drugs. As a toddler, the blood-brain barrier is still developing, so we wouldn't want to give a young child something that could cross the barrier but really shouldn't.

Regarding sex, males and females have differing response rates. Males typically metabolize drugs faster than females, especially if the drugs are injected into muscles. The reason is because females tend to have proportionately more fat than males do, and fat is not as metabolically active as muscle. Body weight also affects drug action, with heavier people requiring greater dosages. Dosages will need to be adjusted for total surface area. Before you get upset by these statements, know that these are generalizations. Fit people metabolize drugs quicker than less-fit people and genes also play a role in drug metabolism. Using alcohol as an example again, beverage alcohol can be enjoyed by both sexes equally. However, if a 150-pound male and a 150-pound female drink equal amounts of beer, the female will likely become drunk before the male. One reason for the sexual differentiation has to do with alcohol dehydrogenase (ADH4), an enzyme that breaks down alcohol. Males tend to have more of the enzyme than females and therefore can metabolize alcohol quicker. Alcohol will be discussed again when we consider the topic of drug abuse.

Circadian cycles can also affect drug actions. Circadian cycles are our natural, biological clocks. You might also know it as our sleep-wake schedule or diurnal body rhythm. We humans have a biological rhythm that cycles about every 24 hours. Some of us have rhythms that are a little longer, and some a little less. Such cycles have been observed in many animal species. Our internal timekeeper is controlled by our brains and light exposure. When light—sunlight or artificial light—reaches the retina in our eyes, a chemical message is relayed to the hypothalamus, another area of the brain involved with biological signals. Fibers from the hypothalamus relay the message to the pineal gland, a pine-coned shaped organ deep in the brain, which makes a hormone called melatonin. During daylight hours, melatonin levels are low and during evening hours, melatonin production increases. This variation is proportional to the length of light exposure, thereby influencing our sleep-wake cycles. Melatonin also has effects on reproduction, sleep disorders, and jet lag. Because increased melatonin levels help us sleep, some people take melatonin supplements as a sleep aid. We also know that artificial blue light from our laptops, tablets, and cell phones can confuse our pineal gland into "thinking" it is daytime and thus decrease our melatonin production, which leads to sleeping problems. For this reason, it is advised to power-down the technology two to three hours before heading off to slumber. Back to circadian cycles and drugs. Sleep-inducing agents should be taken at night, while other drugs that stimulate the body should be taken during the day.

We know *how* drugs get into our bodies, but how do they get *out*? Like most things that enter our bodies, after being metabolized, they are excreted in some form or another. As for drugs, most are excreted via the kidneys, which is why urine is used for drug testing. Because the kidneys filter blood, drug metabolites will show up in the urine. Monitoring drug dosage in people with kidney disease is quite challenging. If the kidneys are not working properly, drugs may stay in circulation a lot longer and exert more effects than intended.

Other routes of excretion include the lungs, sweat, feces, and saliva. If you've ever been talking with a person who has been drinking alcohol, you can definitely smell it because the lungs are an important site for alcohol excretion. Mammary glands are really modified sweat glands, so breast milk contains anything the woman has put into her body. Another example showing how sweat is used as an excretory route involves the common antibiotic, rifampin, which is used to treat tuberculosis (TB). Rifampin gives skin a reddish color as it is excreted in sweat through skin pores. Although not a drug, beets can affect the color of urine and feces, too. Beeturia is passing red-colored or pink-colored urine after eating beets. The color is the result of the pigment betanin, which gives beets their characteristic red color.

In addition to all the good that a particular drug can do, drugs can also be toxic (poisonous). Drug toxicity depends on dosage. Alcohol is a good example. In most cases, a little alcohol does not poison the body. However, alcohol toxicity, also known as alcohol poisoning, can occur if too much beverage alcohol is consumed in a short period of time. In the United States, a "standard drink" is defined as either 12 ounces of beer (5 percent alcohol), 8 ounces of malt liquor (7 percent alcohol), 5 ounces of wine (12 percent alcohol), 3 to 4 ounces of fortified wine like port or sherry (17 percent alcohol), 2 to 3 ounces of liqueur (24 percent alcohol), or 1.5 ounces of distilled spirits/hard liquor (40 percent alcohol). We know that having food in the stomach slows the absorption of alcohol and the liver metabolizes alcohol at a rate of about 15 mL per hour. This 15 mL equates to one standard drink per hour. Per the *Dietary Guidelines for Americans*, a publication of the USDA (United States Department of Agriculture) and HHS (Health and Human Services), women should drink no more than one drink per day and men should drink no more than two drinks per day. Alcohol toxicity depresses breathing, decreases blood pressure, and can cause vascular collapse. Any one of these can lead to death.

An overdose is too much of a drug which causes harm. Overdoses occur through medication errors, poor judgment, confusion, or in cases of attempted suicides. Too much of a drug can act like a poison. Sometimes, there is an antidote. Naloxone (brand name, Narcan) can temporarily reverse an opioid overdose. We'll discuss this a little later when we take into account the most common drugs of abuse.

Drug Tolerance

Drug tolerance occurs when a person develops resistance to a drug's effects. When tolerance happens, more drug must be given to get the desired effect. Tolerance is typically a sign of drug dependence. Drugs that commonly produce tolerance are alcohol, opiates like morphine, and tobacco. How does tolerance actually occur? That is, what happens physiologically that causes a diminished response to a drug so that more must be given to achieve the same thing? Tolerance to a drug can only happen if the drug is used repeatedly. In the case of alcohol or morphine, larger and larger doses must be given over time to produce the same effect that initial, smaller doses did.

Two major players are involved here: enzymes and receptors. Enzymes are catalysts that induce chemical changes. With time and dosage, liver enzymes become more active to the specific drug, thereby causing an increased metabolism of the drug. Receptors are structural proteins on a cell's surface or within the cell's semi-liquid cytoplasm that respond specifically to a substance. Think of a receptor as a little stud along the cell exterior or within the fluid of the cell. These receptors are three-dimensional structures that allow only specific molecules to attach. It works very much like a lock and key. You may have many keys on your keychain, but only one is going to fit in your front door's lock. Cells have many receptors, but only certain molecules can bind to them. Receptors bind to specific factors, such as drugs, hormones, or antigens. Drugs bind to receptors to produce results. Some drugs can also act like a master key and attach to several types of receptors. Tolerance also develops because the number of cell receptor sites that the drug attaches to decrease or the strength of the bond between the receptor and the drug decreases.

Drug tolerance also develops because with repeated exposure to the drug, changes within the cell's organelles also develop. An organelle is a little, specialized structure within a living cell. In this case, the endoplasmic reticulum changes. The endoplasmic reticulum is involved with many functions, but a major function is the detoxification of drugs. It does this through the cytochrome P450 system, which are detoxification enzymes. Drugs increase the activity of the P450 system, which in turn causes proliferation (rapid reproduction) of the endoplasmic reticulum. As the P450 system gets better at metabolizing a drug, the proliferation leads to drug tolerance. Here's an analogy: Pretend that peanut butter is a drug and a slice of bread is the endoplasmic reticulum. You spread two tablespoons of peanut butter on one slice of bread and the tasty goodness covers that one slice just fine and dandy. But, the one slice of bread gets used to two tablespoons of peanut butter and grows larger, say to double its size. If you spread two tablespoons onto what is now essentially two slices of bread, the peanut butter isn't going to cover it so well, and more is needed in order to cover the bread in the manner in which it was accustomed and had the same level of tasty goodness. This is what happens at the cellular level.

It's important to note that tolerance does not develop for all drugs. Antacids, for example, do not attach to cell receptors. You've probably taken an antacid at some point in your life for an upset stomach or indigestion. Antacids work because they decrease stomach acid. But they do this through a simple chemical reaction by interacting with acids already there, altering the stomach's pH, and neutralizing the hydrochloric acid (HCl) in the stomach.

Drug Cumulative Effect

Just as dangerous as tolerance is a cumulative effect. If a person has liver disease or kidney disease, two conditions that affect drug metabolism and drug excretion respectively, more intact drug remains in circulation. When the body cannot completely metabolize or excrete one drug before the next dose is given, an increased response to the drug occurs. Thus, more drug accumulates in the blood and body tissues. Heavy metal poisoning such as lead poisoning and mercury poisoning are examples of cumulative effects. Lead poisoning occurs over months or years as lead in contaminated air, water, and soil is breathed in, drunk, or enters the food supply and accumulates in body tissues. Lead can affect mental and physical development in children; in adults, signs and symptoms can include high blood pressure, joint and muscle pain, reduced sperm count, and miscarriage. High accumulated levels can also cause death.

Mercury poisoning results through breathing mercury vapors, ingesting mercury, and absorbing mercury through the skin. Mercury was also discussed in Chapter 6 in regard to thimerosal in vaccines. Signs and symptoms include impaired motor skills, speaking and understanding language difficulties, physical tremors, and neurological disturbances. Mercury is concentrated in the food chain and fish that swim in contaminated waters accumulate mercury in their bodies. Fish eat both vegetation and smaller fish contaminated with mercury and that mercury becomes biomagnified in their flesh. The protein in fish binds the mercury so tightly that cooking—even deep-frying, pan frying, and boiling—cannot remove the accumulated mercury. For this reason, we must limit eating fish that are high on the food chain such as tuna, pike, walleye, perch, and bass.

Drug Synergism

Synergism is an important action to know because it produces a greater effect than expected when two or more drugs are taken together. Said another way, the interaction of two drugs produces a result that is larger than if each drug were acting separately. Sometimes a synergistic effect is beneficial, as in the case of taking two antibiotics, sulfamethoxazole + trimethoprim—the combination is packaged as Bactrim or Sulfatrim—to treat ear infections, urinary tract infections, and traveler's diarrhea. Other times, the effect is harmful, as in drinking alcohol while also taking the sedative diazepam (Valium). Mixing alcohol with Valium causes excessive sedation and possible overdose. Drug mixing is also problematic in older people because many older people are already taking several prescribed drugs and over-the-counter supplements simultaneously. This is referred to as *polypharmacy*, and it dramatically increases the risk of adverse reactions.

Drug metabolism varies from person to person. As previously described when discussing the P450 system, various enzymes are involved with drug metabolism and it's not always easy to predict how something will affect one person or another. An interesting example is how grapefruit can interfere with certain medications, such as statins to lower cholesterol, hypertension drugs to treat high blood pressure, organ-transplant rejection drugs, corticosteroids to treat ulcerative colitis, and some drugs that treat heart arrhythmia (abnormal heartbeat).

Interestingly, grapefruit doesn't affect all these drugs in all people. But eating grapefruit or drinking grapefruit juice may affect you if you are taking one of these drugs. How? Our intestines make an enzyme called CYP3A4. Some people make more than others. This enzyme is important for metabolizing these aforementioned drugs. Grapefruit can block the action of CYP3A4, preventing drug break down, and causing more drug to remain in the bloodstream longer. This results in too much drug circulating throughout the body. The effects vary from person to person and from drug to drug. For instance, you and your friend may be taking Lipitor to lower your cholesterol and also drinking grapefruit juice. Due to individual variability in the amount of CYP3A4 enzyme secretion, you and your friend could have totally different reactions to the same drug. This was an example of grapefruit interfering with an enzyme causing too much drug to remain in the body. However, the opposite effect can also happen, as occurs with fexofenadine (Allegra), a common drug used to treat allergies. In this case, grapefruit causes less of the drug to enter the bloodstream because it affects proteins called drug transporters. As their name suggests, drug transporters move the drug into our cells for absorption. If the drug transporters are affected, less drug reaches its target. Thus, metabolism isn't altered, rather proteins are altered. Here's a quick synopsis:

drug ingested → grapefruit eaten → enzyme blocked → too much drug in the body
OR
drug ingested → grapefruit eaten → transporter blocked → too little drug in the body

Drug Development

So far, we've covered pharmacology, drug tolerance, cumulative effects, and synergism. But how do we end up with the cornucopia of drugs? We use many over-the-counter drugs to treat common ailments while medical professionals dispense drugs on a daily basis for the purpose of treating acute or chronic conditions. To administer any drug safely, medical practitioners and the average person should know the usual dosage, route of administration, purpose, contraindications, adverse reactions, and major drug interactions. That's a lot to take in! Listening to television drug ads that list the possible side effects could terrify a person. For the most part, we really just want to know the therapeutic effects, which are the positive outcomes. It's the reason we take a drug. We want it to treat our malady. But, like the peanut allergy, harmful effects known as adverse reactions, can also happen.

Before we explore other topics, it's valuable to also understand how prescription drugs are developed. Why? Because there is a public perception that drug companies are just trying to make money at all cost. There is some truth to that statement, at least in the United States: Americans pay more per year for prescription drugs than any other developed country. Consumers in the United States face high out-of-pocket costs for several reasons including a large population of uninsured people and cost-sharing requirements for those with coverage. Compared with other developed countries, it appears that U.S. costs could be reduced if there were universal health coverage, price control strategies, and centralized price negotiations.

"Big Pharma" is the term that is often used to describe the pharmaceutical industry. While drug companies do indeed make lots of money, and some have questionable practices (recall the EpiPen discussion), it is also true that they have to spend lots of money in research, development, and testing just to bring a drug to market. For every success story, there is also a failure— or a hundred failures. The intent here is not to wave a banner for big drug companies; there are background stories worth knowing. In order to be fully-informed, it's good to know how prescription and non-prescription drugs make it to the store shelves.

New drug development can take 7–12 years. Yes, sometimes it's less and sometimes it's more, but 7–12 years is a nice ballpark figure. The United States Food and Drug Administration (FDA) is the government agency in charge of approving new drugs and monitoring them for adverse reactions or toxic effects. Before any drug hits the market, it must be studied and researched before it begins phase trials. Preclinical research is the first step in developing a new drug. During this phase, researchers from universities, pharmaceutical companies, and government labs are consulting published scientific information and databases to gather as much information as possible. Once the mechanisms of a disease are studied, new drugs are developed and analyzed for effectiveness and side effects. To do this work, cell cultures and animal models are both used. It is important to use animals because drugs behave differently when introduced into living creatures. As an example, the first antibiotic, prontosil, had no effects on bacteria in culture, but when the drug was given to mice, the mice livers metabolized the drug into another drug, sulfanilamide. This was the beginning of the new sulfa drug, which proved beneficial in treating infections such as pneumonia. Before this sulfa drug, pneumonia was fatal. Another example doesn't have such a happy ending. In 1933, more than a dozen women were blinded and another died as a result of using the permanent mascara Lash Lure. At this time, there were no drug testing protocols to ensure public safety, and a chemical, p-phenylenediamine, caused blisters, abscesses, and ulcers on the face and eyelids. These led to blindness in some women and another woman died from a bacterial infection induced by the ulcers.

Phases are the stages of pre–FDA approval status. A new drug that hasn't received FDA approval is called an *investigational new drug* (IND), and it must go through clinical testing, which is done in phases. Phase I trials run about four to six weeks when the IND is tested on patients who do not have the disease for which the drug is intended to treat. We call these patients "normal." Phase I trials generally involve 20–80 people and these studies focus on safety and side effects. If this phase goes well, then the drug enters phase II trials.

During phase II trials, the drug is tested on humans who have the disease for which the drug is intended to treat. Phase II trials study about 300 people and look at drug effectiveness. Although this might seem easy, it's actually complicated because another element is figured into the scenario. This element is double-blind placebo-controlled study with powerful statistics. These are scientific studies in which a placebo (sugar pill) and the actual drug that should have therapeutic effect are used, and any information that may influence the behavior of the tester or the subject is withheld until after the testing is completed. Double-blind placebo-controlled research is part and parcel for conducting science and ensures to the best of our ability that drugs are safe and effective. But the story doesn't stop here. We're still only at phase II. Next, the IND enters phase III. During this phase, which generally lasts up to five years, large samples of people are tested and feedback on efficacy and adverse effects is still collected. Several hundred to 3,000 people are involved with phase III studies. If the IND is found to be both safe and effective, the researchers complete a new drug application (NDA). From this point—keeping in mind we're already up to about five years—it may take another two years once the application is made. It's a lot of study. But even so, once a drug is approved, it may be studied further.

After approval, drugs are monitored during phase IV for continued safety issues. While 3,000 people might be a sizable sample size, unpredicted issues could arise when a drug is given to a broader population.

The FDA also monitors drug risks and adverse events through its MedWatch system. Doctors and patients report adverse events through this system so that they may be added to the drug's label. In some cases, too many adverse events may cause a drug to be withdrawn from the market. You may have heard of some of these. For example, Vioxx, a COX-2 selective non-steroidal anti-inflammatory drug (NSAID), did a great job alleviating joint pain and inflammation. After being on the market, however, it showed that people taking the drug were at increased risk for heart attacks and strokes, so it was withdrawn in 2004.

Some drugs, such as those used to treat cancer, may go through several phases of clinical study, culminating in phase IV clinical trials. Phase IV clinical trials are also called post-marketing surveillance trials because these trials study the side effects over time of an approved treatment that is already on the market. These are important studies because they look at thousands of people over a longer period of time who take a particular drug. Here's an example: Cancer Drug A and Cancer Drug B have been on the market for a long period of time. Doctors have used both, but they would really like to know if one is more effective than the other for treating Cancer X. Cancer X could be a lung cancer, kidney cancer, or any other cancer for which Drugs A and B have proven effective against. Doctors involved in such studies generally collaborate with other physicians and researchers who want to know the same thing. Think big cancer research centers like the Cleveland Clinic, Johns Hopkins, Memorial Sloan Kettering Cancer Center, Roswell Park Cancer Institute, University of Texas M.D. Anderson Cancer Center, Dana Farber Cancer Institute, and an exhaustive list of other research institutions across the globe collaborating with pharmaceutical companies. It is quite common to have worldwide involvement in Phase IV clinical trials for cancer since its prevalence is so widespread and its effects so devastating.

The point is that Phase IV clinical trials take all the clinical trials a step farther and may discover something that wasn't known before. For example, in an April 2018 study in *The New England Journal of Medicine*, researchers and medical professionals who were part of the Check-Mate 214 Clinical Trials, found that using two already-on-the market cancer drugs together was better than using this other standard treatment. The real drugs were nivolumab (Opdivo), ipilimumab (Yervoy), and sunitinib (Sutent). It was found that using Opdivo plus Yervoy increased overall survival (OS) and objective response rates better than using Sutent in patients with metastatic renal cell carcinoma (kidney cancer).

Dr. George Smith developed bacteriophage display technology, a method of identifying unknown genes for particular proteins. This technology led to the creation of adalimumab (Humira), Opdivo, and Yervoy. Dr. Smith was a researcher at the University of Missouri in Columbia who earned the 2018 Nobel Prize in Chemistry for this breakthrough. Such technology is enabling development of other drugs utilizing antibody therapies for the treatment of cancer, rheumatoid arthritis, and multiple sclerosis. Basically, immunotherapy is being used instead of chemotherapy to treat disease. These drugs unleash the immune response to fight the cancer. In the past, cancer treatment has revolved around killing the cancer with chemotherapy. But the downside of chemotherapy is that it killed many more cells than just cancer cells, so it wasn't very selective. It was like killing everything in its path versus just killing the bad cancer.

It's also very likely that you've heard of using the immune system to fight cancer if you followed the treatment and recovery of former President Jimmy Carter's melanoma. In 2015, Jimmy Carter was diagnosed with advanced melanoma, an aggressive skin cancer that had spread to his liver and to his brain. The disease progression was so advanced that it prompted him to write letters to his family in case he died. He did receive radiation therapy, and he was also put on the drug pembrolizumab (Keytruda), which is another immunotherapy drug. Fortunately, the drug worked.

The immune system is on constant surveillance, looking for what belongs and what doesn't belong. But sometimes, cancer cells escape the immune response. Some cancer cells can actually "hide" from the immune system by expressing a specific protein that turns off the immune response. Immunotherapy drugs can target this specific protein, namely PD-1, and allow the immune system to attack the cancer. Opdivo is a newer drug than Keytruda and it works the same way by targeting the PD-1 protein. As of 2021, Jimmy Carter's Stage IV cancer has disappeared, and he is currently showing no signs of cancer and actively building homes with Habitat for Humanity. This is great news because over the past three decades, the incidence of melanoma has increased greatly.

Some drugs end up treating another condition than the one for which it was intended. Case in point: Viagra, the little blue pill. We know now that Viagra is a drug used to treat erectile dysfunction (ED), a term that became a household phrase once marketing of Viagra began. What you may not know, though, is that researchers at the drug company Pfizer were trying to find a drug to treat high blood pressure (hypertension). In 1989, chemists at Pfizer in South East England concocted sildenafil citrate, a compound they thought would treat hypertension. By June of 1993, the drug was still in development, but researchers weren't finding any benefit in its treatment of hypertension. Then, in a turn of fate, just months before the company was going to scrap development, researchers discovered an unexpected finding in a group of miners in South Wales. These miners reported having more nighttime erections than usual. With that tidbit of information, researchers thought they might be on to something: a possible treatment for erectile dysfunction, known by many as impotence. What they discovered instead of lowering blood pressure was that sildenafil citrate inhibits an enzyme that plays a role in penis erections. Sildenafil, the active ingredient, relaxes the muscles in the penis, resulting in greater blood flow and erection. New trials were conducted along this other line of research and by March 27, 1998, the world had a new drug. According to an article in Bloomberg, more than 300,000 prescriptions were filled in one week in May 1998. Penis rigidity is measured by a little device called the Rigi-Scan which monitors penis circumference.

Viagra isn't the only "accidental" drug. Another, much more common medicine is aspirin. The aspirin we know today dates back to the late 1890s and is associated with another common household name, Bayer. Felix Hoffmann, a chemist at Bayer in Germany, first used aspirin to treat his father's rheumatism, a disorder marked by joint and muscle inflammation. The history of aspirin is fascinating and really predates Hoffmann and Bayer. Aspirin's key ingredient, acetylsalicylic acid (ASA), can be found in spirea, jasmine, beans, peas, clover, some grasses, and certain trees. Ancient Egyptians extracted the compound from willow bark, and Hippocrates

(460–377 BCE) wrote that willow leaves and bark relieved pain and fever. Noting that aspirin helped his father, Hoffmann thought he was on to something, and Bayer began distributing the white powder to physicians to give to their patients in 1899. In 1915, aspirin was sold as an over-the-counter medication.

Today, many people take 81 mg of aspirin daily for prophylactic measures to prevent deep venous thrombosis (blood clots in deep veins), thromboembolism (moving blood clot), and generalized inflammation reduction among other things. This same dosage is typically found in baby aspirin. Have you ever been curious about the dosage level? Why 81? Why not 80? The answer harkens back to a time when the apothecary system of weights and measures was used. Apothecaries were the forerunners of today's pharmacies and the apothecary system was a system of weights based on the weight of a grain of barleycorn and abbreviated gr. It had been used for centuries in weighing medicines, but the system has been superseded by the metric system, which is based on grams. Some drugs that have been available for long periods of time, like aspirin, are rooted in this history. Standard dosing was 5 grains, which is equivalent to 325 mg. Low-dose aspirin was ¼ of 325, which is 1.25 grains, which is converted to 81 mg.

Aspirin has been called the "wonder drug" because it seems to help so many ailments, and more and more benefits of taking the drug continue to be discovered. Its mechanism of action is to block the production of prostaglandins, by inhibiting COX-1 and COX-2. COX stands for cyclooxygenase, which is an enzyme necessary for prostaglandin synthesis. Prostaglandins are tissue hormones involved in blood clot formation, pain, fever, and inflammation. At least we *think* that's what aspirin does.

Aspirin also blocks thromboxane synthesis. Thromboxanes are biochemically related to prostaglandins and formed from them by COX. Thromboxanes play a role in blood clotting. If you take one baby aspirin every day for 10 days, your platelets—the cellular fragments that play a role in blood clotting—will not contain thromboxane. Your blood will still clot because it has other clotting factors, but it will take longer because without thromboxane the platelets will not clump as easily and will require a little more time. For this reason, if you are about to undergo surgery, you'll be asked to stop taking aspirin for at least 10 days prior to surgery. Other anti-inflammatory drugs such as ibuprofen (Motrin or Advil) and naproxen (Aleve) inhibit thromboxane production, too, but the effect lasts only a few hours. Acetaminophen (Tylenol) has no effect on thromboxane. New research on aspirin also suggests that aspirin's role in cardiovascular health is much broader because it has effects on thromboxane and platelets and inflammation through pathways that may not be fully understood.

Which brings up another point about medicine: we don't always know *how* or *why* medicines work, but that they *do* work. Like Viagra, some drugs come to fruition through serendipity. In many cases, drug development follows progressive steps, but in other cases, we learn more and more about the drug after the fact when thousands of people use the particular drug, or a drug is used for another purpose than originally intended. This is known as prescribing *off label*.

Thalidomide was a drug used between 1957 and 1962 to treat morning sickness. It worked well in this regard. But it was withdrawn from the market when infants were born with hands and feet attached close to the body, resembling the flippers of a seal. The medical term for the condition was *phocomelia*, derived from the Greek work *phoke* meaning *seal* and *melos* meaning *extremity*. Thalidomide has seen a resurgence as a drug used to treat cancer and Hansen disease (leprosy).

Another common drug is penicillin. Even though penicillin has been around for over 90 years, scientists are still finding out new things about the drug. It was first thought that penicillin was effective against certain strains of bacteria because the drug interfered with the bacterium's ability to build a cell wall. Without a cell wall, the contents of the bacterium spilled out, rendering it unable to reproduce or basically killing it. However, new findings suggest that the bacteria get caught in a cycle of building up and tearing down cell wall structures it cannot use. Either way, the drug works.

Taking a course in pharmacology would expose you to many other medicine mysteries. One other worth noting is anesthesia, the total loss of sensation that results from pharmacologic depression of the nervous system. This is a good thing if you are having surgery because temporarily losing consciousness as we "go under" allows painful operations to take place. Early anesthesiology has its roots in 1846 when a doctor at Massachusetts General Hospital held an ether flask near a patient's nose until he lost consciousness. Today, anesthesia is a relatively safe procedure with a fatality percentage of 0.0005 percent. That said, the mechanism of action of many drugs used in anesthesia is not known. General anesthetics may act on specific proteins that affect nerve cell excitability and nerve impulse transmission. With the correct dosage, the unconsciousness is reversible, and you awaken just beautifully.

Propofol is a familiar anesthesia drug because it made headlines in 2009 pertaining to the death of pop star Michael Jackson. Propofol was approved for use in Missouri to execute prisoners, but it was not actually used because drug manufacturers in the European Union threatened to limit the drug's export to the United States if it were going to be used to kill people. Although the mechanism of action is not fully known, we do know that it produces general anesthesia or sedation and is commonly used for surgery. Anesthesia is a different state than normal, restful sleep, so it should never be used to treat sleeplessness, as it was used in the case of Jackson.

Drug names can be unwieldy. Why are there so many names for the same drug? Why are the names so difficult? These are excellent questions. Drug names generally have four different names:

1. There's the *chemical (scientific) name* that describes the drug's atomic and molecular structure.
2. Then, there's the *generic name* that is usually an abbreviation of the chemical name.
3. That's followed by the *official name*, which is listed in the United States Pharmacopoeia and National Formulary.
4. Lastly, there is the *trade (brand) name*, which is selected by the pharmaceutical company and protected with copyright.

Let's use a common example found in our home medicine cabinet: ibuprofen. Ibuprofen is a non-steroidal anti-inflammatory drug (NSAID) taken for pain. Its molecular formula is $C_{13}H_{18}O_2$. (As a comparison, the molecular formula for water is H_2O.)

chemical (scientific) name = ibuprofen
generic name = non-steroidal inflammatory drug (NSAID)
official name = 2-[4-(2-methylpropyl)phenyl]propanoic acid
trade (brand) name = Motrin, Advil, Nuprin, and others

Yes, it's complicated; but chemists and drug developers need this sort of information.

Special Drug Programs

There are special FDA programs to help develop drugs when there is no reasonable expectation that the cost of developing and distributing the drug will ever be recovered by the drug's sale. Think of rare ("orphan") diseases that affect a small population yet are devastating to the person and their loved ones. Amyotrophic lateral sclerosis (ALS), commonly known as Lou Gehrig's Disease, is an example. When a disease is named for a person it is called an eponym. Lou Gehrig (1903–1941) was a New York Yankees baseball player who brought worldwide recognition to the disorder when he abruptly retired after being diagnosed. The disease was first described by French neurologist Jean-Martin Charcot, in 1869. You might also remember that in the Introduction, Jim Obergefell's husband John Arthur died of this dreadful disorder. According to the ALS Association, as many as 30,000 American currently have the disease. Astrophysicist Stephen Hawking died in 2018 of ALS, a disease that totally paralyzes a person while keeping their mind intact and fully functioning. The average life expectancy from diagnosis to death is two to five years. While some do live longer, only about 20 percent will live longer than five years. It is a horrible disease robbing the person of movement while the person is fully aware of what is happening.

The Orphan Drug Act of 1983 exists for such diseases to facilitate drug development for rare diseases. Basically, the FDA allows drug development acceleration and offers "provisional approval" for life-threatening diseases with the goal of possibly saving a person's life. Special FDA programs such as this allow compassion access to unapproved drugs, some of which are free to the person receiving them. The National Organization for Rare Disorders (NORD) lists information on clinical trials and research studies so patients and physicians can see what options are currently available.

Drug Addiction

Addiction is a complex disease. Addiction is not a crime. By definition, it is the habitual psychological or physiological dependence on a substance that is beyond voluntary control. Many people misunderstand drug addiction and how people become addicted to drugs. A common misperception is that people who use drugs are "second-class citizens" or that they simply lack the necessary willpower to quit. As an academic discipline, addiction science studies drug abuse; discovers how drugs affect the brain and other organs; and investigates the mechanisms for drug dependence, tolerance, and recovery. Despite robust scientific evidence surrounding drugs, the brain disease model of addiction still remains controversial. That it's a controversial issue harkens back to an earlier time when mental illness was viewed by society in such light.

Drug addiction is a serious situation currently sweeping the United States. According to the 2017 National Survey on Drug Use and Health (NSDUH), 18.7 million (1 in 12) American adults had a substance use disorder (SUD) and of that number, 46.6 million (1 in 5) had a mental illness. More than 8.5 million had both a substance abuse disorder and a mental illness. The cost to American society is more than $200 billion a year when healthcare, criminal justice, legal, and lost workplace production is figured into the dollar amount. The federal government under past and current administrations is directing billions of dollars in grants to support treatments for the drug addiction crisis. These grants send funds to individual states to finance programs that fight and treat drug addiction. For the fiscal year 2021, the U.S. Senate recommended budgeting $1.8 billion for the Substance Abuse Prevention and Treatment (SAPT) Block Grant.

Much more is known today about drug addiction, but we still have a long way to go. Research has shown that 50 percent of the time, addiction is heritable and other research has shown that both genetics and the environment may play equal roles in addiction development. The question arises, "What is the difference between genetics and heredity?" A hereditary disease can be passed down from one generation to the next. A genetic disease is the result of individual genes that make up a person; and genetic diseases may or may not be hereditary.

Stages of human development also play a role in addition. For example, if drug abuse or alcohol abuse occurs before the brain is fully developed, there is also an increased risk of addiction later in life. This drug addiction risk is in addition to the brain damage that could also occur during development. Full brain development takes 18–25 years and drugs have the potential to change the architecture of growing brains.

Substance abuse experts created a scale to measure the harm and potential misuse of different drugs to come up with an addiction scale. The scale averages three factors: (1) pleasure, (2) psychological dependence, and (3) physical dependence. Scoring was on scale of 0–3; 0 was the least addictive while 3 was the most addictive. From most addictive to least addictive, here are the top five with their scores: 1. Heroin—3; 2. Cocaine—2.4; 3. Nicotine—2.2; 4. Barbiturates—2; and 5. Alcohol—1.9.

Addiction cuts across all population demographics. This includes all ages, sexes, ethnicities, criminal justice history, employment status, and socioeconomic backgrounds. Drug addiction is an equal opportunity malady.

Alcohol

Alcohol is a legal, widely abused, addictive drug. It is produced by the natural fermentation of sugars and has an intoxicating effect. Its effects and metabolism are dependent on agents besides alcohol dehydrogenase, which was briefly discussed at the beginning of this chapter. Many other factors and enzymes—aldehyde dehydrogenase (ALDH), cytochrome P450 and catalase—also influence alcohol metabolism and consequently blood alcohol concentration (BAC). Moreover, variation in genes for these enzymes influence how much alcohol is consumed, the extent of alcohol-related tissue damage, and alcohol dependence. The metabolic pathways utilizing these enzymes are quite complicated. Dehydrogenase enzymes play a role in cellular respiration, which is the process by which our living cells produce energy through complex biochemical pathways. The message is that alcohol metabolism occurs at the cellular level.

Environmental factors such as how quickly alcohol is consumed, how much food is in the stomach, and the type of alcoholic beverage have to be considered. The liver is the main organ for metabolizing alcohol. Again, this is where first-pass metabolism comes into play. Other organs, however, such as the brain, do not contain alcohol dehydrogenase. In this case cytochrome P450 (found in the endoplasmic reticulum of a cell) and catalase (found in peroxisomes of cells) metabolize alcohol. Now, things are getting very complicated. Suffice it to say, those cellular organelles that you zoned out on in school or when reading this book, are very important. Science rocks!

Genetic factors also play a role in alcohol metabolism. As discussed earlier, sex plays a part in how much alcohol dehydrogenase is produced. However, the effects of equivalent amounts of beverage alcohol also varies among populations. Asian and Native American populations have a lower alcohol tolerance than European populations as a result of genetic differences related to alcohol dehydrogenase enzyme production.

To understand this, we once again need to look at evolution. Scientists refer to beverage alcohol as dietary ethanol. As a substance, it is a toxin. We don't need it, yet we can drink it in very small amounts without getting sick. Too much, however, and we're drunk and sick. Alcohol dehydrogenase (ADH4) is produced in our stomach. All primates have ADH4, but not all primates can metabolize alcohol equally. Paleogeneticists sequenced genes from modern day primates and analyzed how they evolved over time. ADH4 was found in primates 50 million years ago. These primates could metabolize small quantities of ethanol slowly. However, about 10 million years ago, a single genetic mutation occurred within our common ancestors, chimpanzees and gorillas. The ADH4 proteins in these primates were much more efficient at metabolizing ethanol. This change occurred at the same time as an abrupt climate change, which also altered the ecosystem and food sources. This "advanced" protein would have enabled early primates to eat rotting, fermenting fruit and survive.

Alcohol affects at least two neurotransmitters: glutamate and gamma-aminobutyric acid (GABA). Glutamate is an excitatory neurotransmitter that nerve cells use to signal other neurons. In biochemistry, this doesn't mean that it's going to excite you, it means that the neuron it acts upon will likely have an action potential. Think of an action potential as a neuron firing; it is completing a circuit much like flipping a switch turns on a light. GABA is an inhibitory neurotransmitter. Inhibitory neurotransmitters decrease the likelihood of neurons generating action potentials. So, glutamate excites neurons and GABA inhibits neurons. With a drink or two, alcohol acts on GABA and you become relaxed. More drinks create a sedative effect. Too many drinks create a next-day hangover. Hangovers are unpleasant experiences that can last for 24 hours or more and are characterized by headache, dry mouth, dizziness, fatigue, upset stomach, diarrhea, vomiting, sweating, nausea, and other ill feelings. Other than knowing that alcohol is the culprit, from a physiological standpoint, the mechanisms behind the hangover are still poorly understood. One hangover theory is linked to alcohol withdrawal: the body experiences a counterbalance with glutamate and GABA and is on overdrive to "reset" the neurotransmitter receptors. Several factors are at play, though, including dehydration, immune system changes, impaired glucose metabolism, sleep deprivation, and others.

Cute sayings have been passed down through the generations as "memory devices" to thwart hangovers. Here are a few: "Wine before liquor, never been sicker." "Liquor before beer, there is no fear." "Mixing drinks gives you a hangover." Is there any truth to these statements in terms of preventing a hangover? The answer is no. It doesn't matter what you drink, but how much you drink that determines whether or not you have a hangover. The only prevention is to not drink alcohol.

What about hangover remedies? The person who invents such a magic pill to cure hangovers would certainly save the world from Mad Dog Malady. However, there is no cure; the treatment plan is rest, nutrient replenishment, fluid intake, and time.

Pervasive inaccurate assumptions regarding drinking also abound. One such assumption revolves around urinating and goes something like this. If you have been out drinking, you should hold your pee as long as possible, because the moment you do decide to go to the bathroom and relieve yourself is the turning point for a night filled with bathroom runs. Not diarrhea runs, but the need to urinate often and with frequency throughout the night. The truth is that those flood gates would open at some point even if you didn't make the mad dash, because there is a point of no return. Once your bladder reaches its over-full capacity, the external urethral sphincter (circular muscle) will relax to cause urination. Normally, you can consciously control this sphincter and thereby control urination. You do this day in and day out and you can even try this exercise at home: While sitting on a toilet, stop and start your flow of urine. You are able to do this by consciously controlling your external urethral sphincter. But, if you've engaged in a night of drinking copious amounts of alcohol, you will wet your pants if you choose to wait too long.

Many people drink alcohol to warm themselves up. You may be reading this and thinking, yes, I've drunk a glass of wine and it definitely warmed me up. Alcohol does indeed have an effect on our blood vessels. At lower levels, alcohol acts as a *vasodilator*, which means the blood vessels relax and expand, which releases heat and you feel warmer. At even higher levels, alcohol now acts as a *vasoconstrictor*, shrinking blood vessels and increasing blood pressure. Research done on the effect of alcohol on vascular function showed that regular or light-to-moderate alcohol intake increases vasorelaxation and lowers blood pressure, but chronic heavy alcohol consumption causes vasoconstriction and increases blood pressure.

People will feel warm when blood vessels dilate, which leads to the false belief that they *are* warm. Heat is better able to escape when blood vessels expand. On a hot day or after exercise, blood vessels dilate to release excess body heat. For this reason, there are many cases of alcohol-induced hypothermia and fatal accidental hypothermia. If you are out ice fishing and tossing back a few beverages, remember to keep your clothes on. People have died from hypothermia because they felt warm and shed their protective outerwear.

Opioids, Marijuana and Tobacco

We are currently in the midst of an opioid epidemic and opioid crisis. Both terms are used interchangeably to describe the widespread increase in the use of prescription and non-prescription opioid drugs across the United States and Canada. First some terms. Opioid and opiate are often used to mean the same thing. From a purist's perspective, an opioid was the original term denoting synthetic narcotics that resembled opiates. Today, opioids refer to both natural and synthetic narcotics. Opiates are preparations derived from the opium plant, commonly known as poppies. When opium is used for nonmedical purposes and sold illegally, it is a narcotic. When used medicinally, it is a drug that relieves pain (analgesic), causes drowsiness (hypnotic), induces sweat (diaphoretic), and treats diarrhea. Compounds from the opium plant include morphine, codeine, and poisons.

As a drug class, opioids are central nervous system (CNS) depressants. This means that they act on the brain and spinal cord, causing relaxation and sedation. Alcohol is a CNS depressant, but other examples include benzodiazepines (Valium and Xanax) and barbiturates (tranquilizers like Amytal and Nembutal). CNS depressants are used in medicine because they are quite effective.

Because they work on the CNS, opioids are narcotic analgesics, meaning the drugs are derived from opium or opium-like compounds and have potent pain-relieving effects. Narcotics, however, also cause significant changes in mood and behavior with an increased potential for dependence and tolerance. Common narcotics include heroin, oxycodone (OxyContin, Roxico-done, and Oxecta), hydrocodone (Vicodin, Lorcet, Lortab), morphine (MS Contin, Oramorph SR, Avinza, and Arymo ER), hydromorphone (Dilaudid and Exalgo), fentanyl (Actiq, Fentora, Duragesic, Subsys, Abstral, Sublimaze, and Lazanda), codeine, methadone (Dolophine and Methadose), meperidine (Demerol), oxymorphone (Opana), tramadol (ConZip, Ultram, and Ryzolt), propoxyphene (Darvon), carfentanil, and buprenorphine. Whew! When taken orally, opioids take effect within 30 minutes and the effects can last 24 hours.

Time for a little anatomy and physiology. Connecting our big brains with our spinal cord is the brainstem. The brainstem is composed of three structures called the midbrain, pons, and medulla oblongata. Sometimes the medulla oblongata is just called the medulla. The midbrain contains areas involved with eye movement and auditory and visual processing. Good to know. The pons and the medulla control the rate and depth of breathing. The pons also controls auto-nomic—the scientific term for automatic—functions such as heartbeat. Here's where it gets very interesting: the pons and medulla oblongata are chock full of opioid receptors, little pro-teins sitting on the cell surfaces. The opioid receptors latch onto opioids. When "docked," the opioids change the way the cells operate, slowing or stopping breathing. This affects automatic, non-voluntary breathing.

You know, however, that you can voluntarily hold your breath. Take a second or two and take a deep breath, hold it, and then let it out. You did this voluntarily. Opioid receptors are also found outside the pons and medulla in areas controlling voluntary breathing. So, when opi-oids attach to those receptors, connections are made, the cell reacts, and breathing is affected in another way. Furthermore, opioid receptors are found in other tissues throughout the body, so they have the propensity to affect various cells in varying ways. Telltale signs that a person has recently taken opioids include: suppressed gag reflex, decreased blood pressure, abnormal heart rhythm, dilated blood vessels in the arms and legs, and constricted eye pupils. Opioids also affect gut motility and gut flora (the microorganisms living in the intestines). However, these effects would not be immediately apparent if you came upon an overdose victim.

Heroin is an illegal highly addictive analgesic drug derived from morphine. Most heroin is injected directly into a vein. As a narcotic, it produces feelings of euphoria—intense excite-ment and happiness. In fact, its name is from the late 19th century German word *Heroin*, which is derived from the Latin word *heros*. *Heros* means *hero*, and heroin gives the user a self-esteem boost, resembling that of a hero. It is a synthetic drug, which means it is made and does not occur naturally in nature. Drug users and drug dealers make it from the resin of poppy plants. First, the sap-like opium is removed from the pod of the poppy flower. The opium is refined to make morphine and then the morphine is refined to make heroin. Like aspirin, heroin came to us originally from the Bayer pharmaceutical company around 1898. Bayer manufactured it as a treatment for tuberculosis (TB) and as a remedy for morphine addition.

Fentanyl is a fast-acting opioid. Once injected, it can cause the diaphragm (dome-shaped muscle of breathing) and chest muscles to "freeze up," causing *wooden chest syndrome*. Chest and abdominal muscles become very rigid, making breathing difficult. High doses of fentanyl during anesthesia can also cause this, making ventilation difficult. Healthcare professionals know to monitor the dosage; street drug users may not be aware. Wooden chest syndrome can be reversed by naloxone (Narcan).

Opioid addiction is a major health crisis. Opioids can take away suffering and make us feel like everything is okay and life is just peachy. Who doesn't want to feel as though everything is always fine and that the sky is always blue? It's like being a glass is half-full kind of person, right? Yes and no. With time, the brain also adapts to opioid use. So, when a person stops taking opioids, the suffering returns, but it seems much greater than it was to begin with. Now the brain itself is producing its own kind of suffering. And the glass is half-empty all the time.

You are perhaps aware of a medication called Narcan. Narcan (naloxone) comes in two forms: an injectable medicine or a nasal spray. Both forms are for the emergency treatment of suspected or known opioid overdose. It begins working within minutes. According to the CDC, life-threatening overdoses from narcotics have reached epidemic levels across the U.S. Signs of overdose are impaired breathing (respiratory depression), extreme sleepiness, and unresponsiveness.

How does Narcan work? Narcan is an overdose-reversal medication that can be inhaled or injected. Once administered, it travels through the bloodstream and attaches to the opioid receptors with a greater affinity than the actual opioid, which means that if Narcan is already "sitting" on the receptor, the opioid can't sit there and render its effect. Several doses may be necessary. Regardless, the person must still receive medical attention after Narcan has been given. Pharmacists in 46 states are able to dispense Narcan without a prescription. Interestingly, the number of naloxone prescriptions doubled from 2017 to 2018, and for every 70 high-dose opioid prescriptions filled, only one naloxone prescription is dispensed. A Narcan kit costs $20 to $40. It is estimated that a single overdose death costs U.S. taxpayers about $30,000; but you can't put a price on a life. On April 19, 2019, the U.S. Food and Drug Administration approved the first generic Narcan.

Addictive narcotics have a history of being trapped in a vicious cycle. In the 1850s the United States had an opium problem. So, morphine was prescribed because it was viewed as less addictive than opium. From here, morphine became the new addictive drug. Heroin, thought to be less addictive than morphine, was then used as a treatment for morphine addiction. Are you following the steps here? First, there was opium, which was replaced by morphine, which was replaced by heroin. At this stage, the addictive drug du jour was heroin. Then, in 1937, German scientists developed methadone as a surgical painkiller. Methadone was introduced into the United States in 1947 as a drug to treat heroin addiction. The problem, like the other "non-addictive substitutes" was that methadone was even more addictive than heroin. Methadone is simply a substitute addiction that takes the place of other opioids.

So why can't somebody just try it once and be done with it? That's an excellent question. Unfortunately, addiction is complicated and there is a genetic component to it. Once individual genes come into play, individual results ensue. While it is true that a one-time introductory experience with heroin won't make you an addict, that first-time use could create such a feeling of euphoria and elation that it leaves you wanting more. And the more you use heroin, the more rapidly your nervous system adjusts to the changes it caused, and now there is another vicious cycle: more drug is needed to achieve the same results. The National Institute on Drug Abuse estimates that roughly 23 percent of people who *try* heroin will become addicted. The moral of the story is obviously to *never try* heroin to begin with.

But what mechanisms are already in place in humans that enable this? The answer lies in the *human* part of our biology. Specifically, we have this powerful neurohormone called dopamine that is secreted naturally. Dopamine is produced in several areas of the brain and released by the hypothalamus, a tiny gland in the brain. This is a very important neurohormone because it helps us move, affects our memory, and influences our mood. These are all good things! But it can also lead to addictive behaviors. Deep in our DNA is coding for seeking positive outcomes.

While dopamine is a feel-good neurohormone, dopamine depletion causes Parkinson's disease. Dopamine is a precursor molecule to norepinephrine and epinephrine, neurotransmitters involved in actions as diverse as increasing or decreasing heart rate and strength of contraction, constricting blood vessels and dilating blood vessels, and relaxing smooth muscles in our bronchial tree and intestines to affecting blood pressure and responding to stress. Epinephrine was discussed earlier in this chapter in our discussion of pseudoephedrine. Dopamine motivates us to do something and plays a role in the reward centers of the brain. Dopamine is also released after exercising giving us the "runner's high," during pregnancy to give the "pregnancy glow," and during other pleasurable activities. Dopamine makes life enjoyable. Previously, it was thought that endorphins caused the so-called runner's high. However, endorphins may be too large to quickly cross the blood brain barrier—a network that acts as a gatekeeper. It is more likely that endorphins dampen pain. Recent research in mice suggests endocannabinoids may also play a role in the runner's high.

Let's circle back to the reward piece. Drugs such as heroin, cocaine, and nicotine (the powerful chemical in tobacco), boost dopamine levels. Cocaine works the opposite of alcohol and is very specific in its actions: Cocaine blocks the recycling of dopamine (and norepinephrine) so that they remain in our bloodstream. Thus, the reward-seeking part of our brains get a spike in dopamine and we feel pretty good. Alas, it also makes us seek out that "thing" that gave us the boost or that adrenaline rush. Consequently, drug cravings begin, leading to addiction.

Trying to stop the addiction is grueling. To illustrate, regular heroin users can experience withdrawal symptoms within 24 hours of their last dose. Withdrawal can be agonizing. So agonizing that the reward centers in the brain "tell us" to avoid such a horrible experience. Heroin withdrawal isn't fatal, but it can be extremely painful. Signs and symptoms include nausea, vomiting, chills, diarrhea, tremors, anxiety, agitation, and cravings. It's like the worst case of flu that you've ever experienced. And then some. Treatment centers can help ease a person through the withdrawal. Once addicted, though, it is a lifelong battle.

How do opioids kill? The most powerful opioids on the streets are heroin, morphine, and fentanyl. These opioids are really good at relieving pain and evoking intense—*intense*—feelings of pleasure. But overdoses are deadly because of their rapid effects. Namely, breathing is affected. Opioids kill people by slowing both the rate and depth of breathing until breathing stops.

How does marijuana fit in and what causes a high? Well, like alcohol, its actions are widespread throughout the body and like cocaine it has very specific actions—communication between neurons. The active ingredient in marijuana is tetrahydrocannabinol (THC), which is a psychoactive agent that enhances communication throughout brain cells. Everything all at once is turned on—you've flipped on all the light switches in the house and everything is better. Brain adaptation does occur and with time, the number of sites activated by THC diminishes.

You're likely thinking, "I've heard a lot about CBD and marijuana." What's the difference between CBD and THC? That's an excellent question! CBD is the abbreviation for cannabidiol. CBD oil is the hot new product in states that have legalized medical marijuana. It is non-intoxicating. Well, then, what's the purpose if I can't get high? It's been touted to treat everything from epileptic seizures and anxiety to inflammation, muscle contractions and stiffness to sleeplessness. Sounds like a wonder drug, doesn't it? It actually just chills you out so it's a nice stress reliever. Don't go running out to your nearest dispensary, though. CBD is not well-regulated, so many products don't contain the promised amounts on the label and some CBD products have been found to also contain THC. The science surrounding both THC and CBD is still murky, too. The reason is that laws have prevented researchers from studying THC and CBD; now, there are more avenues to study marijuana. Stay tuned for more information as the scientific method goes into action.

We've already seen how addiction causes both physiological and psychological dependence while the drug changes the architecture of our cells and body. Did you know that nicotine is *just as addictive* as heroin, cocaine, and alcohol? Nicotine is found in all tobacco products and nicotine withdrawal looks very much like withdrawal from other drugs. Since nicotine is found in tobacco, tobacco in all forms is addictive. Regarding the chemistry behind nicotine, nicotine is a poisonous volatile alkaloid. As a volatile substance, it is easily evaporated at normal temperatures. As an alkaloid, it is a nitrogenous organic compound of plant origin that has profound physiological actions. Many drugs, such as morphine, are alkaloids. Nicotine found in e-cigarettes/vaping may be more addicting than nicotine in ordinary cigarettes.

When nicotine is inhaled in tobacco smoke or touches the mucous membranes of the mouth in smokeless tobacco, it enters the bloodstream within seconds. This immediately increases the heart rate, the amount of blood ejected by the heart ventricular chambers, and the amount of oxygen used by the heart muscle. It also causes euphoria, heightened alertness, and paradoxically a sense of relaxation. Acting on the brain like other addictive drugs, it floods the brain's reward circuits with dopamine, our feel-good chemical. When the good feelings go away, we want more; thus, it is highly addictive. Over time, tolerance develops, meaning that more nicotine is needed to achieve the same results. Tobacco users then require a certain nicotine level to keep them comfortable; this is nicotine dependency. Nicotine withdrawal causes restlessness, irritability, anxiety, difficulty concentrating, and nicotine cravings. This addiction is the primary reason for most tobacco use and a major cause of morbidity and mortality.

Quitting isn't so easy. There may be other chemicals in tobacco that make quitting difficult. Studies in animals show that tobacco smoke causes chemical changes that cannot be explained by nicotine alone. Quitting smoking may be harder than quitting cocaine or heroin. Yes, you read that correctly. There are 28 different studies that have shown that in people trying to quit their particular addiction, 18 percent were able to quit alcohol use, 40 percent were able to quit opiates or cocaine, but only 8 percent were able to quit smoking. Talk about triggering the brain's hot spots.

If you're still thinking that it's easy to quit, try this experiment: Choose something that you absolutely love, whether it is a thing or a person you're in love with. Then, try to go a day or a week or a month without that thing or person. Or, just try to fast—go without eating—for one day. Or, the coup de grace: Put down your cell phone for an entire day. In each of these scenarios, it would be rough. You think about the thing you can't have or the person you can't see a lot. The Robert Palmer song "Addicted to Love" might be playing in your mind right about now.

> You like to think that you're immune to the stuff, oh yeah
> It's closer to the truth to say you can't get enough
> You know you're gonna have to face it, you're addicted to love.

The same is true for a drug addiction, only much, much worse.

The opioid crisis is real. But make no mistake: the opioid epidemic is a homegrown problem. We have an opioid epidemic in the United States because drug companies made the product and then convinced doctors—who are the only people who can prescribe the drugs—that they weren't as addictive as other opioids. First, shame on the drug company. Second, shame on doctors for not remembering their basic pharmacology. Opioids are controlled substances. Controlled substances are drugs that are subject to the U.S. Controlled Substances Act of 1970, which regulates the prescribing and dispensing, as well as the manufacturing, storage, sale, or distribution of substances assigned to five schedules according to their (1) potential for or evidence of abuse, (2) potential for psychic or physiologic dependence, (3) contribution to a public health risk, (4) harmful pharmacologic effect, or (5) role as a precursor of other controlled substances.

The Federal Drug Administration (FDA) oversees drugs in the United States and they released the drug classifications, also called drug schedules, under the Controlled Substance Act. These controlled substances are categorized one through five, using Roman numerals, I, II, III, IV, and V, with the lower numbers assigned to drugs with the most risk toward abuse and dependency. Here's an accounting of the drugs:

- Schedule I controlled substances have no currently accepted medical use. Examples are heroin and lysergic acid diethylamide (LSD).
- Schedule II controlled substances have a high potential for abuse and may lead to severe psychological or physical dependence. Examples are morphine, opium, codeine, oxycodone (OxyContin, Oxycodone, and Percocet) and fentanyl (Sublimaze, Duragesic).
- Schedule III controlled substances have potential for abuse and may lead to moderate or low physical dependence, but it's less than schedules I and II. Examples include combination products containing 15 milligrams of hydrocodone per dosage unit (Vicodin), not more than 90 milligrams of codeine per dosage (Tylenol with codeine), ketamine, and anabolic steroids such as Depo-Testosterone.
- Schedule IV controlled substances have a low potential for abuse relative to Schedule III. Examples include alprazolam (Xanax), diazepam (Valium), and lorazepam (Ativan).
- Schedule V controlled substances have a low potential for abuse relative to substances listed in Schedule IV. Examples include cough syrups containing not more than 200 milligrams of codeine per 100 milliliters or per 100 grams (Robitussin AC and Phenergan with codeine).

Let's return to those Schedule II drugs. A border wall isn't going to keep opioids from entering the United States. Opioids come through airports, ships, trucks, and legal points of entry. The COVID-19 pandemic curtailed shipments. However, opioids also come from right here in the U.S. of A. Currently, more people in the United States die from opioid overdose than from car accidents.

One physician from Florida, Dr. Barry M. Schultz, is currently serving 157 years in prison for fueling the opioid epidemic. In one 16-month period, he prescribed 800,000 opioid pills from his office pharmacy. He even prescribed patients 60 Oxycodone tablets per day! Pharmaceutical companies own part of the problem, too. Purdue Pharma and Insys Therapeutics are very much like drug cartels because they also bribed physicians (with money, expensive dinners, strip club visits) to prescribe the drugs.

The Sackler Family is behind Purdue Pharma, and in 2018 the company admitted in the

United States District Court for the Western District of Virginia that they had misled doctors about just how addictive Oxycontin could be, despite knowing the addiction risks. Between 2008 and 2016, the Sackler Family paid itself more than $4 billion in opioid profits. As a defense, the Sackler Family blamed the abusers, stating they were reckless criminals for becoming addicted. Blaming the consumer! The Sackler Family is among the richest families in America: You'll find Sackler Wings at the Guggenheim Museum, Metropolitan Museum of Art, and the American Museum of National History. They also have their name on buildings of higher education, including Cornell, Columbia, Tufts, George Washington, and McGill. You'd likely think these are good people. They profited from creating an epidemic and just want to pretend they are good people. We're smarter than that. And it's a Schedule II drug; doctors should have been smarter, too. In March 2019, New York State Attorney General Letitia James filed a civil lawsuit against eight members of the Sackler family. In August 2019, Purdue Pharma offered $10–12 billion to settle the opioid claims, and in September 2019, the company declared bankruptcy. After filing for bankruptcy, Purdue Pharma launched a $23.8 million ad campaign for the claims program in an effort to identify the nearly 3,000 lawsuits against the company. In October 2020, Purdue Pharma pled guilty to three federal charges for its role in America's opioid crisis. Purdue Pharma and its owners, the Sackler family, agreed to the largest penalties against a pharmaceutical company ever levied: a criminal fine of $3.544 billion and $2 billion in criminal forfeiture. Purdue also agreed to a civil unsecured bankruptcy claim for $2.8 billion. The Sackler family also agreed to personally pay $225 million in damages. Furthermore, the company agreed to emerge from bankruptcy in an entirely new form, that of a public benefit company (PBC) whereby company proceeds will go to the benefit of state and local opioid abatement programs.

War on Drugs

In June 1971, President Nixon declared a "war on drugs." While in office, he increased the number of federal drug control agencies and enacted no-knock warrants and mandatory sentencing for offenders. (The no-knock warrant is the same legislation Louisville police officers used to enter Breonna Taylor's apartment, ultimately killing her. Unfortunately, police officers were at the wrong apartment for a suspect who was already in custody.)

Rockefeller Drug Laws were enacted in May 1973 in New York. Named after governor Nelson Rockefeller, who wanted to appear "tough on crime," they were harsh laws. Rockefeller used to see drugs as a social problem rather than a criminal one. However, he had presidential aspirations and didn't want to appear too soft, thus Rockefeller Drug Laws were the toughest in the country. The state of Michigan enacted similar laws in 1978. For example, the 650-Lifer Law called for life imprisonment without parole for the sale, manufacture, or possession of at least 650 grams (1.43 pounds) of cocaine or opiates. People were being imprisoned for nonviolent crimes.

In reality, the war was on black people and immigrants. Marijuana, opium, and psychedelic drugs have been used for thousands of years. However, during the 1870s, anti-opium laws were aimed at Chinese immigrants. During the early 1900s, anti-cocaine laws were aimed at black men in the South. In the 1910s and 1920s, anti-marijuana laws were directed at Mexican migrants. Fast forward to today and a disproportionate number of Latino and black people are serving prison sentences related to drugs.

In 2009, President Obama talked about the problem of Nixon's slogan, declared it not useful, and stated his administration would not use it. Today, we are also experiencing a shift in legalizing marijuana and decriminalizing drug addicts. Managing opioid addiction is beneficial, as demonstrated by France, which has reduced heroin overdose by 80 percent. In the United States, approximately 88,000 people (62,000 men and 26,000 women) die from alcohol-related causes annually. Worldwide that total is 3.3 million deaths or 5.9 percent of all global deaths. Contrast that with the number of deaths attributed to marijuana. That number is zero. However, according to an article in the June 7, 2019, issue of *Newsweek*, a Louisiana woman may have died from vaping THC (tetrahydrocannabinol), the active ingredient in marijuana. She would be the first marijuana death ever recorded in the United States. Vaped THC is likely to blame, but more information is needed to tease out the facts. As of the writing of this book, vaping-related deaths are on the rise.

Changing Health Behavior

Changing health behavior is extremely difficult and involves forming habits. How long does it take to form a habit? Some research suggests 21 days. However, other research suggests 18 to 254 days! In practice, habit formation is complicated. Ask any ex-smoker if they still crave cigarettes, and you'll find the answer is likely "yes." And, if they decided to pick up a pack after 20 years of non-smoking, they very likely would form the smoking habit again in under three weeks. It is hard to change. It is very common for people not to do what is best or healthy for themselves. Research has also shown that 38 percent of patients do not follow short term regimens, 43 percent do not follow long term regimens, and 75 percent do not follow life-style changes. Changing behaviors requires work and conscientious effort.

Behavioral change is a process whereby one experiences successes and failures. Knowing that, people who are trying to change a health behavior should expect relapse and failure. Thus, preparations can be made for such an event. Behavioral therapists suggest that individuals use more techniques longer and replace *unwanted* behaviors with *wanted* behaviors. People strive for perfection, which means they might not attempt a change for fear they can't fully accomplish their goal. This isn't very good either because it leads to immobilization and no-change results. Furthermore, it is critical to remember that anything accomplished is better than nothing done perfectly. In order to ensure success, individual goals need to be reasonable because failure to set realistic goals leads to failure. Recording successes, as well as failures, also leads toward progress. Patience is another key component to behavioral change. Too many people quit too soon. Therefore, it is important to employ several techniques to institute change. People are more likely to be successful changers if they are given a choice of options and techniques.

Health can be linked to behavior just as disease is also related to genetics, social circumstances, and environmental pressure. Consider what behaviors people should be held personally accountable for. Did you come up with any? The main issue here is justice. Is it fair? We have a health crisis that is costing money and lives and yet this country does not want a universal health care system. We're totally okay with criminals being given a right to a lawyer, but we're not okay with citizens having the right to a doctor? The Sixth Amendment of the U.S. constitution guarantees the rights of criminal defendants, including the right to a public trial without unnecessary delay, the right to a lawyer, the right to an impartial jury, and the right to know who your accusers are and the nature of the charges and evidence against you. There is no amendment for the right to healthcare. The July 4, 1776, Declaration of Independence does, however, state that "We hold these truths to be self-evident, that all men are created equal, that they are endowed by their Creator with certain unalienable Rights, that among these are Life, Liberty and the pursuit of Happiness."

Pervasive within any change model are persistent myths. To begin, it is a myth to think that change is simple and that changing the desired behavior—whether it is quitting drugs, not drinking alcohol, or limiting your time on the cellphone—simply requires willpower. The phrase "I've tried everything, and nothing works" is also commonly heard. The problem here is that the individual did not use enough techniques at the right time for long enough to institute altered behavior. Another myth is that people do not really change. In actuality, they often do, they just often fail first. If we weren't capable of change, therapists, counselors, psychologists, psychiatrists, social workers, mental health professionals, and myriad other professions would cease to exist. As biological organisms, if we didn't change, we, too would cease to exist.

TAKE-HOME MESSAGE

Drug tolerance and addiction have a basis in individual biology. Some drugs, like alcohol and tobacco, are socially acceptable yet highly addictive. Others, like controlled substances, are so potent that they have to be monitored very closely.

> *Penalties against drug use should not be more damaging to an individual than the use of the drug itself.*
>
> —Jimmy Carter (1924–); 39th president of
> the United States, philanthropist

9

The Final Chapter

Death and Dying

U.S. Landmark Case: Cruzan v. Missouri Department of Health, 497 U.S. 261 (1990); (5–4 decision)

In January 1983, Nancy Cruzan was in a car accident in which she lost control and was thrown from the car. She landed face-down in a ditch filled with water. She was found with no vital signs but was resuscitated and hospitalized. She spent the next three weeks in a coma and was diagnosed as being in a persistent vegetative state (PVS), and a feeding tube was inserted. Five years later her parents asked doctors to remove the feeding tube. The hospital required the Cruzans to get a court order before they would remove the tube as that removal would cause Nancy's death. The Cruzans stated to the trial court that their daughter had told her roommate that she did not want to be kept alive in such circumstances. The local trial court agreed to allow the tube to be removed. On appeal by the state of Missouri, the Missouri Supreme Court reversed the trial court and held that no one could refuse treatment for another person absent a living will or clear, convincing and reliable evidence, which was not the case here.

The Cruzans then appealed to the U.S. Supreme Court, which ruled on the case in 1990. The Court found that a *competent* person has the constitutionally protected right to request or refuse lifesaving treatment. On the other hand, an *incompetent* person cannot make an informed and voluntary choice and must rely on a surrogate to make that choice for them. The Court found that Missouri's requirements of a living will, or clear and convincing evidence of the incompetent person's wishes was not a violation of that person's constitutional rights. Further, the Court stated that Missouri's requirements protected against any abuse of an incompetent person in such circumstances and the state can maintain life support as long as its standards for doing so are reasonable.

Following the U.S. Supreme Court decision, the Cruzan family gathered more evidence of their daughter's wishes about being kept alive in a vegetative state. The state of Missouri had withdrawn from the case in September 1990 since its law had been upheld. With no opposition from the state, the local probate judge found the new evidence compelling and reliable and ordered Nancy's feeding tube removed. The tube was removed on December 14, 1990, and Nancy died December 26, 1990.

Defining Death and Characteristics of Death

The Cruzan v. Missouri Department of Health landmark case is tragic for many reasons and for many family members surrounding the case. It is a reminder that end-of-life documentation is important, even if none of us wants to think about it. Death is profoundly personal, yet often-times we're left without control over it. Death is life's final chapter and this book's final topic. It is a topic fraught with controversy as many people think everyone should die naturally regardless of their personal circumstances. Through the lens of science, defining death is difficult, but euthanasia is an option.

What is death? Sounds like a fairly easy term to define. Nothing is more definite than death. Yet, defining death, much less determining when it has actually occurred, is difficult. What makes it so especially challenging is the number of ambiguous terms. Depending on perspective, varying groups have varying definitions. Medicine, law, science, religion, and the media use death to mean different things. There are the common phrases like passing away, passing on, kicking the bucket, meeting one's maker, pushing up posies, and knocking on heaven's door. Then there are terms like coma, brain death, whole brain death, permanent unconsciousness, persistent vegetative state, somatic death, and biological death.

Picking one definition for death has become increasingly difficult as biomedical devices and bioengineering advances can maintain life processes without conscious control. Through technological medicine, machines keep people alive by maintaining breathing and blood flow even when the brain has permanently stopped functioning. Clinical death is the term used when resuscitation may be achieved. The person looks dead, but there's a good chance of survival through medical intervention.

Just as the right-to-life issue is a polarizing topic, so is the right-to-die issue. Of course, there are myriad related topics in between. For example, is it right to save a life if it means that person is destined to a life of incapacitation? Is it right to bring somebody "back from the dead," if the "post-death" state resembles zombiehood? How should prisoners destined to die behind bars be treated? How should the penal system react to an aged, death-row inmate with dementia who doesn't even know why he is slated for government-induced death? Does the death penalty itself satisfy the requirement for not being cruel and unusual? Is solitary confinement—a kind of personal death—moral? What about life without parole? Is the death penalty acceptable in a nation where so many are opposed to abortion? What is ethical in each of these scenarios? Answers to these questions are complicated. Understanding the science behind death and dying can help shape opinions. This chapter deals with our own mortality. From the cradle to the grave, we're all on a continuum.

Take the East Asian concept of yin and yang, the philosophy that life cannot exist without death. Yin and yang are contrasting opposites. Yin is the principle of the universe associated with earth, dark, and cold. Yang is associated with heaven, light, and heat. Individuals subscribing to yin-yang philosophy use death as a daily reminder to enjoy life. It's a reminder to stop and smell the roses. Sounds simple enough, but still, what is death?

Characteristics of death are complex. In humans, death is manifested by the loss of heartbeat, the absence of spontaneous breathing, and no cerebral activity. Each of these *also has definitions*, so it's no wonder that defining death is an arduous task. One reason for such ambiguity is that there are no clear distinctions between being alive and being dead. In fact, if you've ever been with somebody as they die, you'll discover that dying is a process. It should also be noted that relative to death and dying, one does not necessarily precede the other.

Medical terminology, which aims to be precise and incontrovertible, has often blurred the distinction between life and death. That's probably because like wondering when life begins, there is no absolute time of death. A person can stop breathing and be dead. Your heart can stop beating, rendering you dead. If you can't suck air and you don't have a beating pulse, you're likely dead. But medical care can intervene to prolong life.

As a task of both scientific and philosophical inquiry, let's discuss the concept of death. There are several reasons to study death. One is to better understand disease processes. This is especially true because diseases often cause death. Many professional websites devoted to chronic illnesses, such as cancer, have a checklist of sorts that identifies the signs and symptoms to look for in a loved one who is nearing death. The sites oftentimes identify signs that a person has died.

Another reason to study death, however, is to understand the social and legal implications and responses to death. This area is huge because religion and culture play a role, and laws in the U.S. vary by state. There is one point that everyone can agree on: There is no single marker of death. Like relationships, death is complicated.

In science, death is often divided into two types: somatic and biological. The two are inter-related. Whereas somatic death is the death of the entire organism (person), biological death refers to death at the cellular level. You can have an interruption in blood supply that caused the death of cells and subsequent tissue, as is the case in frostbite, but you are still alive. When we die—as in there are no brain, lung, or heart functions—some of our native body cells or microbes living in and on us can still be alive.

As a biological organism, the reason we breathe is to provide oxygen to our cells so that they can generate adenosine triphosphate (ATP)—the fuel that drives our cells, our metabolism, and enables bodily functions. If you can't breathe, oxygen attainment stops and ATP production ceases. The heart needs ATP to beat. No ATP, no heartbeat. No heartbeat, no ability to pump oxygen-rich, nutrient-dense blood to our brain and other tissues. No blood to our tissues, organs die. For example, if some of your heart cells are deprived of oxygen, areas of the heart can die because of oxygen starvation. (Too much oxygen starvation leads to whole heart death.) The areas of heart tissue that die are not coming back because heart cellular damage cannot be reversed. Evidence of past heart tissue damage can be seen by medical professionals on specific exams. This is the reason why your Uncle Leon can tell you that he had a heart attack and didn't even know it until it was shown in tests taken of his heart. There are very subtle differences between somatic and biological death. Too much biological death leads to somatic death because you simply cannot survive if too much of a vital organ's tissue has died off. Common signs of death with which most of us are probably familiar are body stiffness, cold temperature, and ashen skin.

First up is body stiffness. After a person dies, why do muscles stiffen? The muscles are stiff because they ran out of ATP. The condition is called *rigor mortis*, derived from the Latin words *rigor* which means *stiffness* and *mortis* which means *death*. Going back to the beating heart. Just like the heart, which is a big muscle, the other body muscles need ATP for contraction. They also need ATP for muscle relaxation. If we're not breathing, we're not making ATP, so the muscles can't contract or relax. In fact, after death and with no ATP production, the muscles get "stuck" in contraction. This generalized stiffening of the body occurs one to seven hours after death. Two muscle proteins, myosinogen and paramyosinogen, coagulate in the muscles, causing the tissues to harden. Rigor mortis generally disappears within six days when decomposition begins. If you're a fan of murder mysteries and TV detective shows, you may have wondered how the smart coroner could arrive on the scene and determine the time of death. The degree of rigor factors into the time of death equation.

Second and third common signs of death are cold body and ashen skin. These two are related. If the heart isn't beating, it can't pump blood to tissues, including the skin, so the skin turns pale gray—the color of ash. Without actively metabolizing tissues, they begin to die, and turn cold. We know there's little to no chance of recovery here.

Brain Death and the Harvard Committee

As early as 1968, the report of a Harvard Medical School ad hoc committee proposed widespread adoption of protocols establishing criteria for dealing with a permanently non-functioning brain. The group's goal was to define irreversible coma as the new criterion for establishing time of death.

According to the report, four primary criteria had to be met in order for a person to be pronounced having a permanently non-functioning brain. These include: (1) unreceptivity and unresponsiveness; (2) no respiratory movements; (3) no reflexes; (4) a flat electroencephalogram (EEG).

Unreceptivity and unresponsiveness mean there is a total unawareness to externally applied painful stimuli. Tests to determine this include using a sharp object against the skin to determine pain and squirting cold water in the ear canal to test for responsiveness. Delivering cold water into the ear causes nystagmus, the term for side-to-side eye movements. You can try this experiment at home, but it's not recommended, because it is incredibly uncomfortable. The temperature variation between your ear canal and the cold water squirted in causes a very unpleasant sensation.

Physicians also determine death by checking for reflexes. No spontaneous breathing (Hering-Brewer reflexes) or response to stimuli such as pain, touch, sound, or light after a period of at least one hour is regarded as not breathing or having reflexes. Using various tests, there must be no evidence of other elicitable reflexes. For example, the person must have no reflexes that include pupil movement (shining a light in the eye and looking for the pupils to constrict), eye movement, blinking, postural (changing position) activity, swallowing, yawning, vocalization, corneal reflexes, or pharyngeal reflexes.

Note the abbreviation EEG above: An EEG measures brain electrical activity, while an EKG (also called an ECG) measures heart electrical activity. The EEG assesses brain activity as waveforms that vary in frequency and amplitude, measured in microvoltage and hertz (Hz). Using electrodes attached to sites on the external scalp, the instrument depicts electrical activity occurring on the brain surface. The activity is recorded as alpha, beta, theta, and delta waves. Alpha wave patterns occur in the frequency band of 8–13 Hz and indicate the rhythm of an awake, relaxed person. Beta wave patterns occur in the frequency band of 18–30 Hz; theta waves occur in the frequency band of 4–7 Hz; and delta waves are seen in the 1.5–4 Hz band.

The EEG is a useful tool for healthcare professionals because the waveform frequency and shapes help the clinician determine normal from abnormal. If no electrical activity exists, it is recorded as a flat even EEG. Just as hearts can flat line in EKG measurements, brains can flat line in EEG measurements. Flat EEG is of the greatest confirmatory value in determining brain death.

The committee also stated that both hypothermia and central nervous system depressants could not be present. Hypothermia is discussed shortly, but a person isn't dead until they are warm and dead, so if the person was cold, they couldn't be counted as dead. Central nervous system (CNS) depressants are substances that reduce the function of the brain and spinal cord. In addition to some drugs like barbiturates, alcohol is a powerful CNS depressant. These last criteria ensure that cold, drunk people aren't mistaken as dead. The committee also recommended that these same tests be repeated 24 hours later.

In 1981, the Harvard Criteria were updated. Their purpose was aimed at eliminating errors in classifying a living individual as dead, allowing fewer errors in classifying a dead body as alive, allowing a determination to be made without unreasonable delay, adapting to a variety of clinical situations, and being explicit and accessible to verification.

In 2018, a special report in commemoration of the 50th anniversary of the *Report of the Ad Hoc Committee of the Harvard Medical School to Examine the Definition of Brain Death* was published. Lasting questions such as "Is death defined in terms of the biological failure of the organism to maintain integrated functioning? Can death be declared on the basis of severe neurological injury even when biological functions remain intact? Is death essentially a social construct that can be defined in different ways, based on human judgment?" are still debated.

As part of the World Brain Death Project—yes, there was such a thing—physicians from professional societies the world over convened to reach consensus on how to identify brain death. They reported their findings in the August 3, 2020, edition of *JAMA*. Their main objective was to provide recommendations on determining brain death/death by neurological criteria, abbreviated as BD/DNC. Using a comprehensive review of the literature and expert opinions across a multidisciplinary, international panel, the group provided a report identifying the minimum clinical standards for determining BD/DNC in adults and children. Here are their recommendations from the paper's abstract:

> Prior to evaluating a patient for BD/DNC, the patient should have an established neurologic diagnosis that can lead to the complete and irreversible loss of all brain function, and conditions that may confound the clinical examination and diseases that may mimic BD/DNC should be excluded. Determination of BD/DNC can be done with a clinical examination that demonstrates coma, brainstem areflexia, and apnea. This is seen when (1) there is no evidence of arousal or awareness to maximal external stimulation, including noxious visual, auditory, and tactile stimulation; (2) pupils are fixed in a midsize or dilated position and are nonreactive to light; (3) corneal, oculocephalic, and oculovestibular reflexes are absent; (4) there is no facial movement to noxious stimulation; (5) the gag reflex is absent to bilateral posterior pharyngeal stimulation; (6) the cough reflex is absent to deep tracheal suctioning; (7) there is no brain-mediated motor response to noxious stimulation of the limbs; and (8) spontaneous respirations are not observed when apnea test targets reach pH <7.30 and $Paco_2 \geq 60$ mm Hg. If the clinical examination cannot be completed, ancillary testing may be considered with blood flow studies or electrophysiologic testing. Special consideration is needed for children, for persons receiving extracorporeal membrane oxygenation, and for those receiving therapeutic hypothermia, as well as for factors such as religious, societal, and cultural perspectives; legal requirements; and resource availability.

As in all matters of life and death, it is a complex issue. In addition to the science, familial, cultural, religious, regional, and legal aspects must be considered.

Perspectives on Death

The difficulty in determining death is a direct result of the various concepts of death, identifying the locus of death, and establishing criteria for death. Philosophical, theological, and moral perspectives must be considered when judging the meaning of death. In terms of finding the locus of death, anatomical structures such as the heart, lungs, or brain could be used depending upon the individual perspective. Therefore, criteria of death are deemed necessary for physicians to determine whether a person is actually dead.

Understanding brain death has not escaped the 21st century. It has been suggested that the term be abandoned because it perpetuates professional and public confusion. In fact, the terminology has created a semantic minefield. It has been argued that whole-brain criteria should be used to define death. However, this is not an acceptable definition because although most neurons have ceased functioning, a few may still continue to fire and synapse. Brain-referenced criteria, however, are the most widely accepted criteria for defining death.

When an individual is in a coma, the eyes are closed, and the person is in a sleep-like unconsciousness. In brain death, the eyes are closed, and the person is in an irreversible coma. Both appear very Sleeping Beauty–ish.

The state of permanent unconsciousness is termed a persistent vegetative state. Characteristics of a persistent vegetative state include brainstem function, total absence of cerebral cortex activity, with alternate cycles of eyes open and eyes closed. It is commonly termed "awake but unaware." The brainstem is still functioning? What's happening? The brainstem is an anatomical structure at the base of the brain consisting of three separate, but integrated, areas known as the medulla oblongata, pons, and midbrain. If you were to look at a human brain, you'd see this large mass with convolutions, then a much smaller area called the brainstem, which continues downward ending at the spinal cord. The brainstem has important jobs like maintaining breathing (rate and depth) and heart rate. Thus, the big part of your brain can be totally non-functioning, but you can be "alive" because the brainstem is maintaining breathing and heartbeat. Personality resides in the big part of the brain, the cerebrum. In essence, there is no "person," just a living corpse. The persistence of the term "brain death" in everyday language demonstrates the confusion surrounding an appropriate definition of death.

Although irreversible coma and persistent vegetative state are terms used to describe "brain death," the medical and health communities continue to misuse the term to indicate other diagnoses such as coma or cerebral death. Confusion in the media is evidenced when reporters refer to "brain death" as literally meaning death. Semantic precision is necessary. We can't police word usage. We can only be sure that we understand all the semantic variations.

Several models to alleviate confusion have been proposed in the recent past: the American Bar Association Model, the American Medical Association Model, and the Uniform Brain Death Act. The American Medical Association proposed provisions for determining death and limiting liability for physicians. The previous models were abolished and a new one adopted: the Uniform Determination of Death Act (UDDA).

The American Bar Association, the American Medical Association, and the National Conference of Commissioners on Uniform State Laws have all endorsed the UDDA. Succinctly, the Act states that "An individual who has sustained either (1) irreversible cessation of circulatory and respiratory function, or (2) irreversible cessation of all functions of the entire brain, including the brainstem, is dead." A determination of death must be made in accordance with accepted medical standards.

The concept of "brain death" has been incorporated into the laws of all states, yet many Americans are confused about the terminology. Despite the fact that the Uniform Determination of Death Act has been proposed, few states have adopted this over their existing statutes.

The confusion doesn't end here. Brain activity has been reported even after cessation of cardiac rhythm, undetected arterial blood pressure, and unreactive pupils. After being pronounced clinically dead, a team from the University of Western Ontario in Canada had evidence of one patient displaying delta wave activity. Delta waves have a very slow frequency, are normal if you are in a very deep sleep, but abnormal if you are awake. In the aforementioned Canadian study, with this one patient, single delta wave bursts persisted for about 10 minutes after the heart rhythm had stopped and there was no blood pressure. We need to be very cautious about what this means because this was a sample size of exactly one, so not much can be extrapolated from the data. That said, researchers have identified *death waves* in rat brains from about one minute after decapitation. These findings suggest that the brain and the heart may "die" at different times. Obviously, this has implications that extend into organ harvesting. Enter the relatively new scientific field of necroneuroscience. The word part *necro-* means death and this branch of brain science explores what happens to our brains near death and right after death. It seems that in rare instances, a person can "flat line"—have no EEG waves—only to have a delta wave re-appear, even if for a very short time. Some neurons seem to be resistant to death and can survive a little longer after the heart has stopped. Once again, like life, death is a continuum.

In conclusion, it appears that the debate over when life begins and when it ends will continue, especially in light of biomedical technological advancements. For this reason, one must continue the education process to remain truly abreast of historical information as well as current literature regarding research in this ripe area.

Postmortem Myths and Facts

Following death, most body functions end, but some continue for minutes, hours, days, and up to weeks depending on the tissue. That said, many postmortem functions are myths. One such myth deals with hair and fingernail growth. Neither your hairs nor your fingernails grow after you've been pronounced dead. They only *appear* to grow because the body starts to dry out as time passes after death. When the skin dries, the skin around the nail beds shrinks, thereby exposing more nail surface, which gives the appearance of fingernail growth. Same holds true with hair: less hydrated skin retracts, exposing more of the hair. Neither grows.

However, skin cells *can* live for a few days after a person is "dead." Cells are living, but you are not. As the body's first line of defense and the largest organ in the body, our skin is accustomed to life on the outside and therefore doesn't rely as heavily on a direct blood supply. Note the phrase "as heavily." The skin can snatch nutrients by diffusion, and diffusion doesn't require a beating heart. Diffusion doesn't require energy, and you've seen it in action countless times if you've ever added a teaspoon of sugar to a cup of coffee: The sugar dispersed until it was evenly distributed in the liquid. This is diffusion. No energy needed. Stirring will speed it up, but stirring is an action that was fueled by you. Skin cells can obtain nutrients the same way because the nutrients are close to the surface.

In addition to being the largest visceral (contained in a body cavity) organ in the body, the liver has some remarkable properties. For example, it is regenerative, meaning that it can replace damaged tissue with new cells. If other organs are damaged, the tissue is replaced with scar tissue. That the liver has regenerative properties has been known quite some time, dating back to Greek mythology when Prometheus, who was tied up by Zeus, had his liver eaten repeatedly by an eagle, only to have it grow back after each insult. Scholars argue that the early Greeks likely did not know of the regenerative powers of the liver, but the mythical story stuck. Yet, the liver does indeed repair itself—up to a point. If the liver is damaged beyond repair, a transplant is needed. In living-donor transplants, a portion of the donor liver is removed and used to replace the non-functioning liver in the recipient. The donor's liver regrows to full size—although it doesn't look like the original liver—but it functions. The recipient's new liver grows to normal size and it, too, functions. Building on the regenerative properties of tissues, a relatively new field of medicine, known as regenerative medicine, has emerged.

But, there is another fascinating piece of information related to the liver. After death, hepatocytes (*hepato-*, liver + *-cytes*, cells) still survive anywhere from 30 minutes to seven hours. Postmortem liver function tests are also done in forensic settings, but their usefulness is still being determined as results have not been consistent. Showing that science is cross discipline, postmortem chemistry, also called death chemistry and necrochemistry, is another field of chemistry that deals with the biochemical analysis of body fluids and tissues after death, playing a significant role in forensic pathology.

Another myth is that the body loses weight immediately following death as the soul departs to another world. It does not lose weight other than any solid or fluid weight from bowel or bladder emptying. The mythical weight number that gets tossed around is 21 grams, a value based on a 1907 flawed study conducted by Dr. Duncan MacDougall of Haverhill, Massachusetts. MacDougall tried to determine if the soul had weight by placing the dying patient along with their bed on an industrial scale and measuring the patient's weight before and after death. His study was flawed because his sample size was six (margin for error is greater with a small sample size), he could only document weight loss in four individuals at the time of death, he only used one of those four in his reporting, the equipment was of poor quality, and he had experimenter bias. He did the same experiment with dogs, although it is presumed that he poisoned them to conduct the experiments. How does one find dying dogs otherwise? The 2003 movie, *21 Grams*, references the idea.

Brain cells die pretty quickly after the heart has stopped. This is where medical technology can make things really messy though. While the heart may no longer be beating, as another vital organ, the brain needs oxygen-rich blood. Without a beating heart, the brain is deprived of precious oxygen and nutrients, and its cells begin dying. If the heart *is* restarted, there's a very good chance that the brain isn't going to recover.

Are we dead yet? Bacteria in our gut can live a really long time after we've expired. Think about the fact that there are billions of microscopic organisms in our digestive tract that thrive on the contents contained therein. These bacteria continue doing what they've always done and make gas. This might give one the false impression that someone is still alive if they pass gas. Bacteria continue doing their thing and breaking down the body, creating a rotting mess with lots of gas buildup.

What about those noises that dead bodies make? Besides farting sounds, dead bodies can also make other sounds like someone is moaning or groaning. We call these sounds *vocalizations*, and they are the result of gas (air) passing across the vocal cords. Healthcare workers report hearing this when moving a dead body. If a dead body is repositioned, gas pockets move, and when the gas moves across stiffened vocal cords in the larynx, the result is sound. Actual *words* will not happen since forming words requires the muscles of the mouth to operate. If the person actually *talks*, it is safe to assume they are still alive.

When it comes to death, it seems that some genes also exhibit postmortem activity. Some scientists at the University of Washington were testing gene activation techniques in zebrafish when they stumbled upon an interesting finding: After the fish were dead, about 1 percent of their genes came back to life. We know that genes are a set of chemical instructions that are made of DNA. When genes get activated, those chemical instructions are transcribed by RNA, which cells copy. Think of the gene as a cookbook and think of activation as writing down the ingredients so that something can eventually get made. The University of Washington researchers were initially skeptical of their findings and wrote them off as an instrumentation fluke.

Repeated tests, however, in fish and mice also showed that some genes were activated hours to days after death. This was an unheard of finding and many in the scientific community were skeptical, until a group of researchers at Barcelona's Centre for Genomic Regulation found postmortem gene activation in humans who had donated their bodies to science.

Why this happens is not known but knowing that it *does* could help medical procedures like organ transplants and forensic science investigations to more accurately determine the time of death. The human cadaver studies showed that different genes activated at different time intervals. These reactivated genes have been dubbed *zombie genes* and figuring out where the breakdown between cellular functioning and whole organism functioning occurs has opened a new avenue of scientific research.

Some genes don't appear to be dead after the organism has died, indicating that genes themselves might have a life cycle of sorts. Scientists are calling these genes *pseudogenes*, because they have nonfunctional DNA segments that resemble functional genes. Pseudogenes are likely vestigial—remnants of something that at one time were functional. For example, house cats have taste receptors and genes for sweet flavors, but because cats are meat eaters, these genes for sweet taste sensations have died. Thus, one supposition is that within the human genome some of our pseudogenes can one day be reactivated.

If we didn't embalm, nature, assisted by maggots and insects, would simply take over and return us to dust. Embalming is a common practice of injecting the body with chemicals to preserve it from decay. The practice gained favor during the Civil War when chemicals were used to preserve the body so that it could be shipped back home, identified, and a proper burial given. Early preservatives included arsenic, an element that forms a number of poisonous compounds. Arsenic was replaced with formaldehyde, which is injected into arteries. Formaldehyde is an excellent preservative that decreases decomposition and decay and makes it easier to restore living features, like a smile. Those features are set and then the chemicals are pumped in to keep them in place. Once the chemical is pumped in, the features cannot be changed.

How does embalming occur? In the back room or basement of funeral homes, bodies are "prepared" using a number of tools and tricks. For starters, a nice big artery is found—usually in the neck—and about 2.5 to 3 gallons of fluid is injected through the arterial system while the blood drains out through the jugular vein simultaneously. Where does this blood go? Down the drain, of course. It empties right into the public water and sewer system. Bodies are also prepared in such a manner for cadaver labs where students dissect preserved humans to study and learn about human anatomy.

Hypothermia and the Diving Reflex

There are situations in which a person can be very cold and yet be alive. Which is why a person is not dead until they are *warm and dead*. You read that correctly. Here's a classic example.

Pause and think about the reason you refrigerate food. Most microbes, like bacteria, yeasts, and mold, prefer warm temperatures. Like us, they favor things to be warm and fuzzy with plenty of food. For example, making bread from scratch requires the dough be placed in a warm place to allow the yeast to ferment, yielding alcohol and carbon dioxide and causing the bread dough to rise. If we want to slow microbial growth in general, it's best to cool the food in a refrigerator. For this reason, refrigerated food lasts longer than food left out on the counter. Although we're not food to be stored and eaten later, the idea is the same. Cold temperatures slow down chemical processes in the body, preserving vital tissues.

There have been documented cases of hypothermic people with abnormally low body temperatures coming back to the land of the living. While Hollywood took some artistic liberties in the movie *Titanic*, some of those freezing people floating in the Atlantic Ocean survived. Near-drowning victims who were submerged in frigid waters also have a chance of survival. In fact, cases such as these have been reported for decades.

What prevents death in near-drowning victims? A few factors come into play including environmental conditions. First, the water has to be cold enough to cause hypothermia, and second, the diving reflex has to kick in. Hypothermia is a body temperature significantly lower than normal body temperature. In fact, it is dangerously low. Hypothermia sets in when our bodies lose heat faster than they can generate it. Losing body heat is actually quite easy to do in water because the thermal conductivity is about 25 times greater than air. This means that you lose body heat much faster in water than in air of the same temperature. Jumping into 70° F swimming pool water will cool you quicker than standing on your 70° F front porch. So, the first criterion is established: hypothermia.

The second criterion is the diving reflex. You've seen it in action if you've ever witnessed diving vertebrates, including ducks, seals, and whales. Or, if you've ever seen a human baby under the age of two "swimming" under water. It's a physiological mechanism that allows air-breathing, water-dwelling animals to remain submerged for long periods of time without surfacing for oxygen. In humans, it happens when the face or body is immersed in cold water. When this happens, blood flow is shunted to the vital organs—heart, lungs, and brain—and diverted away from the extremities (arms and legs) and gut. This is an evolutionary feature to preserve vital organs at the expense of non-vital structures. Peripheral arteries constrict, so blood flow to the fingers and toes slows way down. After all, you can survive without a toe or two or three, but not without a heart, lung, or brain. The heart rate slows down, too, causing a condition called bradycardia—a word that means "slow heart." While the heart rate may slow, oxygen supply to the heart and brain is increased. Keep in mind that the reason for needing oxygen is to make ATP. Another nifty thing also happens: To protect against water inhalation, the larynx (voice box) spasms to prevent water from entering the trachea (windpipe). The cold water slows metabolism and reduces the need for oxygen in other body tissues. In warm water, the decreased oxygen level (hypoxia) would lead to irreversible brain damage within about five minutes. In cold water, victims have survived hypoxia for about two hours.

This isn't the end of the story. Once the drowning victim has been recovered, they have to be rewarmed. This rewarming has to be done gradually by drinking warm liquids (not alcohol) if they are able and taking a warm water bath around 104° F. Warm water re-establishes normal body temperature faster than rewarming in air. Alcohol should be avoided because it causes vasodilation (expanded blood vessels) and increases blood flow to the periphery, including the skin, which would speed up heat loss to the environment. This makes sense: on a hot day, you lose a lot of body heat to the air through your skin. Drinking alcohol makes you feel warm because blood flow is diverted to your skin—in fact, you even flush a little.

Cardiovascular Disease

Cardiovascular disease refers to a number of conditions that affect the heart (cardio) and blood vessels (vascular) throughout the body and is a leading cause of morbidity (having disease) and mortality (death) worldwide. In the United States alone, heart disease is the leading cause of death, killing one out of four people. About 600,000 people per year in the U.S. die of a cardiac event. Diseased blood vessels can lead to atherosclerosis (hardening of the arteries), heart attacks, and strokes. Traditional risk factors are age, sex, lipid levels, smoking, hypertension, and diabetes. Taken in total, these seem to predict 75–80 percent of an individual's risk of cardiovascular disease. Another factor should also be considered: birth weight. Numerous robust studies dating back to the 1970s and continuing today show that there is an inverse association between birth weight and cardiovascular disease and stroke risk that persists despite controlling for other known risk factors. Decades-long studies show that babies who weighed less than six pounds at birth have a significantly increased risk of cardiovascular disease. Findings such as these support the role of the prenatal environment in contributing to the development of chronic disease.

Although the reasons are not well understood, the low birth weight risk factor results have been supported by epidemiological evidence for quite some time. It may be that inadequate intra-uterine environment shapes fetal organ systems such that some systems receive more nutrients than others. Low birth weight is consistently associated with increased risk for type 2 diabetes and metabolic syndrome. Maternal obesity is linked to an increased incidence of large-for-gestational age (LGA) babies, and LGA infants are also at increased risk of adult chronic disease. All good information, especially as we work toward increasing maternal health, which positively affects infant birth weight.

Let's pause for a terminology lesson. Heart attacks are also called myocardial infarctions, or MIs for short. *Myocardial* comes from the word parts *myo* meaning *muscle* and *cardial* referring to the *heart*. An infarction means that an area of the heart tissue has been robbed of its blood supply and the tissue has died. So, myocardial infarction = heart attack. Clinicians use the term *necrosis* for tissue death. Disease, injury, or blockage are common causes for blood supply interruption to tissues. This is important to know because another devastating condition, *cardiac arrest*, can also affect the heart and it sounds a lot like a myocardial infarction. With cardiac arrest, heart activity stops. The electrical signals that keep the heart beating may have stopped, the heart muscle itself may have stopped contracting, or both could have happened.

About 14 percent of people who have a heart attack will die from it, thus it's important to talk about resuscitation. Resuscitation methods include cardiopulmonary resuscitation (CPR), defibrillation, and epinephrine injections. (Epinephrine was discussed in detail in Chapter 8.) CPR restores heart and lung function by using artificial breathing and manual closed-chest compression. When not in a hospital, CPR is something a bystander can do to revive a person. Even if you are not trained, doing *something* to try and help is better than doing *nothing*. The American Heart Association has suggested guidelines for what to do if you come upon a person whose heart appears to have stopped. If you have no CPR training, your CPR skills are rusty, or there is a viral pandemic, then drop to your knees beside the distressed person and use hands-only chest compressions at a rate of 100–120 per minute. Do this until help arrives. It will be tiring. Very tiring. Just try pumping your hands on the tabletop now to get an idea.

What does CPR actually do? It keeps oxygenated blood flowing to vital organs, like the heart and brain. Without oxygen, a person can die within 8 to 10 minutes. Even with CPR, there's a good chance the person will still die. But you have to try. According to the American Heart Association, about 45 percent of out-of-hospital heart attack victims survived if a bystander administered CPR. That still leaves 55 percent who do not survive. If you are trained and ready to go, then you already know what to do: check for pulse and breathing; if neither exists, start with 30 chest compressions followed by two rescue breaths and continue the drill until help arrives. Research is currently showing that conventional CPR with breaths may have better outcomes than chest compression only. Regardless of the method used, the next step is key: the person has to get to a hospital. If they don't, the chances of survival drop precipitously.

Defibrillation is another method to counter the physiological events causing a heart attack. However, it only works for ventricular fibrillation or pulseless ventricular tachycardia. This is just a fancy way of saying that the heart's lower chambers are beating irregularly (fibrillation) or rapidly (tachycardia). Both are common causes of heart attack. The defibrillator uses an electrical shock to "reset" the heart so that it reverts to normal functioning by its own pacemaker cells. You've seen these little battery-powered machines hanging on walls in public places. They are commonly known as AEDs, for automated external defibrillators. Studies have shown that immediate use of AEDs can improve survival rates.

Epinephrine is another treatment. Epinephrine is a chemical that can be given to kick start the heart. The more common term for epinephrine is adrenaline. Why are there such disparate terms for the same chemical? The answer lies in the term's origin. If you break apart the term *epinephrine*, you get *epi-*, meaning *above* and *nephros* meaning *kidney*. Putting the term together now tells you the location of the gland that makes epinephrine. Which brings us to the term *adrenaline*, which is derived from the word parts *ad-* meaning *above* and *renal* meaning kidney. So, the adrenal gland is an organ located just above the kidney that secretes the chemical adrenaline, which is also known as epinephrine. Learning anatomy and physiology can be a challenge: Many structures and chemicals have two (or more) different names!

What does adrenaline/epinephrine actually *do* to start the heart? It turns out that it has many roles in the body, like increasing blood flow, increasing breathing rate, and metabolizing carbohydrates with the aim of preparing the muscles for activity. The heart is a big muscle, so a little shot of adrenaline/epinephrine will give it a boost. You've experienced an adrenaline rush every time you've felt anxious or were placed in a stressful situation. Here's a typical example: You're driving your car along a nice paved neighborhood street when a dog darts in front of you. Naturally, you slam on the brakes. Thank your sympathetic nervous system, which prepares you for such situations and squirts a shot of epinephrine into your bloodstream. At that same time, you felt your heart race and you might have been a little shaken. When you realize that you stopped the car before hitting the dog and the dog walks safely to the sidewalk, you stop shaking and your heart rate returns to normal. Thank your parasympathetic nervous system, which returns your body to a resting state. This is our "fight or flight," "freeze for action," and "rest and digest" opposing nervous systems. These instinctive physiological responses prepare us for threatening or stressful situations and ready us for action. We can't stay on heightened alert all the time, though. So, once we're safe, the internal situation resolves via an influx of norepinephrine, which counters the effects of epinephrine.

Unfortunately, there's a caveat to using epinephrine to restart the heart. Research is currently showing that it does so at the expense of the brain. People who had cardiac arrest outside of a hospital and were administered epinephrine by paramedics had slightly improved survival, but they also had diminished neurological function. Basically, they lived longer but were unable to take care of themselves.

Even *if* the heart is restarted in time to preserve brain function, there's still another wrinkle: inflammation. Following a heart attack, the body's immune system goes to work trying to heal the damaged heart. Sounds good, right? Unfortunately, the immune cells (specifically B cell lymphocytes) trigger inflammation, which damages the heart further. Fortunately, there is a drug that can help. This drug is pirfenidone; however, it, too, has side effects, including nausea and vomiting. Of course, what's a little tummy upset and upchuck if it'll save you…

Stopping the heart is usually a very bad thing. But cardiac arrest can be induced for therapeutic reasons. One good reason to stop the heart is to work on it. You wouldn't work on a car's engine while it was running. During open heart surgery, the heart must be stopped. Cardiac arrest is *not* the same thing as a heart attack. Cardiac arrest can happen after a heart attack, but cardiac arrest can also happen on its own. Either situation is not good.

Sometimes, therapeutic hypothermia can be used to preserve heart function in both heart attack and cardiac arrest. With therapeutic hypothermia, cooling devices are used to drop the body temperature to a cool 89° F to 93° F (32° C to 34° C). Keep in mind that normal body temperature hovers around 98.6° F (37° C). Like applying ice to a swollen ankle to reduce swelling, lowering the entire body temperature likely slows down adverse chemical reactions and inflammation. Recall that you're not dead until you are *warm* and dead.

Psychology of Death

It has only been in the last few decades that the psychology of death has begun to appear in the psychological literature, despite the unanimity about its importance to the living. The human concerns and preoccupation that have been voiced about "death and dying" first became a popular area of psychological study in the 1970s, and became the topic of many causes, courses, and research projects. Obviously, the immense interest in dying and the progression towards death is what we really expected and could have foretold, since the end of life is truly an important concern not just for older people, but for everyone, of all ages. For that matter, it is said, "even from the minute of birth, we are beginning to die, if even a cell at a time." One needs to carefully look at this preoccupation and concern (anxiety) with dying and the factors affecting this fear. Moreover, the major aspect of death that is amenable to human control, the psychological conditions under which we die, merit further study.

Contrary to common beliefs and assumptions, older people are not more preoccupied with death than any other group. Older people tend to view death more positively than younger adults, and when asked directly, only a small percentage of older people appear to actually say they fear death. Many older people feel that death is not robbing them of their years to live, like it might a younger person, and thus have less to fear. However, people of all ages think about the subject quite frequently. Developmental psychologist Erik Erikson (1902–1994) felt that since the last developmental stage in life is "integrity," this involves accepting the prospect of death. Nevertheless, as the years begin to roll by, seemingly even more rapidly as we age than when we were young, many do truly have a good deal of underlying death anxiety and fear, particularly on the deepest level of consciousness.

Many influences are prevalent that might increase an older person's anxiety or intensity of death fear. A few of the most common include: being in the stage of integrity, being temporarily close to death, and being invested in the future. Surprisingly enough, the first two influences do not seem to play as an important role as the third, the interest in the future. It appears that older persons who have goals and plans and programs for the future are more fearful of death.

When a person has been diagnosed with a terminal illness, they typically progress through a series of stages in reacting to this knowledge. The stages that have been most often suggested after much clinical research are the following:

1. Denial ("it's not true; it cannot be happening to me" stage). Denial is a universal reaction. One should allow this stage to go on as long as is necessary. It is usually replaced by partial acceptance.

2. Anger ("why me?" stage). Anger may even be projected to family members or the health provider staff. It is hard to deal with, yet one must understand the reason for anger and not necessarily take it personally.

3. Bargaining ("I need to live until my grandchild is born, dear God," stage). With bargaining, the fact of death is accepted, but the person tries to prolong the process by "bargaining" with God or the health care staff. It is generally associated with some guilt and should not be taken too lightly.

4. Depression (stage may be induced by loss already suffered, i.e., hysterectomy or by losses which are pending). Depression is often suffered in moody silence and should not be discouraged, but rather the person should be encouraged to express their sorrow and suffering and should not be treated by telling them to "look at the bright side of things."

5. Acceptance (final stage). Acceptance usually takes place just before actual death occurs. It should not be disturbed by ill-advised, last-ditch, surgical procedures or chemotherapy in order to prolong the "agony" a little bit longer. This stage is not necessarily a happy stage and is often characterized by sleep and a decrease in communication as well as a decrease in feelings. Individuals want to be left alone.

However, the stage theory is not universally accurate. Hope was the most common emotion manifested during the course of a life-threatening illness, and a number of people do not decide that once-and-for-all they are dying. The person may undergo "middle knowledge" where cycling of awareness (knowing) and denial (not knowing) may occur at different times.

There may be another pattern that appears in response to one's death, which may or may not form as a progression of predictable stages. It involves adaptation as a result of previous personality. The way one acts may be correlated with survival. People who maintain a good relationship with others and are assertive in expressing their needs during the illness, tend to live longer than might be expected.

Generally, most people tend to mentally and verbally avoid the "dying." Avoidance and direct expressions of anxiety and intense concern all demonstrate a high level of emotional involvement and may require different types of care. Actual observations of behavior in medical settings do demonstrate that avoidance or distancing is a predominant response, and personnel tend to avoid patients near death. Additionally, the location of dying patients in the care facility is often distant from other patients and house staff areas.

Palliative Care and Hospice

There has been a national movement to provide better terminal and psychologically-sensitive care. An alternative to traditional hospital treatment is palliative care and hospice. Often times these terms connote the end of life, but both can ensure a more comfortable journey to the finish.

Palliative care involves treatment that alleviates symptoms without curing the disease. Hospice care is a centralized program of palliative and supportive services to the dying person and their family. It's not all about medicine. Physical, psychological, social, and spiritual care and services are provided by an interdisciplinary team of professionals and volunteers skilled in end-of-life issues. Hospice normally involves both in-patient and at-home settings. It is a place and program devoted to caring for the dying. Some are affiliated with a religious organization while others are secular. The focus here is on pain control and providing a more supportive psychological environment for the ill individual to spend their last days. Hospice can admit patients for the very end stages of their terminal care, or they often organize domiciliary services for support at home. These specially-skilled volunteers, social workers, nurses and physicians are trained in the management of the dying patient.

Home care is often preferred as surveys have shown that a home death was preferred by a great majority of people even if facilities were available for their care, and if they did not feel they were being a burden to their relatives. Older people favored home death about 4 to 1 over spending one's last days in a hospital. There's simply no place like home.

While we are still healthy and of sound mind, it is good to get our affairs in order. Recall the court case that opened this chapter. Had end-of-life documents been filled out and filed, dying wishes could have been fulfilled. In addition to filling and filing, be sure that copies are made (paper and digital) and that friends and family are fully aware of your personal intentions. Here is a list of such end-of-life documents:

- Living will
- Physicians' Orders for Life-Sustaining Treatment (POLST)
- Power of Attorney (POA) for healthcare/healthcare proxy
- Durable Power of Attorney (POA)
- Do Not Resuscitate (DNR) / Do Not Intubate (DNI) orders
- Diminishing capacity letters
- Organ donor designation
- Life insurance
- Personal property memorandum
- Digital assets memorandum
- Will and testament
- Medical information release = Health Insurance Portability and Accountability Act (HIPAA) waiver authorization
- Authorized user on investment and bank accounts

- Trust
- Letter of intent to loved ones

One might also consider the possibility of modifying hospitals to develop certain standards of humane care and educate staff to be psychologically sensitive to dying people. Doctors must also be trained in medical schools, and there should be education related to the psychology of death and dying in order to operate in this special care area.

Experiencing Challenging Circumstances and Terminal Lucidity

According to the German philosopher Friedrich Nietzsche, "That which does not kill us, makes us stronger." Is this true? If you've read the previous paragraphs, the best answer would be "not necessarily so." Think of the people who have survived a heart attack or some other illness, yet they have a host of resultant other conditions. Some studies do show Nietzsche's phrase to be true from a psychological perspective. Some trauma survivors report post-traumatic growth (PTG), which means that after a horrific event, their mood is enhanced, and they experience positive changes and personal growth. History is replete with examples, including noted psychiatrist Victor Frankl, who survived the Holocaust. He spent three years in four different labor camps and lost his parents, brother, and pregnant wife. His book is often cited as a work that has made a difference in the lives of its readers. Frankl's premise is that humans cannot avoid suffering, but we can choose how we deal with it, find meaning in life, and move on. Basically, we cannot change reality, but we can change how we respond to it.

People who have had a brush with death often report that they no longer fear death and experience personal growth. Many times, they reveal that they no longer care about material possessions or accumulating wealth. They also report finding joy in life's simpler things. Conversely, because people who have knocked on death's door return to the living with a greater sense of what truly matters, if their spouse isn't on board, they also experience a high rate of divorce. An outstanding person from history is Nelson Mandela, who spent nearly 27 years in prison and after his release in 1990 went on to become the first black president of South Africa, winning the Nobel Peace Prize in 1993.

We can hold Frankl and Mandela in high regard but know that not everybody has the mental strength or fortitude to survive. We also must keep in mind that while the mind is strong, the body may not be. As mortals, this is difficult to wrap our brains around. An interesting phenomenon in which dying people are entirely lucid in the hours before death is known as *terminal lucidity*. This term, coined by German biologist Michael Nahm in 2008, explains the totally lucid events of end-of-life people with various brain and mental illnesses who behave as though they are disease free. The classic case is of Anna Katharina Ehmer, a German woman who was presumably unable to hear or talk as a result of contracting meningitis and who spent her life in a mental institution. In the half hour before she died, she sang a song about death. This woman who had allegedly never uttered words in her life was singing "Where does the soul find its home, its peace? Peace, peace, heavenly peace!"

Healthcare professionals who are with dying patients have reported such clarity in many people regardless of illness. It's often referred to as the "end-of-life rally." Palliative care and hospice workers report that it is quite common for dying people to perk up right before death. This event is often disheartening for the surrounding loved ones because they falsely assume that a miracle has occurred, and the dying person is better. In fact, it seems like a cruel trick. Just when you think the end is near, the dying person rebounds and asks for favorite foods, remembers events of the past, talks as you remember them, and then at some point (hours to weeks) slips back into the pre-death stage and dies.

The science behind these physiological and psychological events is still not known, but it has been reported often enough that we know it to be true. In order to scientifically study these terminal lucid periods, researchers would have to engage in practices that are contrary to the premise of hospice and palliative medicine. At the end of life, the toxic chemotherapy, pricking, and prodding are supposed to stop. This is to be a time of allowing death to come in as peaceful a manner as possible. Having machines, monitors, and blood drawn for analysis just wouldn't be right.

Dying of a Broken Heart

Just as elusive as trying to figure out what is happening with the rally at the finish line is the oft cited dying of a broken heart. You know the scenario: Moms and Pops have been married for 65 years, spent every waking hour together in their retirement, and then suddenly Pops dies. Within days, Moms dies too. Is this just anecdotal? Or, is the event real?

The notion of dying of a broken heart has been around for centuries, sparking the genesis of an American folksong, *Barbara Allen*, which began as a traditional Scottish ballad. The story goes something like this: A servant who is in love with Barbara Allen is dying and asks that she be summoned to his death bed. After the servant dies, Barbara Allen hears the funeral bells ringing within the town. She is grief stricken and asks that her grave be dug, too. She dies and the two are buried next to each other. As time goes on, a rose grows out of his grave while a briar grows from hers, eventually intertwining forming a lover's knot to last forever. The earliest known record of the tale dates to 1666. Artists such as Joan Baez, Bob Dylan, and Dolly Parton have recorded versions of the folksong.

Songs about such traditional lore have immortalized the concept. There is, however, statistical evidence supporting death by broken-heartedness. An article published in 1969 showed that there was an increased mortality among widowers aged 44 and older. The researchers followed 4,486 men for nine years following the deaths of their wives and found that during the first six months of bereavement, 213 men died. In practical terms, this meant that the deaths of these men were 40 percent higher than the expected rate for men of the same age. After six months, the death rates were not different than in the general population. So, if you are a man, and your spouse dies, if you can live beyond the first six months postmortem, you'll die just like any other man who has not lost a spouse.

In another article published in March 1987, epidemiologists studied 95,647 widowed persons and discovered that when an older person dies, the widowed spouse had a higher risk of also dying immediately after bereavement. Mortality during the first week after death was more than two-fold compared to expected rates.

The Japanese people have a term for it: Takotsubo cardiomyopathy. Physicians recognized the condition on coronary angiograms and called it *takotsubo*, because the diagnostic image of the heart shape shows that it looks like the Japanese "takotsubo," a pot used to trap octopus. It is also known as stress-induced cardiomyopathy or broken-heart syndrome and the left ventricle weakens as a result of stress. As you might imagine, the list of stressors associated with broken-heart syndrome is long, but the loss of a loved one or pet makes the top ten. The precise mechanism is not known, but a surge in stress hormones, like adrenaline (epinephrine), is thought to play a role. The sudden rush of hormones stuns the heart muscle, specifically the left ventricle and coronary blood vessels, disabling heart pumping. It looks like a heart attack on an electrocardiogram (ECG/EKG). It is usually temporary and only the left ventricle seems to be affected. Before the event, the person experiences chest pain and shortness of breath—the same symptoms of a heart attack.

Recently, researchers at Rice University discovered that pro-inflammatory agents were higher in the blood of recently widowed persons. People who reported higher levels of grief had higher levels of inflammatory markers. The researchers also found that for individuals who reported high levels of depression, there were higher levels of inflammatory markers. Once again, we see how inflammation plays a role in our health.

Death Penalty

Suffering in life is unavoidable. And some of it is needless. Case in point is lethal injection, the government sponsored administration of drugs for the purpose of carrying out capital punishment. Capital punishment is the same as the death penalty. Like euthanasia in pets, the person receiving the drug concoction is supposed to lose consciousness and then stop breathing and lose heart function. This doesn't always happen, and many lethal injections go terribly wrong.

Any idea which country developed this method of execution? What country topped your list? Was it the United States? It should have been. Yes, the U.S. developed it as a method that was supposed to be better and cheaper than electrocution, gas inhalation, hanging, and death by a firing squad. New York physician Julius Mount Bleyer proposed the idea in 1888. The idea was further advanced in Nazi Germany as a way of annihilating unworthy lives, known by the German term, *Lebensunwertes Leben*. Nice. A practice still in vogue today can be compared to Nazi Germany. The Nazis were a mass movement from 1933 to 1945 led by Adolf Hitler. The party is still alive and well in the United States, calls itself the American Nazi Party, and is headquartered in Arlington, Virginia. Let's be clear: capital punishment is equated with Nazism.

At the writing of this book, 14 countries still have the death penalty. In addition to the United States, the countries are Bangladesh, China, Egypt, Ethiopia, India, Indonesia, Iran, Japan, Nigeria, Saudi Arabia, South Korea, Sri Lanka, and Taiwan. The practice has been abolished in every European country. Can you imagine if Germany still had the gas chamber as a way to be put to death by the government?

In February 2019, a bill to repeal Wyoming's death penalty law failed in the state senate, thanks, in part, to the vote of Republican State Senator Lynn Hutchings. Here's her quote: "The greatest man who ever lived died via the death penalty for you and me. I'm grateful to him for our future hope because of this. Governments were instituted to execute justice. If it wasn't for Jesus dying via the death penalty, we would all have no hope." Keep in mind, if you are a religious, single-issue voter, that there were 10 commandments.

Returning to the drug concoction: Many drug manufacturers don't want to be associated with the practice of lethal injection. To that end, 20 American and European pharmaceutical companies, including big Pharma conglomerates like Pfizer, blocked the sale of their drugs for use in lethal injection. You see, for lethal injection to occur in the U.S., the drugs used for the concoction have to be FDA approved. After all, you want the best drugs for the procedure. In August 14, 2018, Carey Dean Moore was executed in Nebraska. His cocktail contained diazepam (an anxiety reliever commonly known as Valium), fentanyl (the narcotic opioid pain reliever receiving so much attention these days), cisatracurium (a skeletal muscle relaxant), and potassium chloride (high amounts cause weakness, paralysis, low blood pressure, irregular heart rhythm, and death). The manufacturer of these drugs was the German pharmaceutical, Fresenius Kabi, who strongly objected to its use.

Pentobarbital, also known as pentobarbitone, is a short-acting barbiturate that causes respiratory arrest in high doses. It is used to execute convicted criminals in the United States. One of the manufacturers of the drug, Lundbeck, forbids its sale to prisons for the purpose of carrying out the death penalty. Nembutal is the trade name of the same drug that is manufactured by Abbott Pharmaceuticals, but it was discontinued in 1999. Known on the streets as "yellow jackets" because the capsule cover was yellow, Nembutal was widely abused. An overdose of Nembutal caused the death of actress Marilyn Monroe in 1962.

Why not just give the drug that put Michael Jackson to sleep forever? In the famed case, the King of Pop's personal physician, Dr. Conrad Murray, told the court he had given Jackson the potent anesthetic propofol, known by the trade name Diprivan. Propofol is used intravenously to induce and maintain general anesthesia. It has a quick onset and a short duration of action, which makes it a nice drug for use during surgeries. Anesthesiologists have cast doubt on the claim that propofol had killed Jackson, because the dosage Dr. Murray claimed to have given him was not sufficient to be lethal. In all likelihood, Jackson received a drug cocktail that contained propofol plus other drugs. These drugs were likely benzodiazepines such as Ativan (lorazepam), Valium (diazepam), or Versed (midazolam). When propofol is mixed with these other drugs, lower dosages are lethal. Both Ativan and Valium can be used before surgery to relieve anxiety while Versed makes you feel relaxed or sleepy before surgery. If you've ever had a colonoscopy, your doctor might have given you Versed just before the procedure. It's short-lived and effective. However, you don't want to mix any of these drugs with other substances, particularly alcohol, because it can slow breathing to the point of death. Not all of these drugs are used for capital punishment, though.

At this point, you might be thinking, so what. These people did terrible, horrific things and should be punished for their crimes. In many cases, that is absolutely true. However, is lethal injection the best way to make these people pay for their transgressions? Especially when innocent people have been executed? For a really morbid exercise, check out www.deathpenaltyinfo. org, the official website of the Death Penalty Information Center (DPIC), a national nonprofit organization founded in 1990 to promote informed discussions related to capital punishment. The DPIC gathers reports from leading researchers and makes statistics available regarding costs associated with capital punishment, the arbitrariness to which it is applied, and the issue of race as it relates to the application of the death penalty. In case this isn't grim enough, one can click on the "upcoming executions" calendar to check out the date, state, and name of the prisoner scheduled to be executed. Failed executions are also tracked. There is even an execution database where you can search by name.

Within the United States, 29 states currently have the death penalty and Texas has executed

108 prisoners since 2010—the most of any state. In July 2019, the U.S. Justice Department announced that it would resume executing death-row inmates. Before 2019, federal-level executions were rare as there was an informal moratorium. There were ten executions by the federal government in 2020: Daniel Lewis Lee on July 14, Wesley Ira Purkey on July 16, Dustin Lee Honken on July 17, Lezmond Charles Mitchell on August 26, Keith Dwayne Nelson on August 28, William Emmett LeCroy, Jr., on September 22, Christopher Andre Vialva, September 24, Orlando Hall, November 20, Brandon Bernard, December 10 and Alfred Bourgeois, December 11. Another execution was planned for December 8, 2020, but did not occur until January 13, 2021. This was for Lisa Montgomery and marked the first time that a woman has been federally executed since 1953. Does this seem normal? It's extremely important to note that studies show no link between the presence or absence of the death penalty and murder rates and some people have been wrongly accused.

Electrocution is another method of enacting the death penalty. It's also another way of accidentally dying. According to the Occupational Safety and Health Administration (OSHA), the federal agency that oversees safe and healthy working conditions, electrocution is the third leading cause of death for construction workers in the United States.

Before we get into the nitty gritty of how electricity kills you, we need to review a little physical science, beginning with electricity followed by voltage (volt), ampere (amps), resistance (ohms), and wattage (watt). Electricity is a form of energy that results from the flow of electrons, those negatively charged particles that surround the nucleus of an atom. It was a stormy night when Benjamin Franklin flew that proverbial kite into the charged sky in Philadelphia and set the groundwork for our understanding of electricity. Ben Franklin had been experimenting with electricity using a Leyden jar for quite some time. What in the world is a Leyden jar? Named after Leyden, a city in the western Netherlands where it was invented in 1745, this object was an early form of capacitor, a device that stores an electric charge. It consisted of a glass jar with layers of metal foil on the outside and the inside. Seven years later came Franklin's most famous experiment. This is when Ben decided to fly a kite during a storm. In June 1752, Franklin harnessed the electricity in the clouds and charged a Leyden jar from the key that was tied to the end of the kite string. This led to the realization that the passage of an electric current through a filament or a pocket of inert gas could illuminate a dark room, sparking the invention of the lightbulb. What does this have to do with the electric chair? We'll get to that, but first we need to know a little bit more about electricity.

A volt is a unit of electromotive force (think current) and is abbreviated V. Batteries come in varying voltages, like a 9-volt battery that operates your household smoke detector. An ampere, abbreviated A, is the unit of electric current. Resistance is impedance or something that slows or stops the effect exerted by one material thing on another. Lastly, wattage (watt) is the operating power of an electrical appliance; it is voltage multiplied by amps (current). For example, 100-watt light bulbs are brighter than 50-watt light bulbs. A plumbing analogy is often helpful: Think of voltage as water pressure, amps (current) as flow rate, and resistance as the diameter of the pipe. There are a couple more terms to add to the mix: direct current (DC) and alternating current (AC). As their names suggest, direct current is electric current that flows in one direction only and alternating current is an electric current that reverses its direction many times a second at regular intervals.

So how does electricity kill you? If as little as 7 milliamps reaches your heart for three seconds, your heart becomes arrhythmic and beats abnormally. However, our skin provides resistance, so electrical shocks don't automatically go to the heart. For a little perspective, lightning is about one billion volts, so if you get struck by lightning, your skin, rubber boots, and anything else will do little by way of resistance. You might think, "Hey, I've heard of people surviving a lightning strike." True. If the path of the lightning bolt doesn't travel through the heart, you could survive. The electricity could also bypass the heart, but hit other vital organs and in essence, cook them. Thus, the best way to survive a jolt of electricity is to make sure it doesn't go through your heart or some other vital organ.

When a person has been *electrocuted*, that means that they have experienced an electric current passing through the body; they may not die. *Electrocution*, on the other hand, is death by an electric current. The terms are *very* similar. Electrocution can be purposeful or accidental. In fact, the term was coined in the United States around 1889, to describe death by the electric chair. The U.S. has now come up with both lethal injection and death by the electric chair.

Now we can discuss AC and DC currents, Thomas Edison, George Westinghouse, General Electric Company, Westinghouse Corporation, and the electric chair. They are all tied together. During the 1880s, hanging was the primary method of carrying out the death penalty. Because this process seemed to be very slow and painful, the New York State government established a commission to study alternate forms of capital punishment. Enter the electric chair.

At the same time, there was a rivalry between Thomas Edison and George Westinghouse. George Westinghouse (1846–1914) was a U.S. engineer, held over 400 patents, developed AC service, and built a huge company known as the Westinghouse Corporation to manufacture his products. Thomas Edison (1847–1931) was a U.S. inventor, held over 1000 patents, developed DC service, and founded the Edison General Electric Company. Westinghouse and Edison were rivals. In a public display to show the perceived superiority of DC over AC, Edison set up a 1000-volt Westinghouse AC generator and killed animals by placing them on an electrified metal plate. Thus, was born the term "electrocution" to mean death by electricity. What a horrific scene to view. At this time, both forms—AC and DC—were now being used to kill people via the electric chair. Edison wanted the AC Westinghouse chair to be used because he thought people wouldn't want this type of current flowing through their homes and he'd win out with his DC type of electricity. The story doesn't end here.

Enter another inventor, Harold Brown (1857–1944). In 1888, Brown wrote a letter to the *New York Post* describing the fatal accident of a boy who died after touching an exposed telegraph wire, which was running on AC current. Okay, this proves that AC kills, but now we have a human involved. While working at the Edison research facility, Brown and an American neurologist, Dr. Frederick Peterson (1859–1938), developed a new electric chair. To aid their design, they used animals as subjects and showed that DC voltage tortured animals but did not kill them while AC voltage killed the animals quickly. Brown and Peterson conducted the so-called "War of Currents" using dogs. On January 1, 1889, the first electrical execution law went into effect. Westinghouse refused to sell AC generators to prisons and funded the appeals for its first would-be victims on the grounds that electrocution was cruel and unusual punishment. Edison and Brown remained proponents of the process. Despite George Westinghouse disapproving of the electric chair and funding appeals, for many years being electrocuted by the electric chair was referred to as being "Westinghoused."

Two ethical issues to note here: (1) Fred Peterson chaired the government committee charged with selecting the best electric chair *while* working for the Edison Company and (2) Animals were tortured in the name of science. Today, we practice scientific research with more scrutiny.

Recall the story of Henrietta Lacks and her immortal cells in Chapter 6. There must be informed consent, total understanding of what is happening, individuals must be voluntary participants, the research must benefit participants and society, no harm must come to the participants, all participants will be treated fairly, and participants have the right to leave the experiment at any time. Many professional scientific organizations also have position statements and animal use policies for educational institutions using animals in teaching or research. Public torture of animals is no longer accepted.

The first person in the world to be executed by electrocution was William Kemmler, a vegetable peddler from Buffalo, New York, who was convicted of killing his common-law wife, Matilda "Tillie" Ziegler, with a hatchet. The execution took place on August 6, 1890, when Kemmler died in the electric chair at Auburn State Prison, part of New York's infamous prison system. It didn't go smoothly. After 700 volts were sent through the body for 17 seconds, the current failed, but Kemmler wasn't dead. So, a second charge of 1,030 volts was delivered for another two minutes. When smoke started billowing from Kemmler's head, viewers declared him dead. While hanging was the method used prior to electrocution, proponents of the electric chair, including dentist Dr. Albert Southwick, viewed this form of death as more humane. After Kemmler's execution, Southwick declared, "We live in a higher civilization from this day on." Westinghouse had a different opinion and stated, "They would have done better with an axe." Figure 9.1 shows the electric chair in use at Auburn State Prison.

Would you like an explicit description of someone executed by an electric chair? Read for yourself the dissent in Glass v. Louisiana, 471 U.S. 1080 (1985); (5–4 decision). The state of Louisiana condemned to death Jimmy Glass. The petitioner, Glass, appealed to the U.S. Supreme Court that "electrocution causes the gratuitous infliction of unnecessary pain and suffering and does not comport with evolving standards of human dignity." The Supreme Court majority found otherwise. However, in his dissent to this decision, Justice William Brennan wrote the following:

Figure 9.1. Electric Chair in Auburn State Prison (Library of Congress).

"This evidence suggests that death by electrical current is extremely violent and inflicts pain and indignities far beyond the mere extinguishment of life. Witnesses routinely report that, when the switch is thrown, the condemned prisoner cringes, leaps, and fights the straps with amazing strength.

The hands turn red, then white, and the cords of the neck stand out like steel bands. The prisoner's limbs, fingers, toes, and face are severely contorted. The force of the electrical current is so powerful that the prisoner's eyeballs sometimes pop out and rest on [his] cheeks. The prisoner often defecates, urinates, and vomits blood and drool.

The body turns bright red as its temperature rises, and the prisoner's flesh swells, and his skin stretches to the point of breaking. Sometimes the prisoner catches on fire, particularly if [he] perspires excessively. Witnesses hear a loud and sustained sound like bacon frying, and the sickly sweet smell of burning flesh permeates the chamber. This smell of frying human flesh in the immediate neighborhood of the chair is sometimes bad enough to nauseate even the Press representatives who are present. In the meantime, the prisoner almost literally boils: the temperature in the brain itself approaches the boiling point of water, and when the post-electrocution autopsy is performed, the liver is so hot that doctors have said that it cannot be touched by the human hand. The body frequently is badly burned and disfigured."

Choice at the End

So, here we are facing death. What if that road to the end is absolutely horrific? It you are facing a painful disease, are in an irreversible coma, or are experiencing a great deal of suffering, doesn't painlessly slipping away toward death sound like a peaceful option? Before making a snap judgment on this, think, and engage in the "have you ever" exercise. Have you ever experienced excruciating pain that would not go away? Have you ever been with a loved one who was in such unrelenting physical agony that they were not able to function? Have you ever experienced someone's death firsthand? Have you ever watched somebody struggling to breathe such that every single breath was a chore? Have you ever seen a once healthy, happy 200-pound man lose 100 pounds in eight months as cancer ravaged his body to the point that he didn't have enough energy to get up to use the bathroom and soiled himself and the bedsheets? Have you ever watched your spouse lose the ability to move as the effects of Lou Gehrig's disease began robbing him of movement beginning in the legs and working its way up to the head such that talking became impossible, eating was out of the question, and choking was a continual, inevitable event—all the while the brain was fully cognizant? Have you ever watched an electrocution of a convicted felon? Keep these thoughts in mind as we discuss euthanasia.

There has been much public discussion and notoriety throughout the world regarding euthanasia. Euthanasia has been called by many monikers such as painless death, assisted death, mercy killing, physician-assisted suicide, active suicide, or the deliberate ending of life for a person suffering an incurable disease. There are slight nuances among the terms, so let's break them down. Euthanasia is *not* the same as physician-assisted suicide. Euthanasia involves a doctor who directly administers a lethal dose of medication to a patient. A common example you may have witnessed is having a pet "put to sleep." Physician-assisted suicide refers to a doctor providing a means for death, such as writing a drug prescription for a patient, thereby providing a lethal means for the person to end their life. Physician-assisted suicide enables the patient to choose the time and removes direct intervention by the physician.

Of course, there are advantages and disadvantages to both practices. At the beginning of their medical careers, physicians usually take a Hippocratic oath, which states the obligations of proper conduct of doctors. Named for the Greek physician Hippocrates, considered the father of Western medicine, the oath establishes the principles of medical confidentiality and non-maleficence. Non-maleficence is just a fancy way of saying, "First, do no harm." Providing the means to end a life or causing the death of a person could be interpreted as doing harm. That said, in 1945, the United Nations was charged with promoting human rights, of which human dignity is seen as a human right. Dying patients want to maintain their human dignity. Imagine if you were totally aware of your surroundings but couldn't move or talk and had to have another adult change your diaper. How degrading is that? The oath is open to interpretation.

In general, the term euthanasia has been replaced by "death with dignity," and it has mixed support. People against it feel that it would be a breach of trust between the physician and patient or the nurse and patient. It may even be perceived as a "threat" by other ill patients, who do not fully understand the situation. At the writing of this book, Washington, D.C., and states with a death with dignity statute include California, Colorado, Hawaii, Maine, New Jersey, Oregon, Vermont, and Washington. Around the world, euthanasia is legal in four countries (Belgium, Colombia, Luxembourg, and the Netherlands) while physician-assisted suicide is legal in seven (Belgium, Canada, Finland, Germany, Luxembourg, the Netherlands, and Switzerland).

Sometimes people are simply too sick to administer lethal doses of drugs to themselves. Recall from Chapter 8 how drugs are delivered and metabolized by the body. If a person cannot swallow or is too weak to give a self-injection or the body has basically shut down and isn't metabolizing much of anything, they will require some other person to assist. In Louisiana, Mississippi, and Nebraska, simply giving a person advice or information on ending life or supplying the means for a person to end her life is considered a criminal act and the physician doing so can be charged with a felony. Information suppression is never good, regardless of your beliefs.

Ideally, euthanasia (not suicide, which is not illegal) is the deliberate ending of life under carefully agreed conditions, with legal agreement. It is usually carried out in the presence of witnesses, with the patient's request, and with agreement of the patient and relatives. Many people still consider it to be "homicide on request."

It should be noted that senicide (killing of the aged) and dementicide (killing of demented people) is a crime. One must understand, too, that euthanasia is not "death" due to the side-effects of drugs used for the relief of intractable pain. It is also not the deliberate withholding of antibiotics or specific surgical procedures, which are temporary measures to prolong a dying person's life.

We can create a situation in which beneficence (doing good) and non-maleficence can happen simultaneously. Such is the case of the *double effect*, a phenomenon in which two different consequences occur as the result of a single action. Let's use end-of-life morphine as an example. As a drug, morphine is an alkaloid of opium and is used as an analgesic (pain reliever), sedative (sleep inducer), and anxiolytic (antianxiety agent). In dying patients, morphine is often given because it eases pain and suffering. This is beneficial. It also shortens the life of the patient by suppressing the respiratory system. This is non-maleficence. Or, is it? Wouldn't you want yourself or your loved one to exit this world as tranquilly as possible, especially when death is inevitable?

Consider the common practice of euthanizing a pet. The euphemism is "putting your pet to sleep." What are the physiological events that occur when the veterinarian begins the process? To begin, many veterinarians will give your furry friend an injection of a sedative, also known as a tranquilizer, to make both the pet and the human comfortable. Yes, I said, human. No, the human is not getting the sedative, but as in humans, sedatives are drugs that quiet nervous excitement. Thus, the vet can give an injection of the sedative to your cat or dog and they will fall asleep. This isn't the end. The pet is literally sleeping, and some may actually snore. This is an important part of the euthanasia process because the injection doesn't hurt the animal and it gives the humans time to be with their furry family member before the lethal injection is given. If the sedative step weren't done, your dog/cat would go from living to dead in a matter of moments—and this is an event that is very hard to watch. Imagine seeing a loved one fainting and never re-awakening. The slumber period before final euthanasia also takes away the fumbling for a vein to use for injecting the final drug.

Acepromazine is a commonly used tranquilizer prior to human anesthesia and surgery, and it is commonly used in other animals. Remember, we humans are animals, too. Many vets will also add another drug such as xylazine, because it also sedates, anesthetizes, and induces muscle relaxation. Other drugs like ketamine, Telazol, Valium, propofol, and medetomidine are also used; the point is that sedation is an important step for easing the transition. After giving this concoction, your dog/cat is sleeping rather peacefully. They no longer feel any pain or anxiety and you can spend time hugging, kissing, and saying goodbye. Barbiturates are then given as the final injection. This drug depresses the central nervous system (brain and spinal cord) and is given in a vein for rapid transport. If the animal is severely dehydrated or the veins have collapsed as often occurs in very sick animals, the drug can be administered via other routes. Generally, the heart stops within one minute after it is administered. Regardless, it is an emotional experience. But it sure beats the experience of having your animal suffer. Since humans and pets are both animals, wouldn't it be nice if we didn't have to suffer?

Anorexia Nervosa and Cachexia

As shown already, there are many ways for a person to die and the physiological aspects have been well described. This chapter will wrap up with two final types because both look identical to the naked eye, yet each has a totally different mechanism, and both will make you think about how the human body responds to seemingly identical circumstances. One is willed; the other cannot be reversed. Both cause death. The afflictions: anorexia nervosa and cachexia. By the strictest definition of the term, *anorexia* means diminished appetite or an aversion to food. For example, you have the flu and are not feeling well and just don't feel like eating. That is anorexia. Or, you have a chronic illness that makes food unpalatable, so you do not eat because you just are not hungry. This is not a hunger strike or a fast. Quick diversion: A person who is fasting chooses not to eat. Fasting people abstain from all or some kinds of food or beverages. Mahatma Gandhi (1869–1948), the father of the nation of India, who was an Indian nationalist and spiritual leader, fasted at different times during India's freedom movement. The practice of fasting has a long history dating back thousands of years. Sometimes fasting is done in protest or civil disobedience; sometimes it is done for religious reasons, and sometimes it is done in preparation for medical tests. Regardless, the purposes are varied. Intermittent fasting, eating every other day, eating five days per week, or just cutting back every other day, may have health benefits such as weight loss and less inflammation.

A type of anorexia is anorexia nervosa. Anorexia and anorexia nervosa are not the same condition, although the terms are often used interchangeably in everyday conversation. *Anorexia nervosa* is a mental disorder. It manifests as an extreme fear of becoming overweight and is accompanied by food aversion. Anorexia nervosa is a compulsive pursuit of thinness at the expense of good health. Much has been written about this disorder, which is characterized by steady weight loss to and below 85 percent of normal weight for height. Individuals with anorexia nervosa achieve profound weight loss through not eating, strenuous exercise, self-induced vomiting to rid the body of recently-eaten food, and laxatives to rid the body of excess water and feces.

From a physiological standpoint, there is profound muscle wasting—a condition called *muscle atrophy*. The fat just below the skin's surface, called subcutaneous fat, diminishes to the point where there is very little to none. This is followed by a slew of other conditions such as anemia (not enough iron in the blood), electrolyte (mineral) imbalance, bradycardia (slow heart rate), hypotension (low blood pressure), asthenia (abnormal physical weakness and lack of energy), exaggerated cold sensitivity, constipation, dry skin with increased pigmentation, and growth of lanugo (fine, soft hair). Women become amenorrheal (failure to menstruate). The risk of osteoporosis (brittle bone disease), fractures, and irreversible bone deformation increases due to lack of minerals such as calcium and phosphorus, and also due to hormonal imbalance. In short, it is a life-threatening eating disorder and is marked by extreme weight loss. If treated in its early stages, anorexia nervosa can be reversed. In 1983, American musician Karen Carpenter died at age 32 as a result of anorexia.

Contrast anorexia nervosa with cachexia. The word *cachexia* comes from the Greek word *kakos* that means *bad* plus the suffix *hexis*, which means *condition of the body*. On the outside, cachexia looks like anorexia nervosa; the difference is that the weight loss and muscle wasting occur as a result of a chronic disease. I tell my students to remember this word by thinking about the abbreviation for cancer, which is CA. The letters C and A are the first two letters of the word *cachexia*. This condition of the body often accompanies cancer. Think of cachexia as cancer anorexia.

People with cachexia generally respond poorly to treatment, have a very short survival time, and suffer greatly. They have no energy, are extremely tired, are fatigued with very little exertion, and have a very low quality of life. It's a horrible condition because the person no longer likes the way they look; they have difficulty interacting with other people, including their own family; eating is very troublesome; and social interactions become increasingly difficult. While writing this book, I had a front-row seat to the ravages of cancer and its concomitant cachexia: Within nine months of diagnosis for kidney cancer, my 205-pound brother lost 105 pounds and died. Before his death, he looked like a skeleton with skin covering his bones. He looked like a walking dead man would look—if he could walk. He could barely walk because he had so little energy. Although cancer drugs (chemotherapy) were beating back the cancer, his cachexia was so far advanced that nothing could save him. He suffered greatly as a result of his cachexia. On the day he passed away, he kept telling me over and over that he wanted to die. His mind was still quite lucid, but his body was ravaged. Imagine experiencing this with a person you cared about deeply.

Enter palliative care. Palliative care is extremely important for people diagnosed with chronic disease or terminal illnesses because it can make the person more comfortable. This is especially true when a person suffers with cachexia. Cachexia has such an effect on a person that studies of people in palliative care settings showed that cachexia was within the top five most troubling symptom clusters. It ranked above pain and difficulty breathing. Diagnostic criteria for cachexia exist; however, once you've seen it, you know it. In a medical setting, basic criteria include a five-pound weight loss in the preceding two months without trying to lose weight or an estimated caloric intake of less than 20 calories per kilogram (2.2 pounds) of body weight.

Putting that into perspective, a 220-pound person weighs 100 kilograms. Twenty calories times 100 kilograms = 2,000 calories. Nutritional counseling and eating more food have relatively little to no effect. People with cachexia simply don't feel like eating, and if they *are* able to eat, the body cannot utilize the calories in a normal fashion.

When comparing cachexia with anorexia, it is important to realize that cachexia is a complex pathophysiological metabolic syndrome and anorexia is a nutritional deficiency with an underlying psychological disorder. Simply put: You can treat anorexia by eating, but eating cannot treat cachexia. In fact, once cachexia has reached a critical point, it cannot even be reversed, despite any type of nutrition or drug therapy. The tissue wasting and multiple organ system involvement are just too profound.

Terminally Ill

End-of-life palliative care is probably the kindest and most humane act that we can impart to our dying loved one. Much is known about terminally ill patients, including how to make them comfortable during their last days along with predicting survival. A scoring system for predicting survival is the palliative prognostic index (PPI). Using assessments such as performance status, clinical symptoms, oral intake, edema, dyspnea (difficulty breathing) at rest, and delirium, the PPI can predict with high accuracy the length of time a person has to live. The scores from each of these prognostic domains are added together and applied to the scale.

- A score of 0 to 2.0 is associated with a median survival of 90 days.
- A score of 2.1–4.0 is associated with a median survival of 61 days.
- A score greater than 4.0 is associated with a median survival of 12 days.

In my personal experience, my brother had a PPI of 2.5–4.0. This value was calculated on August 17; he died on October 16. Do the math. Since he died at 11:45 p.m., he was a mere 15 minutes shy of the 61-day mark. This scale is highly predictive of survival and is routinely used in palliative medicine.

I frequently lecture my students on end-of-life decision making along with the role of electrolytes. These are the same electrolytes discussed a few paragraphs ago. You see, if you have too much potassium (K^+) in your blood, it will shut down your heart. Since I teach the future healthcare workers of the world, I literally preach that if it looks like I'm not going to make it, pretty please just give me a little potassium bolus in my IV line. Too much potassium stops the heart and I'll drift off into perpetual slumber. Along those same lines, I also tell them that if I do wake up from my coma and see a former student standing there, I want to be able to say or think, "Thank God it's you!" instead of "Sweet Jesus take me!"

There is a great advantage if continuity of care can be maintained for both the dying patient and the surviving relatives. Surviving relatives experience bereavement. To be bereaved means that you are deprived of a loved one because he or she has died. It is the ultimate absence; and the period of grieving and the expression of grief afterward is known as bereavement. Doctors and health care personnel should make sure that bereaved relatives have the chance to go through the stages of the grieving process. These stages are, generally, numbness, weeping and sobbing, and then depression. One should permit the bereaved person to live through these stages and not short-circuit the process unless the depression appears to be interminable (endless), which may be the case in old age. If endless, treatment may be necessary. Treatment may involve simple encouragement to socialize or the use of anti-depressant drugs accompanied with psychiatric management. Research has shown that antidepressants work best when accompanied by talk therapy.

In addition to the general bereavement, the death of a spouse of an older person may also include other hardships. For example, the loss of pension may leave the survivor less well-off. Decisions include deciding whether the older person can remain alone or should move in with relatives, or if the person needs to apply for some form of care. These decisions should not be made immediately and suddenly, but after careful consideration. For certain, no irrevocable decisions should be acted upon until the main grieving period has come to an end. Even for younger people who lose a loved one, the law of permanency sets in around year two after the death. Year one might be a year of shock, disbelief, and myriad other feelings. Year two is when you realize the loved one really isn't coming back. Grief is something you live with forever.

End-of-Life Decision-Making Organizations

Exit International is familiar with Nembutal, the aforementioned drug that killed Marilyn Monroe. According to the Australian organization's website, Exit International is a leading end-of-life choices, information, and advocacy organization. It was founded in 1997 by Dr. Philip Nitschke. Their foundation is based on the premise that dying is not a medical process and as such it does not always—if ever—need medical intervention. They aim to ensure that all rational adults have access to the best available information so that people have absolute control over when and how they die. Where does Nembutal intersect with this organization? Scam internet sites are using Dr. Philip Nitschke's image to sell the banned drug Nembutal. The organization has published *The Peaceful Pill Handbook* both in print and online that provides research and information on voluntary euthanasia and assisted suicide for older people, those who are seriously ill, and family and friends of seriously ill people.

Another organization, Final Exit Network, originally organized in 1980 as the Hemlock Society, recognizes the need for compassionate support and death with dignity education in all states. Hemlock is a highly poisonous European plant of the parsley family. Extracts from the plant are used as a sedative, antispasmodic, and pain reliever. It's also used to make poisonous potions; such a potion is believed to have poisoned the ancient Athenian philosopher Socrates. Final Exit's website also offers information on living wills, advance directives, exit guide services, and helpful information and resources on death and dying, including the book *Final Exit*.

On a Final Note

Death and dying are not very upbeat topics on which to end a book. However, giving suggestions for decreasing premature mortality is an optimistic way to conclude. As of September 2, 2021, the world's oldest person is Kane Tanaka from Fukuoka, Japan. She was 118 years, 243 days old.

Want to live a long life? Eat well, get plenty of rest, exercise regularly, have a strong social network, and increase the number of neurons (nerve cells) in your brain. Recent research is linking neuron number in the cerebral cortex to longevity. The cerebral cortex is the outer layer of the cerebrum, the brain's biggest part. It is composed of folded gray matter and plays an important role in consciousness. Whatever your body size, animals with more neurons in their cerebral cortex have a better shot at living longer.

How do we grow new neurons? The fancy term for such growth is neurogenesis, a word that literally means *formation of new neurons*. When I began my teaching career over two decades ago, I lectured on nature versus nurture and the work of Michael Kaplan. In the 1970s, Kaplan reported that enriched environments enhance the growth of new neurons. Basically, if you compare the brains of animals kept in isolation with the brains of animals who were given interesting toys, the interesting toy group formed more neural connections and had more neurons. His work was largely poo-pooed, he left research science, and had a productive career as a rehabilitation doctor.

Kaplan's work has been revisited and new work on neurogenesis is enlightening. What we're learning today is that physical exercise will increase neuron number. Physical exercise increases neurogenesis, while stress and depression hamper new neuron growth. Neurogenesis doesn't occur everywhere in the brain, but it does occur in the hippocampus, olfactory bulbs, and cerebral cortex. The hippocampus is a brain part involved with emotion and memory while the olfactory bulbs are involved with smell. We do know that smell is a strong trigger for memory. You've experienced this: you smell something and immediately it reminds you of something. Chocolate chip cookies fresh out of the oven may remind you of your beloved grandma. The cerebral cortex (*cortex* means *bark*) is the outer layer of the cerebrum (the brain's biggest part) that plays an important role in consciousness.

Anyhow, it was once thought that we couldn't form new brain cells. We had to preserve what we had. Yet, we know we can learn new things as we age, referred to as brain plasticity. Thus, it is advisable to keep challenging ourselves, exploring new things, exercising, and building new neurons to live a longer life. Yet, what happens when that long life is mired in health issues?

Of the almost 200 countries in the world, the United States is one of the wealthiest. Using data from the International Monetary Fund, world wealth is ranked on gross domestic product (GDP) per capital based on purchasing power parity. Basically, it compares the currencies of nations in relation to the cost of goods, thereby determining its economy. Note that the U.S. *is one of the wealthiest*. It is not the wealthiest. That title is held by Qatar. The U.S. is #12, sandwiched between #11 San Marino and #13 Saudi Arabia. Guess where the United States does rank #1? The U.S. was #1 in the world for health expenditure as a percent of the GDP. Astonishing, isn't it? The life expectancy is 78.6 years in the United States. However, Americans have a shorter life expectancy than many other countries (the shortest among most "first-world" countries), ranking 46th in the world. The top five countries with the longest life expectancy are Hong Kong, Japan, Macao, Switzerland, and Singapore. Our neighbors to the north, Canada, ranks 16th. The United States also ranks 33rd out of 36 countries in infant mortality, with a rate of 5.9 deaths per 1,000 live infant births. Moreover, advances in medicine have caused health expenditures to rise and life expectancy to increase in most affluent countries, but not in the United

States. What we do know is that lifestyle, access to healthcare, prenatal care, and good nutrition play a role in life expectancy. Moreover, adopting five healthy lifestyle habits can reduce premature mortality and prolong life expectancy.

Need a little more information, like, how many years longer? A great study that used over 34 years of follow-up data found that female Americans age 50 lived 14.0 years longer and males age 50 lived 12.2 years longer when they adhered to a life that embraced five low-risk factors. Here's the lowdown on the low-risk factors:

Risk Factor 1: Never smoking.
Risk Factor 2: Body Mass Index (BMI) of 18.5 to 24.9 kg/m²
Risk Factor 3: Thirty or more minutes of moderate to vigorous physical activity per day
Risk Factor 4: Moderate alcohol intake (5 to 15 g/day, or roughly one standard drink, for women and 5 to 30 g/day, roughly two standard drinks, for men)
Risk Factor 5: High-quality diet rich in fruits and vegetables

The moral of the story? Prevention of disease and enhancement of lifestyle choices should be part of our daily lives.

TAKE-HOME MESSAGE

Defining death is difficult and dying is a life-long process. Religious and personal beliefs preclude people from accepting euthanasia as a viable option. Some diseases and illnesses cause so much pain and human distress that euthanasia can be the most humane option we can offer to suffering people.

> *No one wants to die. Even people who want to go to heaven don't want to die to get there. And yet death is the destination we all share. No one has ever escaped it. And that is as it should be, because death is very likely the single best invention of life. It is life's change agent. It clears out the old to make way for the new.*
> —Steve Jobs (1955–2011); co-founder of Apple Inc.

Bibliography

Reputable General Websites

Ad Fontes Media: www.adfontesmedia.com

Agency for Healthcare Research and Quality: www.ahrq.gov

American Cancer Society: www.cancer.org

American Council on Science and Health: www.acsh.org

American Diabetes Association: www.diabetes.org

American Heart Association: www.heart.org

American Journal of Clinical Nutrition: www.ajcn.org

American Medical Association: www.ama-assn.org

Animal Research Info: www.animalresearch.info/en

Big Think: www.bigthink.com

BioEthics Education Project: www.beep.ac.uk/content/1.0.html

Brookings Institution: www.brookings.edu

Center for Reproductive Rights Retrieved: www.reproductiverights.org/about-us

Centers for Disease Control and Prevention: www.cdc.gov

Cleveland Clinic: www.my.clevelandclinic.org

Clinical Correlations: www.clinicalcorrelations.org/about

Cochrane: www.cochrane.org

The Conversation: www.theconversation.com/us

Fact Check.Org A Project of the Annenberg Public Policy Center: www.factcheck.org

Federal Citizen Information Center: www.gsa.gov/portal

Go Ask Alice!: www.goaskalice.columbia.edu

Guttmacher Institute: www.guttmacher.org

Health On the Net: www.hon.ch/en

Johns Hopkins Medicine: www.hopkinsmedicine.org

Live Science: www.livescience.com

Mayo Clinic: www.mayoclinic.org

The Media Bias Chart: www.adfontesmedia.com

Medscape: www.medscape.com/today

Merck Manuals: www.merckmanuals.co

New England Journal of Medicine: www.nejm.org

Office of Dietary Supplements (ODS): www.ods.nih.gov

The People's Pharmacy: www.peoplespharmacy.com

Pew Research Center: www.pewresearch.org

ProCon.org—Pros and Cons of Controversial Issues: www.procon.org

ProPublica: www.propublica.org

PubMed—NCBI: www.ncbi.nlm.nih.gov

Quackwatch: www.quackwatch.org

Rand Corporation: www.rand.org

Reuters News Agency: www.reuters.com

Rewire News Group: www.rewirenewsgroup.com

Science-Based Medicine: www.sciencebasedmedicine.org

Science News Magazine: www.sciencenews.org

Snopes.com: www.snopes.com

TheSkimm: www.theskimm.com

The Straight Dope: www.straightdope.com

ThoughtCo: www.thoughtco.com

Union of Concerned Scientists: www.ucsusa.org

USAFacts: www.usafacts.org

The Logic of Science: www.thelogicofscience.com

Preface

Christensen, A. (n.d.). *Washington and Lee University LibGuides: Ruth Bader Ginsburg: A Reading List: Arguments before the Supreme Court*. Retrieved from https://libguides.wlu.edu/law/RBG/arguments.

Introduction

AllSides | Balanced news via media bias ratings for an unbiased news perspective. (2019, May 27). AllSides. https://www.allsides.com/unbiased-balanced-news.

Ambady, N., & Rosenthal, R. (1993). Half a minute: Predicting teacher evaluations from thin slices of nonverbal behavior and physical attractiveness. *Journal of Personality and Social Psychology*, 64(3), 431–441. https://doi.org/10.1037/0022–3514.64.3.431.

Are Coffee and Alcohol the Fountain of Youth? (2018, March 5). Retrieved from The People's Pharmacy website: https://www.peoplespharmacy.com/2018/03/05/are-coffee-and-alcohol-the-fountain-of-youth/.

Avalanches, explained. (2019, July 19). Retrieved from National Geographic website: https://www.nationalgeographic.com/environment/natural-disasters/avalanches/.

Bansal, A., Garg, C., Pakhare, A., & Gupta, S. (2018).

Selfies: A boon or bane? *Journal of Family Medicine and Primary Care*, 7(4), 828. https://doi.org/10.4103/jfmpc.jfmpc_109_18.

Bratsberg, B., & Rogeberg, O. (2018). Flynn effect and its reversal are both environmentally caused. *Proceedings of the National Academy of Sciences*, 115(26), 6674–6678. https://doi.org/10.1073/pnas.1718793115.

Campoy, A. (n.d.). *US-born Americans commit more rape and murder than immigrants, Texas data show*. Quartz. Retrieved from https://qz.com/1227461-trumps-immigration-claims-debunked-texas-data-show-us-born-americans-commit-more-rape-and-murder/.

Carpenter, S. K., Wilford, M. M., Kornell, N., & Mullaney, K. M. (2013). Appearances can be deceiving: instructor fluency increases perceptions of learning without increasing actual learning. Psychonomic Bulletin & Review, 20(6), 1350–1356. https://doi.org/10.3758/s13423-013-0442-z.

Cenziper, D. O., Jim. (2016). *Love Wins: the Lovers and Lawyers Who Fought the Landmark Case for Marriage Equality*. HarperCollins.

Corsi, J. R. (2008). *The Obama nation: Leftist politics and the cult of personality* (1st Threshold Editions hardcover ed.). Threshold Editions/Simon & Schuster.

Gorman, S. E., & Gorman, J. M. (2017). *Denying to the grave: Why we ignore the facts that will save us*. Oxford ; New York: Oxford University Press.

Hamzelou, J. (n.d.). Newly-discovered human organ may help explain how cancer spreads. Retrieved from New Scientist website: https://www.newscientist.com/article/2164903-newly-discovered-human-organ-may-help-explain-how-cancer-spreads/.

Huang, Y. C., & Lin, S. H. (n.d.). Assessment of Charisma as a Factor in Effective Teaching. *Educational Technology and Society*, 17, 284–295.

Infographic: Report: Facebook Poses A Major Threat To Public Health. (n.d.). Statista Infographics. Retrieved from https://www.statista.com/chart/22660/health-misinformation-on-facebook/.

Interactive Media Bias Chart—Ad Fontes Media. (n.d.). Retrieved from https://www.adfontesmedia.com/interactive-media-bias-chart/.

Isaacson, R. L., McKeachie, W. J., & Milholland, J. E. (1963). Correlation of teacher personality variables and student ratings. *Journal of Educational Psychology*, 54(2), 110–117. https://doi.org/10.1037/h0048797.

Loeb, S., Sengupta, S., Butaney, M., Macaluso, J. N., Czarniecki, S. W., Robbins, R., … Langford, A. (2019). Dissemination of Misinformative and Biased Information about Prostate Cancer on YouTube. European Urology, 75(4), 564–567. https://doi.org/10.1016/j.eururo.2018.10.056.

Mankiw, N. G. (2013). Defending the One Percent. *Journal of Economic Perspectives*, 27(3), 21–34.

Mercier, H., & Sperber, D. (2019). *Enigma of reason.*

Cambridge, Massachusetts: Harvard University Press.

Murray, H. G., Rushton, J. P., & Paunonen, S. V. (1990). Teacher personality traits and student instructional ratings in six types of university courses. *Journal of Educational Psychology*, 82(2), 250–261. https://doi.org/10.1037/0022-0663.82.2.250.

Naftulin, D. H., Ware, J. E., & Donnelly, F. A. (1973). The Doctor Fox Lecture: A Paradigm of Educational Seduction. *Journal of Medical Education*, 48(7), 630–635.

O'Neill, J. E., & Corsi, J. R. (2004). *Unfit for command: Swift boat veterans speak out against John Kerry*. Regnery Pub.; Distributed to the trade by National Book Network.

Patrick, C. L. (2011). Student evaluations of teaching: effects of the Big Five personality traits, grades and the validity hypothesis. *Assessment & Evaluation in Higher Education*, 36(2), 239–249. https://doi.org/10.1080/02602930903308258.

Peter, L. J., & Hull, R. (2009). *The Peter principle: why things always go wrong* (1st Collins Business ed.). New York: Collins Business.

Pew Research. (2015, September 10). What Americans know and don't know about science. Retrieved from Pew Research Center Science & Society website: https://www.pewresearch.org/science/2015/09/10/what-the-public-knows-and-does-not-know-about-science/.

Piketty, T., & Saez, E. (2014). Inequality in the long run. *Science*, 344(6186), 838–843. https://doi.org/10.1126/science.1251936.

Schuld, J., Slotta, J. E., Schuld, S., Kollmar, O., Schilling, M. K., & Richter, S. (2011). Popular Belief Meets Surgical Reality: Impact of Lunar Phases, Friday the 13th and Zodiac Signs on Emergency Operations and Intraoperative Blood Loss. *World Journal of Surgery*, 35(9), 1945–1949. https://doi.org/10.1007/s00268-011-1166-8.

Selby, L. (2015, October 27). "Full moon madness" in the ER: Myth or reality? *The DO*. Retrieved from https://thedo.osteopathic.org/2015/10/full-moon-madness-in-the-er-myth-or-reality/.

Selfie-Taking Woman Apologizes After Jaguar Attack At Arizona Zoo. (n.d.). Retrieved from NPR.org website: https://www.npr.org/2019/03/11/702386685/selfie-taking-woman-apologizes-after-jaguar-attack-at-arizona-zoo.

Sloman, S. A., & Fernbach, P. (2017). *The knowledge illusion: Why we never think alone*. New York: Riverhead Books.

Sommeiller, E., & Price, M. (2018, July 19). The new gilded age. Income inequality in the U.S. by state, metropolitan area, and county. *Economic Policy Institute*. Retrieved from https://www.epi.org/publication/the-new-gilded-age-income-inequality-in-the-u-s-by-state-metropolitan-area-and-county/.

Supreme Court Landmarks. (n.d.). Retrieved from United States Courts website: https://www.

uscourts.gov/about-federal-courts/educational-resources/supreme-court-landmarks.

Sutton, R. I. (2007). *The no asshole rule: building a civilized workplace and surviving one that isn't* (1st ed.). New York: Warner Business Books.

The ALS Association. (n.d.). Retrieved from http://www.alsa.org/about-als/.

Topic: Social media. Statista. Retrieved from https://www.statista.com/topics/1164/social-networks/.

Trump may owe his 2016 victory to "fake news," new study suggests. (2018, February 15). *The Conversation.* Retrieved from http://theconversation.com/trump-may-owe-his-2016-victory-to-fake-news-new-study-suggests-91538.

U.S. Reports: New York Times Co. v. Sullivan, 376 U.S. 254 (1964). [Image]. (n.d.). Retrieved from Library of Congress, Washington, D.C. 20540 USA website: https://www.loc.gov/item/usrep376254/.

Williams, R. G., & Ware, J. E. (1976). Validity of student ratings of instruction under different incentive conditions: A further study of the Dr. Fox effect. *Journal of Educational Psychology*, 68(1), 48–56. https://doi.org/10.1037/0022–0663.68.1.48.

Chapter 1

Aviles, J. M., Whelan, S. E., Hernke, D. A., Williams, B. A., Kenny, K. E., O'Fallon, W. M., & Kopecky, S. L. (2001). Intercessory Prayer and Cardiovascular Disease Progression in a Coronary Care Unit Population: A Randomized Controlled Trial. *Mayo Clinic Proceedings*, 76(12), 1192–1198. https://doi.org/10.4065/76.12.1192.

Bacon, F. (2012). *Novum organum "new method."* Place of publication not identified: Bottom Of The Hill Publishing.

Blitz, M. (2016, February 23). The amazing true story of how the microwave was invented by accident. *Popular Mechanics.* Retrieved from https://www.popularmechanics.com/technology/gadgets/a19567/how-the-microwave-was-invented-by-accident/.

Chalmers University of Technology. (2011, October 26). New weapon against cancer: Microwaves can be used to create medical images. *ScienceDaily.* Retrieved from www.sciencedaily.com/releases/2011/10/111025090349.htm.

The Condition of Education—Preprimary, Elementary, and Secondary Education—High School Completion—Public High School Graduation Rates—Indicator May (2019). (n.d.). Retrieved from https://nces.ed.gov/programs/coe/indicator_coi.asp.

Desilver, D. (2017, February 15). U.S. students' academic achievement still lags that of their peers in many other countries. Pew Research Center. Retrieved from http://pewrsr.ch/2kLfozD.

Gauchat, G. (2012). Politicization of Science in the Public Sphere: A Study of Public Trust in the United States, 1974 to 2010. *American Sociological Review*, 77(2), 167–187. https://doi.org/10.1177/0003122412438225.

Hambrick, D. Z., & Marquardt, M. (2018, February 6). *Cognitive ability and vulnerability to fake news.* Retrieved from https://www.scientificamerican.com/article/cognitive-ability-and-vulnerability-to-fake-news/.

Hodge, D. R. (2007). A Systematic Review of the Empirical Literature on Intercessory Prayer. Research on Social Work Practice, 17(2), 174–187. https://doi.org/10.1177/1049731506296170.

Jukic, A. M., Baird, D. D., Weinberg, C. R., McConnaughey, D. R., & Wilcox, A. J. (2013). Length of human pregnancy and contributors to its natural variation. *Human Reproduction*, 28(10), 2848–2855. https://doi.org/10.1093/humrep/det297.

Kahan, D. M., Landrum, A., Carpenter, K., Helft, L., & Jamieson, K. H. (2017). Science Curiosity and Political Information Processing. *Political Psychology*, 38(S1), 179–199. https://doi.org/10.1111/pops.12396.

Kennedy, B., & Hefferon, M. (2019, March 28). What Americans Know About Science. Retrieved from Pew Research Center Science & Society website: https://www.pewresearch.org/science/2019/03/28/what-americans-know-about-science/.

Martini, F., Nath, J. L., Bartholomew, E. F., Ober, W. C., Ober, C. E., & Hutchings, R. T. (2018). *Fundamentals of anatomy & physiology.* Retrieved from http://www.myilibrary.com?id=1040174.

Michigan State University. (2007, February 27). Scientific Literacy: How Do Americans Stack Up?. *ScienceDaily.* Retrieved from www.sciencedaily.com/releases/2007/02/070218134322.htm.

Patton, K. T., & Thibodeau, G. A. (2016). *Structure & function of the body* (15th edition). St. Louis, Missouri: Elsevier.

Phillips, C., & Axelrod, A. (2005). *Encyclopedia of wars.* In *Facts on File Library of World History.* New York: Facts on File, Inc.

Rogers, A. (n.d.). Americans Trust Scientists, Until Politics Gets in the Way | WIRED. Retrieved from https://www.wired.com/story/americans-trust-scientists-until-politics-gets-in-the-way/.

Shuster, M. (2017). *Biology for a changing world* (3rd edition). New York, NY: W. H. Freeman.

Smith, J., & Kolbaba, G. (2017). *The impossible: The miraculous story of a mother's faith and her child's resurrection* (First edition). New York: FaithWords, Hachette Book Group.

Stoet, G., & Geary, D. C. (2017). Students in countries with higher levels of religiosity perform lower in science and mathematics. Intelligence, 62, 71–78. https://doi.org/10.1016/j.intell.2017.03.001.

Taylor, M. R., Simon, E. J., Dickey, J., & Hogan, K. A. (Eds.). (2017). *Campbell biology: concepts & connections* (Ninth edition). New York, NY: Pearson.

TCFAQ C5c) Why don't we try to destroy tropical

cyclones by nuking. (n.d.). Retrieved from https://www.aoml.noaa.gov/hrd/tcfaq/C5c.html.

Tharoor, S. (2018). *Why I am a Hindu*. Scribe US.

Trump suggested dropping nuclear bombs into hurricanes to stop them from hitting the U.S. (n.d.). Retrieved from Axios website: https://www.axios.com/trump-nuclear-bombs-hurricanes-97231f38-2394-4120-a3fa-8c9cf0e3f51c.html.

U.S. Reports: Edwards, Governor of Louisiana, et al. v. Aguillard et al., 482 U.S. 578 (1987). [Image]. (n.d.). Retrieved from Library of Congress, Washington, D.C. 20540 USA website: https://www.loc.gov/item/usrep482578/.

Verghese, A. (2010). *Cutting for stone* (1st Vintage Books ed.). New York: Vintage Books.

Chapter 2

Animal Research.Info: The global resource for scientific evidence in animal research. (n.d.). Retrieved from ari.info website: http://www.animalresearch.info/en/.

Assoc. for Molecular Pathology v. Myriad Genetics, Inc., 569 U.S. 576 (2013). (n.d.). Retrieved from Justia Law website: https://supreme.justia.com/cases/federal/us/569/576/.

Blackburn, E. H., & Epel, E. (2017). *The telomere effect: a revolutionary approach to living younger, healthier, longer*.

Carlos, C., C, A., & Nm, B. (2017). The Importance of Dietary Behavior to the Health of Monozygotic Twins. *Journal of Community Medicine & Health Education*, *07*(03). https://doi.org/10.4172/2161-0711.1000526.

Cloning Fact Sheet. (n.d.). Retrieved from Genome.gov website: https://www.genome.gov/about-genomics/fact-sheets/Cloning-Fact-Sheet.

Cooney, A. L., Abou Alaiwa, M. H., Shah, V. S., Bouzek, D. C., Stroik, M. R., Powers, L. S., ... McCray, P. B. (2016). Lentiviral-mediated phenotypic correction of cystic fibrosis pigs. *JCI Insight*, *1*(14). https://doi.org/10.1172/jci.insight.88730.

Cornell Alliance for Science. (n.d.). Retrieved from Alliance for Science website: https://allianceforscience.cornell.edu/.

Dedrick, R. M., Guerrero-Bustamante, C. A., Garlena, R. A., Russell, D. A., Ford, K., Harris, K., ... Spencer, H. (2019). Engineered bacteriophages for treatment of a patient with a disseminated drug-resistant *Mycobacterium abscessus*. *Nature Medicine*, *25*(5), 730. https://doi.org/10.1038/s41591-019-0437-z.

Delerue, F., & Ittner, L. M. (2017). Generation of Genetically Modified Mice through the Microinjection of Oocytes. *Journal of Visualized Experiments* (124). https://doi.org/10.3791/55765.

Diabetes | Social Media | Resource Center | Diabetes | CDC. (2020, February 13). https://www.cdc.gov/diabetes/library/socialmedia/infographics/diabetes.html.

Gabbett, M. T., Laporte, J., Sekar, R., Nandini, A.,

McGrath, P., Sapkota, Y., ... Fisk, N. M. (2019). Molecular Support for Heterogonesis Resulting in Sesquizygotic Twinning. *New England Journal of Medicine*, *380*(9), 842–849. https://doi.org/10.1056/NEJMoa1701313.

Genetic Disease Foundation: Hope Through Knowledge. (n.d.). Retrieved from http://www.geneticdiseasefoundation.org/.

Hesman Saey, T. (2018, October 13). A recount of human genes ups the number to at least 46,831. *Science News*, *194*(7), 5.

Hoekstra, C., Zhao, Z. Z., Lambalk, C. B., Willemsen, G., Martin, N. G., Boomsma, D. I., & Montgomery, G. W. (2008). Dizygotic twinning. *Human Reproduction Update*, *14*(1), 37–47. https://doi.org/10.1093/humupd/dmm036.

How did Public Opinion About Entering World War II Change Between 1939 and 1941? Americans—United States Holocaust Memorial Museum. (n.d.). Retrieved from https://exhibitions.ushmm.org/americans-and-the-holocaust/us-public-opinion-world-war-II-1939-1941.

Lee, H. (2014). Genetically Engineered Mouse Models for Drug Development and Preclinical Trials. *Biomolecules & Therapeutics*, *22*(4), 267–274. https://doi.org/10.4062/biomolther.2014.074.

Louise Joy Brown. (n.d.). Retrieved from Louise Joy Brown website: https://www.louisejoybrown.com/.

Machin, G. (2009). Familial monozygotic twinning: A report of seven pedigrees. *American Journal of Medical Genetics Part C: Seminars in Medical Genetics*, *151C*(2), 152–154. https://doi.org/10.1002/ajmg.c.30211.

Molteni, M. (2019, July 30). The World Health Organization Says No More Gene-Edited Babies. *Wired*. Retrieved from https://www.wired.com/story/the-world-health-organization-says-no-more-gene-edited-babies/.

The Nobel Prize in Chemistry 2020. (n.d.). NobelPrize.Org. Retrieved from https://www.nobelprize.org/prizes/chemistry/2020/summary/.

Nylander, P. P. S. (1979). The Twinning Incidence in Nigeria. *Acta Geneticae Medicae et Gemellologiae: Twin Research*, *28*(04), 261–263. https://doi.org/10.1017/S0001566000008746.

Reference, G. H. (n.d.). Genetics Home Reference, Your Guide to Understanding Genetic Conditions. Retrieved from Genetics Home Reference website: https://ghr.nlm.nih.gov/.

Salzberg, S. L. (2018a). Open questions: How many genes do we have? *BMC Biology*, *16*(1). https://doi.org/10.1186/s12915-018-0564-x.

Salzberg, S. L. (2018b). Open questions: How many genes do we have? *BMC Biology*, *16*(1). https://doi.org/10.1186/s12915-018-0564-x.

Shur, N. (2009). The genetics of twinning: From splitting eggs to breaking paradigms. *American Journal of Medical Genetics Part C: Seminars in Medical Genetics*, *151C*(2), 105–109. https://doi.org/10.1002/ajmg.c.30204.

Standord at The Tech: Understanding Genetics. (n.d.). Retrieved from https://genetics.thetech.org/.

Stem Cell Basics III. | stemcells.nih.gov. (n.d.). Retrieved from https://stemcells.nih.gov/info/basics/3.htm.

Stern, A. (2005). *Eugenic nation: Faults and frontiers of better breeding in modern America.* Berkeley: University of California Press.

Tharps, L. L. (2016). *Same family, different colors: confronting colorism in America's diverse families.* Boston: Beacon Press.

University of Iowa. (2016, September 20). Gene therapy for cystic fibrosis lung disease. Retrieved from ScienceDaily website: https://www.sciencedaily.com/releases/2016/09/160920093723.htm.

What Are Mirror Twins? Here's Everything You Want to Know. (2020, August 21). Healthline. https://www.healthline.com/health/pregnancy/mirror-twins.

Chapter 3

About the Creation Museum. (n.d.). Retrieved from Creation Museum website: https://creationmuseum.org/about/.

The Belmont Report. (2010, January 28). [Text]. HHS. Gov. https://www.hhs.gov/ohrp/regulations-and-policy/belmont-report/index.html.

Brown, J. (n.d.). *John Brown, fl. 1854. Slave Life in Georgia: A Narrative of the Life, Sufferings, and Escape of John Brown, a Fugitive Slave, Now in England.* Retrieved from https://docsouth.unc.edu/neh/jbrown/jbrown.html.

Chaplin, G. (2007). Geographic Distribution of Environmental Factors Influencing Human Skin Coloration. *American Journal of Physical Anthropology, 125,* 292–302.

Dao, A. H., & Netsky, M. G. (1984). Human tails and pseudotails. *Human Pathology, 15*(5), 449–453. https://doi.org/10.1016/s0046-8177(84)80079-9.

Escaped slave Gordon, also known as "Whipped Peter," showing his scarred back at a medical examination, Baton Rouge, Louisiana. (n.d.). [Image]. Library of Congress, Washington, D.C. 20540 USA. Retrieved from https://www.loc.gov/resource/ppmsca.54375/.

Health Equity Now | ADA. (n.d.). Retrieved from https://www.diabetes.org/healthequitynow.

Herron, J. C., & Freeman, S. (2013). *Evolutionary analysis* (5th edition). San Francisco, CA: Pearson Education.

History Engine: Tools for Collaborative Education and Research | Search. (n.d.). Retrieved from https://historyengine.richmond.edu/search/citation/78978.

Hoffman, K. M., Trawalter, S., Axt, J. R., & Oliver, M. N. (2016). Racial bias in pain assessment and treatment recommendations, and false beliefs about biological differences between blacks and whites. *Proceedings of the National Academy of Sciences, 113*(16), 4296–4301. https://doi.org/10.1073/pnas.1516047113.

Hogarth, R. A. (2019). The Myth of Innate Racial Differences Between White and Black People's Bodies: Lessons From the 1793 Yellow Fever Epidemic in Philadelphia, Pennsylvania. *American Journal of Public Health, 109*(10), 1339–1341. https://doi.org/10.2105/AJPH.2019.305245.

McPherson & Oliver, photographer. (1863) *Escaped slave Gordon, also known as "Whipped Peter," showing his scarred back at a medical examination, Baton Rouge, Louisiana.* Baton Rouge, Louisiana, United States, 1863. [Baton Rouge, La.: Publisher not identified, 2 April] [Photograph] Retrieved from the Library of Congress, https://www.loc.gov/item/2018648117/.

National Archives at Atlanta. (n.d.). Item470-full23.jpg. Retrieved from https://www.archives.gov/atlanta/exhibits/item470-full.html.

Nauright, J., & Wiggins, D. K. (2019). *Routledge handbook of sport, race and ethnicity.*

Perry, G. H., Foll, M., Grenier, J.-C., Patin, E., Nedelec, Y., Pacis, A., Barakatt, M., Gravel, S., Zhou, X., Nsobya, S. L., Excoffier, L., Quintana-Murci, L., Dominy, N. J., & Barreiro, L. B. (2014). Adaptive, convergent origins of the pygmy phenotype in African rainforest hunter-gatherers. *Proceedings of the National Academy of Sciences, 111*(35), E3596–E3603. https://doi.org/10.1073/pnas.1402875111.

Presser, L. (2020, May 19). The Black American Amputation Epidemic. *ProPublica.* https://features.propublica.org/diabetes-amputations/black-american-amputation-epidemic/.

Presser, L. (n.d.). *Black Diabetics Lose Limbs at Triple the Rate of Others. Here's How Health Care Leaders Are Starting to Act.* ProPublica. Retrieved from https://www.propublica.org/article/black-diabetics-lose-limbs-at-triple-the-rate-of-others-heres-how-health-care-leaders-are-starting-to-act?token=Gg58888u2U5db3W3CsuKrD0LD_VQJReQ.

Quinn, S. (2018, April 27). The Shameful History of Human Zoos. *HistoryCollection.Com.* https://historycollection.com/the-shameful-history-of-human-zoos/.

Reactions to creation "museum." (2007, May 25). Retrieved from NCSE website: https://ncse.com/news/2007/05/reactions-to-creation-museum-001073.

Savitt, T. L. (1982). The Use of Blacks for Medical Experimentation and Demonstration in the Old South. *The Journal of Southern History, 48*(3), 331. https://doi.org/10.2307/2207450.

Wilber, R. L., & Pitsiladis, Y. P. (2012). Kenyan and Ethiopian Distance Runners: What Makes Them so Good? *International Journal of Sports Physiology and Performance, 7*(2), 92–102. https://doi.org/10.1123/ijspp.7.2.92.

Chapter 4

Article Information | Albany Government Law Review. (n.d.). Retrieved from http://www.albany governmentlawreview.org/archives/Pages/article-information.aspx?volume=5&issue=2&page=440.

Beck, A. (2013). *Sexual Victimization in Prisons and Jails Reported by Inmates, 2011–12*. 108.

Bollinger, A. (2020, September 23). *GOP Senator introduces bill that could require genital exams for girls competing in school sports*. LGBTQ Nation. https://www.lgbtqnation.com/2020/09/gop-senator-introduces-bill-require-genital-exams-girls-competing-school-sports/.

Butler, J. (2004). *Undoing gender*. New York; London: Routledge.

Conley, G. (2016). *Boy erased: a memoir*. New York, New York: Riverhead Books.

Diamond, M., & Beh, H. G. (2009). *David Reimer's Legacy: Limiting Parental Discretion* (SSRN Scholarly Paper No. ID 1446966). Retrieved from Social Science Research Network website: https://papers.ssrn.com/abstract=1446966.

Drescher, J. (2015). Out of DSM: Depathologizing Homosexuality. *Behavioral Sciences, 5*(4), 565–575. https://doi.org/10.3390/bs5040565.

The Gay/Trans Panic Defense: What It is, and How to End It. (n.d.). Retrieved from https://www.americanbar.org/groups/crsj/publications/-member-features/gay-trans-panic-defense/.

Heide, A. (2018, January 29). Intersexuality: Hello, I Am the Third Option. *Die Zeit*. Retrieved from https://www.zeit.de/zeit-magazin/leben/2018-01/intersexuality-gender-identity?wt_zmc=sm.ext. zonaudev.mail.ref.zeitde.share.link.x.

Human Rights Campaign Glossary of Terms. (n.d.). HRC. Retrieved from https://www.hrc.org/resources/glossary-of-terms.

Intersex Society of North America | A world free of shame, secrecy, and unwanted genital surgery. (n.d.). Retrieved from http://www.isna.org/.

Karyotype (Normal): Image Details—NCI Visuals Online. (n.d.). Retrieved from https://visualsonline.cancer.gov/details.cfm?imageid=2721.

Khazan, O. (2016, April 27). Why Men With Older Brothers Are More Likely to be Gay. Retrieved from The Atlantic website: https://www.theatlantic.com/health/archive/2016/04/gay-brothers/480117/.

Lambert, J. (2019). No 'gay gene': Massive study homes in on genetic basis of human sexuality. *Nature, 573*(7772), 14–15. https://doi.org/10.1038/d41586-019-02585-6.

Law Lecture: Finding Friendship in a Contentious Place. (2020, February 13). https://www.toledoalumni.org/events/law-finding-friendship.html.

Legislative Tracker: Keep Tabs on All Legislation Related to LGBTQ Discrimination in 2019. (n.d.). Retrieved from Freedom for All Americans website: https://www.freedomforallamericans.org/2019-legislative-tracker/.

Markey, E. J. (2018, July 10). *Text—S.3188–115th Congress (2017–2018): Gay and Trans Panic Defense Prohibition Act of 2018 (2017/2018)* [Webpage]. https://www.congress.gov/bill/115th-congress/senate-bill/3188/text.

Martin, J. (2020, February 27). *LBGTQ groups: South Carolina law is putting students at risk*. AP NEWS. https://apnews.com/cac1092d96c5675aae79287 af64370fd.

Military & Veterans. (n.d.). Retrieved from National Center for Transgender Equality website: https://transequality.org/issues/military-veterans.

Ming, D. (2019, May 1). Female runners with high testosterone must take hormone suppressants to compete, sports court rules. Retrieved from Vice News website: https://news.vice.com/en_us/article/wjvda4/female-runners-with-high-testosterone-must-take-hormone-blockers-to-compete-sports-court-rules.

Minter, S., Esq., & Director, N. L. (n.d.). National Center for Lesbian Rights. Retrieved from National Center for Lesbian Rights website: http://www.nclrights.org/.

North, A. (2019, June 11). Alabama's law forcing sex offenders to get chemically castrated, explained. Retrieved July 31, 2019, from Vox website: https://www.vox.com/identities/2019/6/11/18661514/-alabama-chemical-castration-bill-kay-ivey-effects.

Obergefell v Hodges—Law School Case Briefs for Class Prep. (n.d.). Retrieved from https://www.lexisnexis.com/lawschool/resources/p/casebrief-obergefell-v-hodges.aspx.

Researchers at Powdermill Nature Reserve observe rare gynandromorph bird containing both male and female characteristics. (2020, September 28). Carnegie Museum of Natural History. https://carnegiemnh.org/press/researchers-at-powdermill-nature-reserve-observe-rare-gynandromorph-bird-containing-both-male-and-female-characteristics/.

Saey, T. H. (2018, October 31). These DNA differences may be linked to having same-sex partners. Retrieved from Science News website: https://www.sciencenews.org/article/genetics-dna-homo sexuality-gay-orientation-attractiveness-straight.

Sanders, A. R., Beecham, G. W., Guo, S., Dawood, K., Rieger, G., Badner, J. A., ... MGS Collaboration. (2017). Genome-Wide Association Study of Male Sexual Orientation. *Scientific Reports, 7*(1), 16950. https://doi.org/10.1038/s41598-017-15736-4.

Soh, D. W. (n.d.). Cross-Cultural Evidence for the Genetics of Homosexuality. Retrieved from Scientific American website: https://www.scientificamerican.com/article/cross-cultural-evidence-for-the-genetics-of-homosexuality/.

Texas SB17 | 2019–2020 | 86th Legislature. (n.d.). Retrieved from LegiScan website: https://legiscan.com/TX/bill/SB17/2019.

The LGBT Bar. (n.d.). Retrieved from The National LGBTBar Association website: https://lgbtbar.org/about/about-us/.

Urrutia, D. E. U. (n.d.). Total Solar Eclipse 2024: Here's What You Need to Know. Retrieved from Space.com website: https://www.space.com/-41552-total-solar-eclipse-2024-guide.html.

Vanderlaan, D. P., Blanchard, R., Zucker, K. J., Massuda, R., Fontanari, A. M. V., Borba, A. O., … Lobato, M. I. R. (2017). Birth order and androphilic male-to-female transsexualism in Brazil. *Journal of Biosocial Science*, 49(4), 527–535. https://doi.org/10.1017/S0021932016000584.

Vatican office blasts gender theory, questions intentions of transgender people. Retrieved from National Catholic Reporter website: https://www.ncronline.org/news/vatican/vatican-office-blasts-gender-theory-questions-intentions-transgender-people.

Wiederman, M. W., Bockting, W. O., Nicolai, K., Bullough, V. L., & Fisher, T. D. (2000). Book reviews. *Journal of Sex Research*, 37(4), 378–386. https://doi.org/10.1080/00224490009552061.

Chapter 5

Abstinence-Only-Until-Marriage Programs Are Ineffective and Harmful to Young People, Expert Review Confirms. (2017, August 14). Retrieved from Guttmacher Institute website: https://www.guttmacher.org/news-release/2017/abstinence-only-until-marriage-programs-are-ineffective-and-harmful-young-people.

Akmal, M., Qadri, J. Q., Al-Waili, N. S., Thangal, S., Haq, A., & Saloom, K. Y. (2006). Improvement in Human Semen Quality After Oral Supplementation of Vitamin C. *Journal of Medicinal Food*, 9(3), 440–442. https://doi.org/10.1089/jmf.2006.9.440.

American Academy of Pediatrics. (n.d.). Retrieved from AAP.org website: http://www.aap.org/en-us/Pages/Default.aspx.

American College of Nurse-Midwives | Home. (n.d.). Retrieved from http://www.midwife.org/.

A pregnant 11-year-old rape victim in Ohio would no longer be allowed to have an abortion under new state law. (n.d.). Retrieved from https://www.cbsnews.com/news/ohio-abortion-heartbeat-bill-pregnant-11-year-old-rape-victim-barred-abortion-after-new-ohio-abortion-bill-2019–05–13/.

Ballard, O., & Morrow, A. L. (2013). Human Milk Composition. *Pediatric Clinics of North America*, 60(1), 49–74. https://doi.org/10.1016/j.pcl.2012.10.002.

Bek, K. M., & Laurberg, S. (1992). Risks of anal incontinence from subsequent vaginal delivery after a complete obstetric anal sphincter tear. *BJOG: An International Journal of Obstetrics and Gynaecology*, 99(9), 724–726. https://doi.org/10.1111/j.1471-0528.1992.tb13870.x.

Bryant-Comstock, K., Bryant, A. G., Narasimhan, S., & Levi, E. E. (2016). Information about Sexual Health on Crisis Pregnancy Center Web Sites: Accurate for Adolescents? *Journal of Pediatric and Adolescent Gynecology*, 29(1), 22–25. https://doi.org/10.1016/j.jpag.2015.05.008.

CDC. (2020, March 19). *Vaccinating Boys and Girls Against HPV*. Centers for Disease Control and Prevention. https://www.cdc.gov/hpv/parents/vaccine.html.

Chesson, H. W., Blandford, J. M., Gift, T. L., Tao, G., & Irwin, K. L. (2004). The estimated direct medical cost of sexually transmitted diseases among American youth, 2000. *Perspectives on Sexual and Reproductive Health*, 36(1), 11–19. https://doi.org/10.1363/psrh.36.11.04.

Christ, G., & Washington, J. (2018, March 28). UH explains how it lost all the embryos in its fertility clinic (photos, video). Retrieved from Cleveland website: https://www.cleveland.com/healthfit/2018/03/uh_explains_how_it_lost_all_th.html.

Cooney, S. (n.d.). States Enacted 60 Restrictions on Women's Reproductive Health | Time. Retrieved from http://time.com/4619534/-reproductive-health-abortion-restrictions-2016/?utm_source=emailshare&utm_medium=email&utm_campaign=email-share-article&utm_content=20190107.

Corey, S. (n.d.). *3 Ways To Lower Abortion Rates That Are More Effective Than Closing Clinics*. Elite Daily. Retrieved from https://www.elitedaily.com/news/politics/lower-abortion-clinics/1290840.

Crew, B. (n.d.). Here's How Many Cells in Your Body Aren't Actually Human. Retrieved from ScienceAlert website: https://www.sciencealert.com/how-many-bacteria-cells-outnumber-human-cells-microbiome-science.

Darran, S., & Scutti, S. (n.d.). DA drops all charges against Alabama's Marshae Jones. Retrieved August 1, 2019, from CNN website: https://www.cnn.com/2019/07/03/us/pregnant-alabama-woman-manslaughter-indictment/index.html.

Dietz, H. P., Gillespie, A. V. L., & Phadke, P. (2007). Avulsion of the pubovisceral muscle associated with large vaginal tear after normal vaginal delivery at term. *Australian and New Zealand Journal of Obstetrics and Gynaecology*, 47(4), 341–344. https://doi.org/10.1111/j.1479-828X.2007.00748.x.

Documentary film AKA Jane Roe on FX. (n.d.). Retrieved from https://www.fxnetworks.com/shows/aka-jane-roe.

Feldman, N., & Pugliese, N. (n.d.). New Lawsuit Reveals More Sexual Abuse Allegations Against Boy Scouts Of America. Retrieved from NPR.org website: https://www.npr.org/2019/08/07/749041591/new-lawsuit-reveals-more-sexual-abuse-allegations-against-boy-scouts-of-america.

Flexmort » CuddleCot™. (n.d.). Retrieved from http://flexmort.com/cuddle-cots/.

For the First Time in North America, a Woman Gives Birth After Uterus Transplant From a Deceased Donor. (2019, July 9). Retrieved from Health Essentials from Cleveland Clinic website: https://

health.clevelandclinic.org/for-the-first-time-in-north-america-woman-gives-birth-after-uterus-transplant-from-deceased-donor/.

Freedman, L. R., Landy, U., & Steinauer, J. (2008a). When There's a Heartbeat: Miscarriage Management in Catholic-Owned Hospitals. *American Journal of Public Health, 98*(10), 1774–1778. https://doi.org/10.2105/AJPH.2007.126730.

Freedman, L. R., Landy, U., & Steinauer, J. (2008b). When There's a Heartbeat: Miscarriage Management in Catholic-Owned Hospitals. *American Journal of Public Health, 98*(10), 1774–1778. https://doi.org/10.2105/AJPH.2007.126730.

Gyhagen, M., Bullarbo, M., Nielsen, T., & Milsom, I. (2013). The prevalence of urinary incontinence 20 years after childbirth: A national cohort study in singleton primiparae after vaginal or caesarean delivery: Urinary incontinence 20 years after childbirth. *BJOG: An International Journal of Obstetrics & Gynaecology, 120*(2), 144–151. https://doi.org/10.1111/j.1471-0528.2012.03301.x.

Hill, J. (2003). Posthumous sperm retrieval. *The Lancet, 361*(9371), 1834. https://doi.org/10.1016/S0140-6736(03)13400-9.

Holmes, M. M., Resnick, H. S., Kilpatrick, D. G., & Best, C. L. (1996). Rape-related pregnancy: estimates and descriptive characteristics from a national sample of women. *American Journal of Obstetrics and Gynecology, 175*(2), 320–324; discussion 324–325.

Jewish Telegraphic Agency. (n.d.). Retrieved from Jewish Telegraphic Agency website: https://www.jta.org/about-us.

Ma, Y., Zhang, P., Wang, F., Yang, J., Yang, Z., & Qin, H. (2010). The relationship between early embryo development and tumourigenesis. *Journal of Cellular and Molecular Medicine, 14*(12), 2697–2701. https://doi.org/10.1111/j.1582-4934.2010.01191.x.

Moore, L. (2012, August 20). Rep. Todd Akin: 'Legitimate Rape' Statement and Reaction. *The New York Times*. Retrieved from https://www.nytimes.com/2012/08/21/us/politics/rep-todd-akin-legitimate-rape-statement-and-reaction.html.

National Intimate Partner and Sexual Violence Survey: 2010 Summary Report. (2010). 124.

National Sexual Violence Resource Center (NSVRC). (n.d.). Retrieved from National Sexual Violence Resource Center website: https://www.nsvrc.org/.

NW, 1615 L. St, Suite 800Washington, & Inquiries, D. 20036USA202–419–4300 | M.-857–8562 | F.-419–4372 | M. (n.d.). Racial, ethnic diversity increases yet again with the 117th Congress. *Pew Research Center*. Retrieved from https://www.pewresearch.org/fact-tank/2021/01/28/racial-ethnic-diversity-increases-yet-again-with-the-117th-congress/.

Ohio Bill Prohibiting Abortion Insurance Coverage (HB 182). (n.d.). Retrieved from Rewire.News website: https://rewire.news/legislative-tracker/law/ohio-bill-prohibiting-abortion-insurance-coverage-hb-182/.

Pink Tax Repeal Act (2019—H.R. 2048). (n.d.). GovTrack.Us. Retrieved from https://www.govtrack.us/congress/bills/116/hr2048.

Planned Parenthood | Official Site. (n.d.). Retrieved from https://www.plannedparenthood.org.

Platenburg, G., & Brice-Saddler, M. (2019, June 29). A bullet, a miscarriage and an unthinkable question: Who's the victim, and who is to blame? *Washington Post*. Retrieved from https://www.washingtonpost.com/nation/2019/06/29/bullet-miscarriage-an-unthinkable-question-whos-victim-who-is-blame/.

Rennison, C. M. (n.d.). Bureau of Justice Statistics (BJS)—Rape and Sexual Assault: Reporting to Police and Medical Attention, 1992–2000. Retrieved from https://www.bjs.gov/index.cfm?ty=pbdetail&iid=1133.

Reporting of Sexual Violence Incidents. (n.d.). Retrieved from National Institute of Justice website: https://www.nij.gov:443/topics/crime/rape-sexual-violence/pages/rape-notification.aspx.

Reproductive Health Access Project. (n.d.). Retrieved from Reproductive Health Access Project website: https://www.reproductiveaccess.org/.

Richards, C., & Peterson, L. (2018). *Make trouble: standing up, speaking out, and finding the courage to lead* (First Touchstone hardcover edition). New York: Touchstone, an imprint of Simon & Schuster, Inc.

Roche, D. (2020, October 12). *Pete Buttigieg's abortion comments to Chris Wallace go viral: "I trust women to draw the line."* Newsweek. https://www.newsweek.com/pete-buttigiegs-abortion-comments-chris-wallace-go-viral-i-trust-women-draw-line-1538231.

Rowlands, S. (2011). Misinformation on abortion. *The European Journal of Contraception & Reproductive Health Care, 16*(4), 233–240. https://doi.org/10.3109/13625187.2011.570883.

Santelli, J. S., Kantor, L. M., Grilo, S. A., Speizer, I. S., Lindberg, L. D., Heitel, J., ... Ott, M. A. (2017). Abstinence-Only-Until-Marriage: An Updated Review of U.S. Policies and Programs and Their Impact. *Journal of Adolescent Health, 61*(3), 273–280. https://doi.org/10.1016/j.jadohealth.2017.05.031.

Services, U. S. D. of H. and H. (n.d.). U.S. Department of Health & Human Services [Text]. Retrieved from HHS.gov website: https://www.hhs.gov/.

Shefi, S., Raviv, G., Eisenberg, M. L., Weissenberg, R., Jalalian, L., Levron, J., ... Madgar, I. (2006). Posthumous sperm retrieval: analysis of time interval to harvest sperm. *Human Reproduction, 21*(11), 2890–2893. https://doi.org/10.1093/humrep/del232.

Speier, J. (2019, April 4). *Actions—H.R.2048–116th Congress (2019–2020): Pink Tax Repeal Act (2019/2020)* [Webpage]. https://www.congress.gov/bill/116th-congress/house-bill/2048/all-actions.

Speier, J. (2018, April 10). H.R.5464–115th Congress (2017–2018): Pink Tax Repeal Act [Webpage].

Retrieved from https://www.congress.gov/bill/ 115th-congress/house-bill/5464.

Stanger-Hall, K. F., & Hall, D. W. (2011). Abstinence-Only Education and Teen Pregnancy Rates: Why We Need Comprehensive Sex Education in the U.S. *PLOS ONE*, *6*(10), e24658. https://doi.org/10.1371/journal.pone.0024658.

The American College of Obstetricians and Gynecologists—ACOG. (n.d.). Retrieved from https://www.acog.org/.

The Hyde Amendment and Coverage for Abortion Services. Jul 30, A. R. P., & 2019. (2019, July 30). Retrieved from The Henry J. Kaiser Family Foundation website: https://www.kff.org/womens-health-policy/issue-brief/the-hyde-amendment-and-coverage-for-abortion-services/.

U.S. Department of Labor—Employment Protection For Workers Who Are Pregnant Or Nursing. (n.d.). Retrieved from https://www.dol.gov/wb/maps/2.htm#Alabama.

U.S. Reports: Roe v. Wade, 410 U.S. 113 (1973). [Image]. (n.d.). Retrieved from Library of Congress, Washington, D.C. 20540 USA website: https://www.loc.gov/item/usrep410113/.

Washington, J. (2018, April 11). Are frozen embryos human life? Fertility treatments present ethical dilemmas. Retrieved from cleveland.com website: https://www.cleveland.com/healthfit/2018/04/are_frozen_embryos_human_life_1.html.

World Health Organization. (n.d.). Retrieved from https://www.who.int.

Chapter 6

American College of Allergy, Asthma, & Immunology. (n.d.). Retrieved from ACAAI Public Website: https://acaai.org/.

Autism Speaks. (n.d.). Retrieved from Autism Speaks website: https://www.autismspeaks.org/.

Autism Spectrum Disorder—NIMH. (n.d.). Retrieved from https://www.nimh.nih.gov/health/topics/-autism-spectrum-disorders-asd/index.shtml.

Branswell, H. (2019, April 23). As calls for measles vaccination rise, it's crickets from the White House. Retrieved from STAT website: https://www.statnews.com/2019/04/23/measles-vaccination-calls-rise-crickets-from-white-house/.

Bruesewitz v. Wyeth LLC, 562 U.S. 223 (2011). (n.d.). Retrieved from Justia Law website: https://supreme.justia.com/cases/federal/us/562/223/.

Carroll, A. E., & Vreeman, R. C. (2009). *Don't swallow your gum! myths, half-truths, and outright lies about your body and health* (1st ed.). New York: St. Martin's Griffin.

CDC. (2019, October 4). Measles Cases and Outbreaks. Retrieved from Centers for Disease Control and Prevention website: https://www.cdc.gov/measles/cases-outbreaks.html.

CDC. (2018, May 7). 1918 Influenza Pandemic. Retrieved from Centers for Disease Control and Prevention website: https://www.cdc.gov/features/1918-flu-pandemic/index.html.

Cornwall, W. (2020, June 30). *Just 50% of Americans plan to get a COVID-19 vaccine. Here's how to win over the rest.* Science | AAAS. https://www.sciencemag.org/news/2020/06/just-50-americans-plan-get-covid-19-vaccine-here-s-how-win-over-rest.

Coronavirus disease (COVID-19): Herd immunity, lockdowns and COVID-19. (n.d.). Retrieved from https://www.who.int/news-room/q-a-detail/-herd-immunity-lockdowns-and-covid-19.

December 09, N. L.-S. W. & 2020. (n.d.). *When will COVID-19 vaccines start to make a difference?* Livescience.Com. Retrieved from https://www.livescience.com/coronavirus-vaccination-rate-model.html.

From biodefence to the DRC: How the Ebola vaccine became one of the fastest vaccines to license in history. (n.d.). Retrieved from https://www.gavi.org/vaccineswork/biodefence-drc-how-ebola-vaccine-became-one-fastest-vaccines-license-history.

Global Measles Outbreaks. https://www.cdc.gov/globalhealth/measles/globalmeaslesoutbreaks.htm.

Gorski, D. (n.d.). Chemical castration of autistic children leads to the downfall of Dr. Mark Geier. Retrieved from https://sciencebasedmedicine.org/-chemical-castration-of-autistic-children-leads-to-the-downfall-of-dr-mark-geier/.

History of Vaccines—A Vaccine History Project of The College of Physicians of Philadelphia | History of Vaccines. (n.d.). Retrieved from file:///Users/judinath/Zotero/storage/XIEYDBQM/www.historyofvaccines.org.html#null.

Iannelli, V., & MD. (2016, October 7). George Bernard Shaw on Vaccines. Retrieved from VAXOPEDIA website: https://vaxopedia.org/2016/10/07/-george-bernard-shaw-on-vaccines/.

Jacobs, J. P., Jones, C. M., & Baille, J. P. (1970). Characteristics of a Human Diploid Cell Designated MRC-5. *Nature*, *227*(5254), 168–170. https://doi.org/10.1038/227168a0.

Jain, A., Marshall, J., Buikema, A., Bancroft, T., Kelly, J. P., & Newschaffer, C. J. (2015). Autism Occurrence by MMR Vaccine Status Among US Children With Older Siblings With and Without Autism. *JAMA*, *313*(15), 1534. https://doi.org/10.1001/jama.2015.3077.

King, B. H. (2015). Promising Forecast for Autism Spectrum Disorders. *JAMA*, *313*(15), 1518. https://doi.org/10.1001/jama.2015.2628.

List of Vaccines | CDC. (2019, April 15). Retrieved from https://www.cdc.gov/vaccines/vpd/-vaccines-list.html.

McCarthy, J. (2008). *Louder than words: A mother's journey in healing autism.* Retrieved from https://archive.org/details/louderthanwordsm00mcca_0.

McCarthy, J., & Kartzinel, J. (2010). *Healing and preventing autism: A complete guide.* New York: Plume, Penguin Group.

Meet The First Person in The US To Get COVID Vaccine. (2020, December 15). Kaiser Health News. https://khn.org/morning-breakout/meet-the-first-person-in-the-us-to-get-covid-vaccine/.

Mina, M. J., Kula, T., Leng, Y., Li, M., Vries, R. D. de, Knip, M., ... Elledge, S. J. (2019). Measles virus infection diminishes preexisting antibodies that offer protection from other pathogens. *Science*, 366(6465), 599–606. https://doi.org/10.1126/science.aay6485.

NA06814. (n.d.). Retrieved from https://www.coriell.org/0/Sections/Search/Sample_Detail.aspx?Ref=AG06814-N&PgId=166.

NG05965. (n.d.). Retrieved from https://www.coriell.org/0/Sections/Search/Sample_Detail.aspx?Ref=AG05965-C&PgId=166.

Parkinson, J. (n.d.). *FDA Fast Tracks 2 COVID-19 Vaccines*. ContagionLive. Retrieved from https://www.contagionlive.com/news/fda-fast-tracks-2-covid19-vaccines.

Patel, M. K., & Orenstein, W. A. (2019). Classification of global measles cases in 2013–17 as due to policy or vaccination failure: a retrospective review of global surveillance data. *The Lancet Global Health*, 7(3), e313–e320. https://doi.org/10.1016/S2214-109X(18)30492-3.

Pfizer and BioNTech Granted FDA Fast Track Designation for Two Investigational mRNA-based Vaccine Candidates Against SARS-CoV-2 | Pfizer. (n.d.). Retrieved from https://www.pfizer.com/news/-press-release/press-release-detail/pfizer-and-biontech-granted-fda-fast-track-designation-two.

Riedel, S. (2005). Edward Jenner and the history of smallpox and vaccination. *Proceedings (Baylor University. Medical Center)*, 18(1), 21–25. https://doi.org/10.1080/08998280.2005.11928028.

Rural and Hindu children have the highest immunization rates in India | Institute for Healthcare Policy & Innovation. (n.d.). Retrieved from https://ihpi.umich.edu/news/rural-hindu-children-have-highest-immunization-rates-india.

Skloot, R. (2010). *The immortal life of Henrietta Lacks*. New York: Crown Publishers.

States With Religious and Philosophical Exemptions From School Immunization Requirements. (n.d.). National Conference of State Legislatures. Retrieved from https://www.ncsl.org/research/health/school-immunization-exemption-state-laws.aspx.

The Influenza Epidemic of 1918. (n.d.). Retrieved from https://www.archives.gov/exhibits/influenza-epidemic/.

Understanding the journey to herd immunity. (n.d.). Retrieved from https://www.path.org/articles/understanding-journey-herd-immunity/?gclid=EAIaIQobChMImK_gv6rO7wIVUPDACh2AzwnJEAAYBCAAEgKOY_D_BwE.

U.S. National Library of Medicine—National Institutes of Health [Other]. (n.d.). from https://www.nlm.nih.gov/.

Vally, H. (n.d.). What is herd immunity and how many people need to be vaccinated to protect a community? Retrieved from The Conversation website: http://theconversation.com/what-is-herd-immunity-and-how-many-people-need-to-be-vaccinated-to-protect-a-community-116355.

Vazquez, M. (n.d.). Trump now says parents must vaccinate children in face of measles outbreak. Retrieved from CNN website: https://www.cnn.com/2019/04/26/politics/donald-trump-measles-vaccines/index.html.

Wakefield, A., Murch, S., Anthony, A., Linnell, J., Casson, D., Malik, M., ... Walker-Smith, J. (1998). RETRACTED: Ileal-lymphoid-nodular hyperplasia, non-specific colitis, and pervasive developmental disorder in children. *The Lancet*, 351(9103), 637–641. https://doi.org/10.1016/S0140-6736(97)11096-0.

Chapter 7

About NCCIH. (n.d.). NCCIH. Retrieved from https://www.nccih.nih.gov/about.

Anti-Vaccination Attitudes within the Chiropractic Profession: Implications for Public Health Ethics. (n.d.). Retrieved from http://www.tihcij.com/Articles/Anti-Vaccination-Attitudes-within-the-Chiropractic-Profession-Implications-for-Public-Health-Ethics.aspx?id=0000377.

Brody, H., & Brody, D. (2000). *The placebo response: How you can release the body's inner pharmacy for better health* (1st ed.). New York: Cliff Street Books.

Cassidy, J. D., Boyle, E., Côté, P., He, Y., Hogg-Johnson, S., Silver, F. L., & Bondy, S. J. (2008). Risk of Vertebrobasilar Stroke and Chiropractic Care: Results of a Population-Based Case-Control and Case-Crossover Study. *Spine*, 33(Supplement), S176–S183. https://doi.org/10.1097/BRS.0b013e3181644600.

Chiropractic: In Depth. (2011, November 21). Retrieved from NCCIH website: https://nccih.nih.gov/health/chiropractic/introduction.htm.

Complementary, Alternative, or Integrative Health: What's In a Name? (2011, November 11). Retrieved from NCCIH website: https://nccih.nih.gov/health/integrative-health.

Davis, M. A., Sirovich, B. E., & Weeks, W. B. (2010). Utilization and expenditures on chiropractic care in the United States from 1997 to 2006. *Health Services Research*, 45(3), 748–761. https://doi.org/10.1111/j.1475-6773.2009.01067.x.

Depression: How effective are antidepressants? In *InformedHealth.org [Internet]*. Institute for Quality and Efficiency in Health Care (IQWiG). (2020) https://www.ncbi.nlm.nih.gov/books/NBK361016/.

Firanescu, C. E., de Vries, J., Lodder, P., Venmans, A., Schoemaker, M. C., Smeets, A. J., ... Lohle, P. N. M. (2018). Vertebroplasty versus sham procedure for painful acute osteoporotic vertebral compression fractures (VERTOS IV): Randomised sham controlled clinical trial. *BMJ*, k1551. https://doi.org/10.1136/bmj.k1551.

Freedman, D. H. (2011, June 7). The Triumph of New-Age Medicine. Retrieved from The Atlantic website: https://www.theatlantic.com/magazine/archive/2011/07/the-triumph-of-new-age-medicine/308554/.

Gleberzon, B., Lameris, M., Schmidt, C., & Ogrady, J. (2013). On Vaccination & Chiropractic: when ideology, history, perception, politics and jurisprudence collide. *The Journal of the Canadian Chiropractic Association*, 57(3), 205–213.

Gouveia, L. O., Castanho, P., & Ferreira, J. J. (2009). Safety of Chiropractic Interventions: A Systematic Review. *Spine*, 34(11), E405–E413. https://doi.org/10.1097/BRS.0b013e3181a16d63.

Huang Lian (Chinese Herb) in Treating Patients With Advanced Solid Tumors—Full Text View—ClinicalTrials.gov. (n.d.). Retrieved from https://clinicaltrials.gov/ct2/show/NCT00014378.

Integrative Medicine | Cleveland Clinic. (n.d.). Retrieved from Cleveland Clinic website: https://my.clevelandclinic.org/pediatrics/departments/integrative-medicine.

Kanodia, A. K., Legedza, A. T. R., Davis, R. B., Eisenberg, D. M., & Phillips, R. S. (2010). Perceived Benefit of Complementary and Alternative Medicine (CAM) for Back Pain: A National Survey. *The Journal of the American Board of Family Medicine*, 23(3), 354–362. https://doi.org/10.3122/jabfm.2010.03.080252.

Licensure of Complementary and Alternative Practitioners. (2011a). *Virtual Mentor*, 13(6), 374–378. https://doi.org/10.1001/virtualmentor.2011.13.6.pfor1–1106.

Licensure of Complementary and Alternative Practitioners. (2011b). *Virtual Mentor*, 13(6), 374–378. https://doi.org/10.1001/virtualmentor.2011.13.6.pfor1–1106.

Louw, A., Diener, I., Fernández-de-las-Peñas, C., & Puentedura, E. J. (2016). Sham Surgery in Orthopedics: A Systematic Review of the Literature. *Pain Medicine*, pnw164. https://doi.org/10.1093/pm/pnw164.

Massage & Bodywork State Licensing Requirements. (2015, June 3). Retrieved from Associated Bodywork & Massage Professionals website: https://www.abmp.com/practitioners/state-requirements.

Nahin, R. L., Barnes, P. M., Stussman, B. J., & Bloom, B. (2009). Costs of complementary and alternative medicine (CAM) and frequency of visits to CAM practitioners: United States, 2007. *National Health Statistics Reports* (18), 1–14.

Naturopathy. (2012, January 12). Retrieved from NCCIH website: https://nccih.nih.gov/health/naturopathy.

Offit, P. A. (2013a). *Do you believe in magic? The sense and nonsense of alternative medicine* (First edition). New York: Harper.

Offit, P. A. (2014). *Do you believe in magic? Vitamins, supplements, and all things natural: A look behind the curtain.* HarperCollins.

Offit, P. A. (2013b, November 11). Washington Post: Alternative medicines are popular, but do any of them really work? Retrieved from Paul Offit website: http://paul-offit.com/washington-post-alternative-medicines-are-popular-but-do-any-of-them-really-work/.

Shelton, J. (2004). *Homeopathy: how it really works.* Amherst, N.Y: Prometheus Books.

Traditional Chinese Medicine: In Depth. (2009, April 1). Retrieved from NCCIH website: https://nccih.nih.gov/health/whatiscam/chinesemed.htm.

U.S. Reports: Dent v. West Virginia, 129 U.S. 114 (1889). [Image]. (n.d.). Retrieved from Library of Congress, Washington, D.C. 20540 USA website: https://www.loc.gov/item/usrep129114/.

Wayback Machine. (2014, April 24). Retrieved from https://web.archive.org/web/20140424011335/http://ahc.memberclicks.net/assets/documents/ChiroHistoryPrimer.pdf.

Chapter 8

A Brief History of the Drug War. (n.d.). Retrieved from Drug Policy Alliance website: http://www.drugpolicy.org/issues/brief-history-drug-war.

Abuse, N. I. on D. (n.d.). National Institute on Drug Abuse (NIDA). Retrieved from https://www.drugabuse.gov/.

Alcohol Facts and Statistics. (2019, April 25). National Institute on Alcohol Abuse and Alcoholism (NIAAA). https://www.niaaa.nih.gov/publications/brochures-and-fact-sheets/alcohol-facts-and-statistics.

ALS Association Greater New York Chapter | New York | New Jersey. Retrieved from The ALS Association Greater New York Chapter website: http://als-ny.org.

American Addiction Centers. (n.d.). Retrieved from https://americanaddictioncenters.org/.

Aronson, S. M. (2014). The ancient vocabulary of medical prescriptions. *Rhode Island Medical Journal (2013)*, 97(9), 67.

Bevilacqua, L., & Goldman, D. (2009). Genes and Addictions. *Clinical Pharmacology & Therapeutics*, 85(4), 359–361. https://doi.org/10.1038/clpt.2009.6.

Bourkaib, M., Polomeni, P., & Schwan, R. (2014). Management of opioid addiction with buprenorphine: French history and current management. *International Journal of General Medicine*, 143. https://doi.org/10.2147/IJGM.S53170.

Callum, P. (2019, June 7). THC overdose: Has first death from marijuana exposure been recorded in the United States? Retrieved from Newsweek website: https://www.newsweek.com/thc-overdose-death-marijuana-exposure-united-states-1442742.

Carrigan, M. A., Uryasev, O., Frye, C. B., Eckman, B. L.,

Myers, C. R., Hurley, T. D., & Benner, S. A. (2015). Hominids adapted to metabolize ethanol long before human-directed fermentation. *Proceedings of the National Academy of Sciences*, 112(2), 458–463. https://doi.org/10.1073/pnas.1404167111.

Commissioner, O. of the. (2019, April 19). FDA approves first generic naloxone nasal spray to treat opioid overdose. Retrieved from FDA website: http://www.fda.gov/news-events/press-announcements/fda-approves-first-generic-naloxone-nasal-spray-treat-opioid-overdose.

Dioscorides, P. (1555). De Materia Medica. Retrieved from https://www.wdl.org/en/item/10632/.

Fuss, J., Steinle, J., Bindila, L., Auer, M. K., Kirchherr, H., Lutz, B., & Gass, P. (2015). A runner's high depends on cannabinoid receptors in mice. *Proceedings of the National Academy of Sciences*, 112(42), 13105–13108. https://doi.org/10.1073/pnas.1514996112.

Gonzales v Raich—Law School Case Briefs for Class Prep. (n.d.). Retrieved from https://www.lexisnexis.com/lawschool/resources/p/casebrief-gonzales-v-raich.aspx.

Grisel, J. (2019). *Never enough: the neuroscience and experience of addiction* (First edition). New York: Doubleday.

Kudo, R., Yuui, K., Kasuda, S., & Hatake, K. (2015). [Effect of alcohol on vascular function]. *Nihon Arukoru Yakubutsu Igakkai Zasshi = Japanese Journal of Alcohol Studies & Drug Dependence*, 50(3), 123–134.

Mann, B. (n.d.). New York Lawsuit Claims Sackler Family Illegally Profited From Opioid Epidemic. Retrieved from NPR.org website: https://www.npr.org/2019/03/28/707722556/new-york-lawsuit-claims-sackler-family-illegally-profited-from-opioid-epidemic.

Motzer, R. J., Tannir, N. M., McDermott, D. F., Arén Frontera, O., Melichar, B., Choueiri, T. K., … Escudier, B. (2018). Nivolumab plus Ipilimumab versus Sunitinib in Advanced Renal-Cell Carcinoma. *New England Journal of Medicine*, 378(14), 1277–1290. https://doi.org/10.1056/NEJMoa1712126.

NARCAN® (naloxone) Nasal Spray. (n.d.). Retrieved from https://www.narcan.com/.

National Research Council (US) Committee to Update Science, M. (2004). *Safety Testing*. Retrieved from https://www.ncbi.nlm.nih.gov/books/NBK24645/.

Neves, F. F., Cupo, P., Muglia, V. F., Elias Junior, J., Nogueira-Barbosa, M. H., & Pazin-Filho, A. (2013). Body packing by rectal insertion of cocaine packets: a case report. *BMC Research Notes*, 6(1). https://doi.org/10.1186/1756-0500-6-178.

Nitschke, P., & Stewart, F. (2016). *The peaceful pill ehandbook*.

Nivolumab Combined With Ipilimumab Versus Sunitinib in Previously Untreated Advanced or Metastatic Renal Cell Carcinoma (CheckMate 214)—Full Text View—ClinicalTrials.gov. (n.d.). Retrieved from https://clinicaltrials.gov/ct2/show/NCT02231749.

Nutt, D., King, L. A., Saulsbury, W., & Blakemore, C. (2007). Development of a rational scale to assess the harm of drugs of potential misuse. *The Lancet*, 369(9566), 1047–1053. https://doi.org/10.1016/S0140-6736(07)60464-4.

Prescription Drug Spending Why Is the U.S. an Outlier? (n.d.). Retrieved from https://www.commonwealthfund.org/publications/issue-briefs/2017/oct/paying-prescription-drugs-around-world-why-us-outlier.

Purdue Opioid Addiction Class Action Settlement. (2020, March 3). Top Class Actions. https://topclassactions.com/lawsuit-settlements/open-lawsuit-settlements/opioids/purdue-opioid-addiction-class-action-settlement/.

Rare Disease Research Studies & Current Clinical Trials. (n.d.). Retrieved from NORD (National Organization for Rare Disorders) website: https://rarediseases.org/for-patients-and-families/-information-resources/info-clinical-trials-and-research-studies/.

RigiScan® Plus. (n.d.). Retrieved from http://gotopmedical.com/rigiscan%C2%AE-plus.html.

Rockefeller, G. N. (n.d.). The Drug Laws That Changed How We Punish. Retrieved from NPR.org website: https://www.npr.org/2013/02/14/171822608/-the-drug-laws-that-changed-how-we-punish.

SAMHSA—Substance Abuse and Mental Health Services Administration [Text]. (n.d.). Retrieved from https://www.samhsa.gov/.

Sanders, L. (n.d.). Opioids kill. Here's how an overdose shuts down your body. Retrieved from Science News website: https://www.sciencenews.org/article/opioid-crisis-overdose-death.

Shields, L. B. E., Rolf, C. M., & Hunsaker, J. C. (2015). Sudden Death Due To Acute Cocaine Toxicity-Excited Delirium in a Body Packer. *Journal of Forensic Sciences*, 60(6), 1647–1651. https://doi.org/10.1111/1556-4029.12860.

Squeglia, L. M., Jacobus, J., & Tapert, S. F. (2009). The Influence of Substance Use on Adolescent Brain Development. *Clinical EEG and Neuroscience*, 40(1), 31–38. https://doi.org/10.1177/155005940904000110.

Tozzi, J., & Hopkins, J. S. (2017, December 11). *A History of Viagra*. Retrieved from https://www.bloomberg.com/news/features/2017-12-11/the-little-blue-pill-an-oral-history-of-viagra.

Why People Start Using Tobacco, and Why It's Hard to Stop. (n.d.). Retrieved from https://www.cancer.org/cancer/cancer-causes/tobacco-and-cancer/-why-people-start-using-tobacco.html.

Zakhari, S. (2006). *Overview: How Is Alcohol Metabolized by the Body?* 29(4), 10.

Zaybak, A., Güneş, Ü. Y., Tamsel, S., Khorshid, L., & Eşer, İ. (2007). Does obesity prevent the needle from reaching muscle in intramuscular injections? *Journal of Advanced Nursing*, 58(6),

552–556. https://doi.org/10.1111/j.1365–2648. 2007.04264.x.

Chapter 9

Adamo, L., Staloch, L. J., Rocha-Resende, C., Matkovich, S. J., Jiang, W., Bajpai, G., … Mann, D. L. (2018). Modulation of subsets of cardiac B lymphocytes improves cardiac function after acute injury. *JCI Insight, 3*(11). https://doi.org/10.1172/jci. insight.120137.

Akashi, Y. J., & Ishihara, M. (2016). Takotsubo Syndrome. *Heart Failure Clinics, 12*(4), 587–595. https://doi.org/10.1016/j.hfc.2016.06.009.

Amnesty International. (n.d.). Retrieved from https:// www.amnesty.org/en/who-we-are/.

Balsamo, M., & Long, C. (2019, July 26). US government will execute inmates for first time since 2003. Retrieved from AP NEWS website: https:// apnews.com/1ab434fcc31e429089eb5e14de0d13 24.

Barker, D. J., Bull, A. R., Osmond, C., & Simmonds, S. J. (1990). Fetal and placental size and risk of hypertension in adult life. *BMJ (Clinical Research Ed.), 301*(6746), 259–262.

Barker, D. J., Osmond, C., Golding, J., Kuh, D., & Wadsworth, M. E. (1989). Growth in utero, blood pressure in childhood and adult life, and mortality from cardiovascular disease. *BMJ (Clinical Research Ed.), 298*(6673), 564–567.

Beham, I. (2018, February 13). Zombie Genes: Study Finds Gene Activity After Death. From ValueWalk website: https://www.valuewalk. com/2018/02/gene-activity-death-study/.

Bellis, M. (2017, April 6). Death, Money, and the History of the Electric Chair. Retrieved from ThoughtCo website: https://www.thoughtco.com/- death-money-and-the-history-of-the-electric- chair-1991890.

Cardiopulmonary resuscitation (CPR): First aid. (n.d.). Retrieved from Mayo Clinic website: https:// www.mayoclinic.org/first-aid/first-aid-cpr/ basics/art-20056600.

Countries With Death Penalty 2020. (n.d.). Retrieved from https://worldpopulationreview.com/- country-rankings/countries-with-death-penalty.

CPR and ECC. (n.d.). Retrieved from https://cpr. heart.org/AHAECC/CPRAndECC/UCM_ 473161_CPR-and-ECC.jsp.

Curhan, G. C., Chertow, G. M., Willett, W. C., Spiegelman, D., Colditz, G. A., Manson, J. E., … Stampfer, M. J. (1996). Birth weight and adult hypertension and obesity in women. *Circulation, 94*(6), 1310–1315.

Death with Dignity. (n.d.). Death With Dignity. Retrieved from https://www.deathwithdignity. org/about/.

Deterrence. (n.d.). Retrieved from Death Penalty Information Center website: https://death penaltyinfo.org/policy-issues/deterrence.

DPIC | Death Penalty Information Center. (n.d.). Retrieved from https://deathpenaltyinfo.org/.

Dying With Dignity | Physician Aided Death | Assisted Suicide States. (n.d.). Retrieved from http://www. finalexitnetwork.org/.

Editors, H. com. (n.d.). *First execution by electric chair.* HISTORY. Retrieved from https://www.history. com/this-day-in-history/first-execution-by- electric-chair.

The electric chair in Auburn State Prison. (n.d.). [Image]. Library of Congress, Washington, D.C. 20540 USA. Retrieved from https://www.loc.gov/resource/ cph.3a36053/.

Euthanasia & Physician-Assisted Suicide (PAS) around the World—Euthanasia—ProCon.org. (n.d.). Euthanasia. Retrieved from https://euthanasia.procon. org/euthanasia-physician-assisted-suicide-pas- around-the-world/.

Exit International. (n.d.). Retrieved from https:// exitinternational.net/.

Fagundes, C. P., Brown, R. L., Chen, M. A., Murdock, K. W., Saucedo, L., LeRoy, A., … Heijnen, C. (2019). Grief, depressive symptoms, and inflammation in the spousally bereaved. *Psychoneuroendocrinology, 100*, 190–197. https://doi.org/10.1016/j. psyneuen.2018.10.006.

Fall, C. H., Osmond, C., Barker, D. J., Clark, P. M., Hales, C. N., Stirling, Y., & Meade, T. W. (1995). Fetal and infant growth and cardiovascular risk factors in women. *BMJ (Clinical Research Ed.), 310*(6977), 428–432.

Farinholt, P., Park, M., Guo, Y., Bruera, E., & Hui, D. (2018). A Comparison of the Accuracy of Clinician Prediction of Survival Versus the Palliative Prognostic Index. *Journal of Pain and Symptom Management, 55*(3), 792–797. https://doi.org/10.1016/j. jpainsymman.2017.11.028.

Fearon, K., Strasser, F., Anker, S. D., Bosaeus, I., Bruera, E., Fainsinger, R. L., … Baracos, V. E. (2011). Definition and classification of cancer cachexia: an international consensus. *The Lancet Oncology, 12*(5), 489–495. https://doi. org/10.1016/S1470–2045(10)70218–7.

Findings International Comparison | 2018 Annual Report. (n.d.). America's Health Rankings. Retrieved from https://www.americashealthrankings. org/learn/reports/2018-annual-report/findings- international-comparison.

Frankl, V. E. (2014). *Man's search for meaning.* Boston: Beacon Press.

Fulton, G. B., & Metress, E. K. (1995). *Perspectives on death and dying.* Boston: Jones and Bartlett.

Fumeaux, L., Scarpelli, M. P., Tettamanti, C., & Palmiere, C. (2018). Usefulness of liver function tests in postmortem samples. *Journal of Forensic and Legal Medicine, 56*, 51–54. https://doi.org/10.1016/j.jflm. 2018.03.011.

Global Health Observatory Life expectancy and Healthy life expectancy—Data by country. (n.d.).

Retrieved from http://apps.who.int/gho/data/node.main.688.

Golembiewski, K. (2019, August 2). After You Die, These Genes Come to Life. from The Crux website: http://blogs.discovermagazine.com/crux/2019/08/02/genes-activate-after-death/.

Greer, D. M., Shemie, S. D., Lewis, A., Torrance, S., Varelas, P., Goldenberg, F. D., Bernat, J. L., Souter, M., Topcuoglu, M. A., Alexandrov, A. W., Baldisseri, M., Bleck, T., Citerio, G., Dawson, R., Hoppe, A., Jacobe, S., Manara, A., Nakagawa, T. A., Pope, T. M., … Sung, G. (2020). Determination of Brain Death/Death by Neurologic Criteria: The World Brain Death Project. *JAMA*. https://doi.org/10.1001/jama.2020.11586.

Gross, C. G. (2000). Neurogenesis in the adult brain: death of a dogma. *Nature Reviews Neuroscience*, *1*(1), 67–73. https://doi.org/10.1038/35036235.

Hartley, V. (2016, July 5). How a Near Death Experience Can Change a Person's Life? Retrieved from Watkins website: https://www.watkinspublishing.com/how-an-nde-can-change-a-persons-life/.

Heart Disease Facts & Statistics | cdc.gov. (2018, October 9). Retrieved from https://www.cdc.gov/heartdisease/facts.htm.

Herculano-Houzel, S. (2018). Longevity and sexual maturity vary across species with number of cortical neurons, and humans are no exception. *Journal of Comparative Neurology*. https://doi.org/10.1002/cne.24564.

Holland, F. J., Stark, O., Ades, A. E., & Peckham, C. S. (1993). Birth weight and body mass index in childhood, adolescence, and adulthood as predictors of blood pressure at age 36. *Journal of Epidemiology and Community Health*, *47*(6), 432–435.

Humphry, D. (2002). *Final exit: the practicalities of self-deliverance and assisted suicide for the dying* (3rd ed.). New York: Delta Trade Paperback.

International Monetary Fund (IMF). (n.d.). Retrieved from https://www.imf.org/external/index.htm.

Johns Hopkins Medicine, based in Baltimore, Maryland. (n.d.). Retrieved from https://www.hopkinsmedicine.org/.

Kaplan, M. S. (2001). Environment complexity stimulates visual cortex neurogenesis: Death of a dogma and a research career. *Trends in Neurosciences*, *24*(10), 617–620. https://doi.org/10.1016/s0166-2236(00)01967-6.

Kaprio, J., Koskenvuo, M., & Rita, H. (1987). Mortality after bereavement: a prospective study of 95,647 widowed persons. *American Journal of Public Health*, *77*(3), 283–287.

Kato, K., Lyon, A. R., Ghadri, J.-R., & Templin, C. (2017). Takotsubo syndrome: aetiology, presentation and treatment. *Heart*, *103*(18), 1461–1469. https://doi.org/10.1136/heartjnl-2016-309783.

Kontis, V., Bennett, J. E., Mathers, C. D., Li, G., Foreman, K., & Ezzati, M. (2017). Future life expectancy in 35 industrialized countries: projections with a Bayesian model ensemble. *The Lancet*, *389*(10076), 1323–1335. https://doi.org/10.1016/S0140-6736(16)32381-9.

Kruszelnicki, K. (2006). *Great mythconceptions: the science behind the myths*. Kansas City, MO: Andrews McMeel Publ.

Law, C. M., de Swiet, M., Osmond, C., Fayers, P. M., Barker, D. J., Cruddas, A. M., & Fall, C. H. (1993). Initiation of hypertension in utero and its amplification throughout life. *BMJ (Clinical Research Ed.)*, *306*(6869), 24–27.

Li, Y., Pan, A., Wang, D. D., Liu, X., Dhana, K., Franco, O. H., … Hu, F. B. (2018). Impact of Healthy Lifestyle Factors on Life Expectancies in the US Population. *Circulation*, *138*(4), 345–355. https://doi.org/10.1161/CIRCULATIONAHA.117.032047.

Lopez, A. L., & Williams, R. T. (2018, December 13). Spending more on health care may not lengthen life the most. Retrieved from Magazine website: https://www.nationalgeographic.com/magazine/2019/01/spending-money-health-care-may-not-extend-life/.

MacDougall, D. (1907). The Soul: Hypothesis Concerning Soul Substance Together with Experimental Evidence of the Existence of Such Substance. *American Medicine*, *2*, 240–243.

Morita, T., Tsunoda, J., Inoue, S., & Chihara, S. (1999). The Palliative Prognostic Index: a scoring system for survival prediction of terminally ill cancer patients. *Supportive Care in Cancer: Official Journal of the Multinational Association of Supportive Care in Cancer*, *7*(3), 128–133.

Nahm, M., & Greyson, B. (2013). The death of Anna Katharina Ehmer: a case study in terminal lucidity. *Omega*, *68*(1), 77–87.

National Research Council (US), & Institute of Medicine (US). (2013). *U.S. Health in International Perspective: Shorter Lives, Poorer Health* (S. H. Woolf & L. Aron, Eds.). Retrieved from http://www.ncbi.nlm.nih.gov/books/NBK115854/.

Nichol, G., Leroux, B., Wang, H., Callaway, C. W., Sopko, G., Weisfeldt, M., … Ornato, J. P. (2015). Trial of Continuous or Interrupted Chest Compressions during CPR. *New England Journal of Medicine*, *373*(23), 2203–2214. https://doi.org/10.1056/NEJMoa1509139.

Norton, L., Gibson, R. M., Gofton, T., Benson, C., Dhanani, S., Shemie, S. D., … Young, G. B. (2017). Electroencephalographic Recordings During Withdrawal of Life-Sustaining Therapy Until 30 Minutes After Declaration of Death. *Canadian Journal of Neurological Sciences / Journal Canadien Des Sciences Neurologiques*, *44*(02), 139–145. https://doi.org/10.1017/cjn.2016.309.

Occupational Safety and Health Administration. (n.d.). Retrieved from https://www.osha.gov/.

Osmond, C., & Barker, D. J. (2000). Fetal, infant, and childhood growth are predictors of coronary heart disease, diabetes, and hypertension in adult men and women. *Environmental Health Perspectives*,

108 Suppl 3, 545–553. https://doi.org/10.1289/ehp.00108s3545.

Panneton, W. M. (2013). The mammalian diving response: An enigmatic reflex to preserve life? *Physiology (Bethesda, Md.)*, *28*(5), 284–297. https://doi.org/10.1152/physiol.00020.2013.

Parkes, C. M., Benjamin, B., & Fitzgerald, R. G. (1969). Broken heart: a statistical study of increased mortality among widowers. *British Medical Journal*, *1*(5646), 740–743.

Progressive Secular Humanist. (n.d.). Retrieved from https://www.patheos.com/about-patheos.

Rich-Edwards, J. W., Stampfer, M. J., Manson, J. E., Rosner, B., Hankinson, S. E., Colditz, G. A., … Willet, W. C. (1997). Birth weight and risk of cardiovascular disease in a cohort of women followed up since 1976. *BMJ*, *315*(7105), 396–400. https://doi.org/10.1136/bmj.315.7105.396.

Roelofs, K. (2017). Freeze for action: neurobiological mechanisms in animal and human freezing. *Philosophical Transactions of the Royal Society B: Biological Sciences*, *372*(1718), 20160206. https://doi.org/10.1098/rstb.2016.0206.

Smith, C. J., Ryckman, K. K., Barnabei, V. M., Howard, B. V., Isasi, C. R., Sarto, G. E., … Robinson, J. G. (2016). The impact of birth weight on cardiovascular disease risk in the Women's Health Initiative. *Nutrition, Metabolism, and Cardiovascular Diseases: NMCD*, *26*(3), 239–245. https://doi.org/10.1016/j.numecd.2015.10.015.

Truog, R. D., Berlinger, N., Zacharias, R. L., & Solomon, M. Z. (2018). Brain Death at Fifty: *Exploring Consensus, Controversy, and Contexts*. *Hastings Center Report*, *48*, S2–S5. https://doi.org/10.1002/hast.942.

The World Federation of Right to Die Societies. (n.d.). Retrieved from https://www.worldrtd.net/.

U.S. Reports: Cruzan v. Director, MDH, 497 U.S. 261 (1990). [Image]. (n.d.). Retrieved from Library of Congress, Washington, D.C. 20540 USA website: https://www.loc.gov/item/usrep497261/.

Valdez, R., Athens, M. A., Thompson, G. H., Bradshaw, B. S., & Stern, M. P. (1994). Birth weight and adult health outcomes in a biethnic population in the USA. *Diabetologia*, *37*(6), 624–631.

World Development Indicators | The World Bank. (n.d.). Retrieved from http://wdi.worldbank.org/table/2.12.

Yu, Q. C., Lipsky, M., Trump, B. F., & Marzella, L. (1988). Response of human hepatocyte lysosomes to postmortem anoxia. *Human Pathology*, *19*(10), 1174–1180. https://doi.org/10.1016/s0046-8177(88)80149-7.

Zhang, X., Zhang, W., Wang, C., Tao, W., Dou, Q., & Yang, Y. (2019). Chest-compression-only versus conventional cardiopulmonary resuscitation by bystanders for children with out-of-hospital cardiac arrest: A systematic review and meta-analysis. *Resuscitation*, *134*, 81–90. https://doi.org/10.1016/j.resuscitation.2018.10.032.

Index

Numbers in *bold italics* indicate pages with illustrations